Cooperative Effects in Matter and Radiation

Cooperative Effects in Matter and Radiation

EDITED BY

Charles M. Bowden and D. W. Howgate

U. S. Army Missile Research and Development Command
Redstone Arsenal, Alabama

AND

Hermann R. Robl

U.S. Army Research Office
Research Triangle Park, North Carolina

PLENUM PRESS • NEW YORK AND LONDON

Library of Congress Cataloging in Publication Data

Main entry under title:

Cooperative effects in matter and radiation.

"Papers presented at the Cooperative Effects Meeting which was held as part of the U. S. Army Sponsored Symposium on New Laser Concepts at Redstone Arsenal, Alabama, November 30-December 2, 1976."

Includes index.

1. Superradiance—Congresses. 2. Superfluorescence—Congresses. 3. Lasers—Congresses. I. Bowden, Charles M. II. Howgate, D.W. III. Robl, Hermann R. IV. U.S.—Army. V. Symposium on New Laser Concepts at Redstone Arsenal, 1976. VI. Title: Matter and Radiation.

QC476.S86C66 539.2 77-21591

ISBN 0-306-31078-3

Papers presented at the Cooperative Effects Meeting which was held as part of the U.S. Army Sponsored Symposium on New Laser Concepts at Redstone Arsenal, Alabama, November 30—December 2, 1976

A Division of Plenum Publishing Corporation
227 West 17th Street, New York, N.Y. 10011

Printed in the United States of America

PREFACE

This volume contains the large majority of the papers presented at the Cooperative Effects Meeting which was held as part of the US Army Sponsored Symposium on New Laser Concepts at Redstone Arsenal, Alabama, from November 30 through December 2, 1976. The motivation for the meeting was to bring together a representative cross-section of research scientists active in related areas of cooperative effects in matter-radiation field interaction and coherent pulse generation and propagation. An emphasis was placed upon the rapidly developing areas of superradiance and superfluorescence, with a balance between theory and experiment in regard to the choice of speakers. This meeting came at a very fortunate time when new experimental results in metal vapors and gases have just recently been realized. Also represented on the program were areas dealing with new laser concepts such as the free electron laser and two photon amplifier.

A few supplemental papers are included in this volume which were authored by participants at the meeting, but were not present on the agenda, primarily due to limited time. These were included because of their relation to the content of papers which were presented and/or were the subject of discussion among attendees.

The meeting consisted of eleven invited papers and two workshop sessions, each with a panel. The order of the papers in this volume generally follows the order of their presentation on the agenda. However, the supplemental papers have been inserted where appropriate. An effort has been made to preserve the content and and flavor of the discussions following each presentation as well as that of the panel discussions. An edited version of these notes is included at the end of the book along with the agenda. The program was planned by a committee consisting of the following:

D. W. Howgate
Laser Science Directorate
High Energy Laser Laboratory
US Army Missile Research and Development Command (MIRADCOM)
Redstone Arsenal, Alabama

C. M. Bowden
Physical Sciences Directorate
Technology Laboratory
US Army Missile Research and Development Command
Redstone Arsenal, Alabama

R. W. Mitchell
Laser Science Directorate
High Energy Laser Laboratory
US Army Missile Research and Development Command
Redstone Arsenal, Alabama

The editors and the program committee would like to express gratitude to the Army High Energy Laser Laboratory of Redstone Arsenal, Alabama, and in particular to its Director, Dr. R. D. Rose, for the sponsorship of the Symposium, and for the vital encouragement provided by Dr. Rose, both for the organization of the Symposium and for the publishing of these Proceedings. We are also particularly indebted to Colonel James M. Kennedy, Jr., Chief of the Physics and Mathematics Branches, US Army Research and Standardization Group (Europe), whose efforts contributed considerably to the success of the meeting.

Finally, we would like to express gratitude to Mrs. Sherry Troglen of the Physical Sciences Directorate, MIRADCOM, Redstone Arsenal, Alabama, and to Miss Ann Hill of the European Office of Research, London, for administrative assistance.

C. M. BOWDEN
D. W. HOWGATE
HERMANN R. ROBL
13 April 1977

CONTENTS

INTRODUCTORY REMARKS

Hopefully, these Proceedings on the Cooperative Effects session of the recent international meeting on New Concepts for High Energy Lasers held at the US Army Missile Command in December 1976, will convey not only the excellent experimental data recently taken on superfluorescence in cesium vapor and superradiance in methyl fluoride but the excitement of the heated theoretical discussions precipitated by this data. Certainly the fervor of the joust between Dr. Rodolfo Bonifacio advocating the mean-field approach and Dr. Michael Feld proponent of the stimulated decay model is difficult to capture. Nevertheless, the essence of their arguments and their interpretation of the recent data should make interesting reading.

I further hope that these proceedings, although only a sampling of the exciting work being done by investigators exploring new concepts for lasers (see Laser Focus Apr 77 for a synopsis of the entire meeting at Redstone Arsenal), will provide a stimulus for continued exchange of information and ideas among those poorly funded scientists exploring new concepts for high energy lasers.

R. D. Rose
Director
High Energy Laser Laboratory
US Army Missile Research and
Development Command

SUPERRADIANCE IN EXPERIMENTALLY RELEVANT REGIMES*

J. C. MacGillivray and M. S. Feld†

Department of Physics and Spectroscopy Laboratory

Massachusetts Institute of Technology

Cambridge, Massachusetts 02139

Abstract: This paper explores the assumptions made in the semiclassical description of superradiance. Simple expressions for observable output parameters in several experimentally relevant regimes are given. Implications of these results to some possible applications of superradiance are discussed.

I. INTRODUCTION

In view of the renewed experimental interest in superradiance,[1-4] a detailed list of specific limiting conditions[5] for the applicability of the simple theoretical model[6,7] which accurately described the results of the initial experiments[8] seems appropriate. In this paper the semiclassical model and its exact solution in the "ideal superradiance" limit will be briefly reviewed. The effect of relaxing each of the assumptions made in obtaining this limit will be described, emphasizing the amount by which the parameters of an actual system can depart from ideality without significantly altering the analytical expressions for the expected output radiation. The changes in these expressions which occur when some of the constraints are further relaxed will then be discussed. Finally, the implications of these results to some potential applications of superradiance will be mentioned.

*Work supported in part by National Science Foundation and US Army Research Office (Durham).

†Alfred P. Sloan Fellow.

These considerations should be of particular interest to those attempting to observe superradiance in other systems, and are relevant to the problems of x-ray laser system design[9] and ultrashort pulse generation.

Superradiance is the spontaneous radiative decay of an assembly of atoms or molecules in the collective mode. It is the optimal process for extracting coherent energy from an inverted system. In this process incoherent emission induces a small macroscopic polarization in an inverted two-level medium which gives rise to a growing electric field and consequently an increasing polarization in space and time. After a long delay, a highly directional pulse is emitted, often accompanied by ringing. The peak output power is proportional to the square of the number of radiators, N.

In our theoretical model[7] the semiclassical approach (classical fields, quantized molecules) has been adopted in order to take propagation effects fully into account. Semiclassical discussions have also been given by Burnham and Chiao,[10] Friedberg and Hartmann,[11] Arecchi and Courtens,[12] and Bullough.[13] In fact, Dicke[14] gave a semiclassical description in his original paper. For a discussion of quantized field models, see Bonifacio and Lugiato[15] and references contained therein.

The coupled Maxwell-Schrödinger equations in the slowly-varying envelope approximation, written in complex form, are[7,16]

$$\partial \mathcal{E}/\partial x = -\kappa \mathcal{E} + 2\pi k \, \Sigma_{v,M} \mathcal{P} \quad , \tag{1a}$$

$$\partial \mathcal{P}/\partial T = -\left(\frac{1}{T_2} - ikv\right)\mathcal{P} + \frac{\mu_z^2}{\hbar} \mathcal{E} n + \Lambda_p \quad , \tag{1b}$$

$$\partial n/\partial T = \Lambda - n/T_1 - (1/\hbar) \, \mathrm{Re} \, (\mathcal{E}\mathcal{P}^*) \quad . \tag{1c}$$

Here $\mathcal{P}(x,T,v,M)$ and $\mathcal{E}(x,T)$ are the complex, slowly varying envelopes of the polarization density per velocity interval dv in degenerate M_J-state M and of the electric field, respectively, at position x and retarded time $T = t-x/c$; $n(x,T,v,M)$ is the inversion density; κ accounts for diffraction or other loss; T_1 is the population decay time; T_2 is the polarization decay time; Λ is a source term describing the rate of production of n; μ_z is the dipole moment component parallel to the direction of polarization; and $\Sigma_{v,M}$ denotes a velocity integral and a sum over degenerate M-states. The remaining notation is the same as in Ref. 7.

Spontaneous emission from the exicted state is simulated in this model by a randomly phased polarization source term Λ_p which describes the rate of production of $\mathcal{P}$. (The expression for Λ_p is given in Ref. 7.) The superradiant process can be initiated by either spontaneous emission or background thermal radiation. However, only spontaneous emission will be considered in this paper, since blackbody radiation (described in detail in Ref. 7) is relatively unimportant at wavelengths shorter than 50 μ, as in the experiments of current interest.

Three basic assumptions are incorporated in Eqs. (1), the implications of which will be discussed below:

1) The semiclassical model with a polarization source term to simulate spontaneous emission is used, instead of a quantized field model.

2) The plane wave approximation is utilized. Thus, effects associated with finite beam diameter are neglected.

3) The interaction of forward and backward travelling waves is ignored.

Computer solutions of Eqs. (1) should be used for precise comparisons with experimental data. However, approximate analytical solutions which are in close agreement with the computer results can be obtained in certain limiting cases. These results are useful in estimating relevant experimental parameters and as an aid to understanding the underlying physical processes.

II. IDEAL SUPERRADIANCE

In this limit an exact solution of the resulting equations can be obtained, with simple expressions for experimentally observable quantities such as output intensity, pulse width, and delay time. These can be useful in determining the feasibility of a proposed superradiant scheme and in optimizing an existing system.

The assumptions made to obtain the "ideal superradiance" limit, in addition to those built into Eqs. (1), are: (4) $1/T_1 = 1/T_2 = 1/T_2^* = 0$, where T_2^* is the dephasing time; (5) $\mathcal{P}(t = 0) = 0$ (no initial polarization at the superradiant transition); (6) $\kappa = 0$; (7) no level generacy (n is summed over degenerate M states and μ_z is averaged over M states); (8) swept excitation (system inverted by a pulse travelling longitudinally through the medium at the speed of light); (9) zero inversion time (system inverted instantaneously); and (10) no feedback. Furthermore, (11) Λ_p is set equal to zero and replaced by an equivalent delta-function input electric field. Each of these assumptions is discussed below.

Given these assumptions, Eqs. (1) become[7]

$$\partial E/\partial x = 2\pi k P \quad , \tag{2a}$$

$$\partial P/\partial T = \mu_z^2 n E/\hbar \quad , \tag{2b}$$

$$\partial n/\partial T = - EP/\hbar \quad , \tag{2c}$$

and n, E, and P are all real. The solution of these equations is $n = n_0 \cos\psi$, $P = \mu_z n_0 \sin\psi$, $n_0 = n(t = 0)$, and

$$d\psi/dT = \mu_z E/\hbar \quad , \tag{3}$$

where

$$\psi(x,T) = \int_{-\infty}^{T} (\mu_z/\hbar) E(x,T')dT' \tag{4}$$

is the partial area of the pulse. (The total area $\theta(x) \equiv \psi(x,\infty)$.) Applying the transformation $w = \sqrt{2xT}$ to Eqs. (2a) and (3) gives the pendulum equation,[17]

$$\psi'' + (1/w)\psi' = \sin\psi/(T_R L) \quad , \tag{5}$$

where $\psi = \psi(w)$ and

$$T_R = \frac{\lambda}{2\pi}\frac{\hbar}{2\pi}\frac{1}{\mu_z^2 n_0 L} \tag{6}$$

is the characteristic radiation damping time of the collective system.

Equations (2b–c) give rise to the familiar Bloch vector picture. As can be seen from the zT dependence of Eq. (5), this system is analogous to a spatial array of coupled pendula, initially tipped at a uniform small angle $\psi(w = 0) = \theta(x = 0) = \theta_0$,[18] which fall as a phased array.

The solution of Eq. (5) is completely determined by two parameters,[10] T_R and the initial tipping angle θ_0. For a given value of θ_0, a single curve relates $T_R^2 I_p$ to T/T_R (see Fig. 4 of Ref. 7), and approximate expressions in terms of $\phi \equiv \ln(2\pi/\theta_0)$ can be derived for the peak output power

$$I_p \approx 4N\hbar\omega/T_R\phi^2 \propto N^2 \quad , \tag{7a}$$

the width of the output pulse

$$T_w \approx T_R\phi \propto N^{-1} \quad , \tag{7b}$$

and the energy contained in the first lobe of emitted radiation

$$E_p \approx 4N\hbar\omega/\phi \propto N \quad . \tag{7c}$$

The delay time from the inversion to I_p is

$$T_D \approx T_R\phi^2/4 \propto N^{-1} \quad , \tag{7d}$$

so that $T_D \approx T_w\phi/4$. Typically, $10 < \phi < 20$.

III. APPLICABILITY OF IDEAL SUPERRADIANCE

The regions of validity of each of the assumptions listed above will now be discussed.

A. Simplifying Assumptions Which Have Little Effect on Output

We first consider those assumptions which can be completely removed without significantly affecting the ideal solution.

a) The semiclassical approach describes the system for $T \gg T_R$, since at T_R there is one photon in each mode of the radiation field.[7] Although, strictly speaking, the semiclassical description breaks down for $T < T_R$, we are only interested in the output at T_D, which is typically 25-100 T_R. Fluctuations in the fields during the first T_R will have little effect on the output at T_D due to the logarithmic dependence of the output on the initial conditions through ϕ. Thus, the randomly phased polarization source, constructed to be consistent with the requirements of thermal equilibrium,[7] should give correct results for $T \gg T_R$.

b) Computer analysis shows that the effect of Λ_p on the evolution of the system is almost identical to that of a delta-function input $\mathcal{E}$ field of appropriate magnitude to give $\phi \approx \ln\left[\sqrt{2\pi N}\,(2\pi\alpha L)^{3/4}\right]$, where αL is the small-signal field gain, so that $\alpha L = T_2'/T_R$, where T_2' is the inverse linewidth. This is understandable since fluctuations at the far end of the medium are amplified over the greatest length and therefore dominate.

c) Computer analysis of the interaction between forward and backward travelling waves shows that this effect is virtually negligible in all swept excitation systems, and it is also negligible in uniformly excited systems for which $L/c \lesssim T_D$. (This latter case is discussed below.) This is so because the forward and backward waves only become sizable in the same region after much of the stored energy has been radiated.[19]

d) Computer calculations show that replacing μ_z^2 by its average value over M states and n by its sum over M states has little effect on the output radiation. Therefore, the influence of level degeneracy is insignificant.

B. Assumptions Which Can Significantly Affect Output

For the remaining assumptions, small deviations from ideality are of little importance but large deviations can cause significant changes in the output. The following conclusions have been verified by computer solutions of Eqs. (1).[5]

a) The lifetimes need not be infinite, which would imply infinite gain:

i) The net gain (gain minus loss) must be large enough so that the total area of the output pulse can grow to π. This leads to the requirement[5]

$$(\alpha - \kappa)L \gtrsim \phi \quad . \tag{10a}$$

In the opposite limit where $(\alpha-\kappa)L < 1$, collective effects can still be important (since $T_R \ll T_{sp}$) but only a small fraction of the energy is radiated coherently (since $T_2' \ll T_R$); this regime, which we refer to as "limited superradiance",[7] includes such familiar effects as free induction decay and echos. In the intermediate regime, where $1 < (\alpha-\kappa)L < \phi$, the peak intensity will be significantly less than that given by Eq. 7(a); analytical results in this regime can be obtained from the linear theory of Crisp.[20]

ii) T_1 must be greater than T_D [Eq. 7(d)], otherwise the population will decay incoherently and reduce the amount of coherent output. This leads to the condition for efficient output

$$\alpha L \gtrsim (\phi^2/4)T_2'/T_1 \quad . \tag{10b}$$

There is no similar requirement on T_2 or T_2^* since effects due to large dephasing or polarization decay rates are offset by high gain.[7]

b) As long as the inversion time τ is less than the observed delay time, a non-zero τ will have little effect on the output other than to increase the observed delay time from T_D to $T_D + \tau/2$. This gives the requirement

$$\tau \lesssim 2T_D \quad . \tag{10c}$$

If the superradiant output occurs before the inversion process terminates, then only the early part of the population inversion can contribute to the first burst of radiation.[21] In the simple case where the inversion density in the absence of superradiant emission is equal to Λt, Λ constant, then[9]

$$(T_D)_{observed} = \sqrt{4\pi T_{sp}\phi^2/\lambda^2 L\Lambda} \quad ,$$

$$I_p = 2h\omega AL\Lambda \quad ,$$

and

$$T_w = 8\sqrt{\pi T_{sp}/\lambda^2 L\Lambda} \quad .$$

Simple forms other than Λt can also be solved analytically, and graphical solutions are also possible.[9]

This same method can be used to provide an approximate solution for the case where $T_1 < T_D$, in violation of requirement (10b) above.

c) Excessive loss κ can diminish superradiance in two ways. It can reduce the net gain, making requirement (10a) harder to fulfill. Superradiant behavior also requires that

$$\int \kappa \, dx \lesssim \phi/4 \quad . \tag{10d}$$

When $\int \kappa \, dx \gtrsim \phi/4$ the pulse stops narrowing and the intensity no longer grows with length. For constant κ, $I_p = N\hbar\omega/4T_R(\kappa L)^2$.[22]

In the case of diffraction of a Gaussian beam, $\int \kappa \, dx = 1/2 \ln[1 + (\lambda L/A)^2]$. This quantity is always small when the Fresnel number is larger than unity.

d) Uniform excitation (entire system inverted simultaneously, in contrast to swept excitation) will have little effect on the output as long as the transit time $T_{tr} = L/c$ is less than the observed delay time, other than to increase the observed delay time to $T_D + T_{tr}/2$. When T_{tr} is longer than the delay time, the system will no longer radiate as a single entity; this places a condition on the length L_c ("cooperation length"):

$$L \lesssim L_c = \sqrt{cT_{sp}\ \phi^2/2n\lambda^2} \quad . \tag{10e}$$

Longer systems will break up into a number of independently radiating segments in a manner described by Arecchi and Courtens.[12] In this limit, the output intensity $I_p \simeq nA\hbar\omega c$ and no longer increases with increasing length. Note that requirement (10e) does not apply to swept excitation.

e) The presence of polarization at the superradiant transition $t = 0$ will have little effect provided that

$$P(t = 0) \ll \mu_z n\theta_0 \quad . \qquad (10f)$$

Larger values of $P(t = 0)$ are equivalent to increasing the initial tipping angle of the Bloch vector, which shortens the delay time and reduces the ringing. This increases the difficulty of completing the inversion process before coherent emission begins. In principle, a pulse of area exactly π could completely invert an initially absorbing medium without residual polarization. An energy conservation argument shows that such a pulse would have to be shorter than T_R in order not to lose area as it traverses the medium; for longer pulses, the effects of self-induced transparency and pulse propagation become relevant.[10,12,16,23,24] As a practical matter, schemes to directly invert two-level systems are probably not feasible due to problems associated with loss, level degeneracy, transverse variations in the electric field associated with beam profile, and the difficulty of generating a pulse of exactly area π.

These problems can be circumvented by using indirect excitation methods such as three level pumping[8] and two photon excitation with a nonresonant intermediate state. All observations of superradiance up to now have employed such schemes. However, the problem remains that when the pump radiation is turned off, a large residual polarization could be left at the pump transition. This can result in superradiance at _this_ transition, which would deplete the population available for superradiance at the desired wavelength. This problem can be overcome by using an incoherent pump pulse, or by choosing a much shorter wavelength for the pump transition (to increase its T_R).

One should also note that in indirect excitation schemes the background emission which initiates superradiance can be modified by the presence of the pump field through multiple quantum transitions. This would increase the effective initial tipping angle, particularly if τ is long.

f) The effect of feedback on the output is negligible as long as the output field due to the initializing spontaneous emission is significantly greater than the additional output which results from the feedback process. Comparing these two quantities gives a condition on the feedback fraction F:

$$F \lesssim 4/(e^{0.35\phi}\,\phi^2) \quad . \qquad (10g)$$

($F < 10^{-4}$ in the HF experiments.[8,25]) In long systems, where transit time is appreciable, the influence of feedback decreases. Computer calculations in this regime have not yet been made.

The effect of significant feedback is to drastically shorten the delay times and reduce the ringing; the effect is analogous to continuing to push a pendulum after it has started to fall. The system acts as if it were subject to a different, larger initial condition.

g) The plane wave approximation breaks down when the solid angle factor μ of Rehler and Eberly[26] (a function of A, L, and λ) falls in the small Fresnel number regime of Fig. 5 of Ref. 26; the break between the two regimes is relatively sharp and occurs near Fresnel number $2A/\lambda L \approx 1/10$.

For small Fresnel number $T_R = T_{sp} 4\pi/nA\lambda$. The output should be independent of length in this regime, but the output intensity should still be proportional to the square of the population inversion density. We have done no computer analysis in the small Fresnel number regime.

To properly account for the spatial variations of $\mathcal{E}$ associated with finite beam diameter, and with focusing and defocusing in a high gain medium, a transverse spatial dependence must be added to Eqs. (1). This aspect of the analysis deserves further attention.

IV. SOME APPLICATIONS OF SUPERRADIANCE

A. Spin-Phonon Superradiance

It may be possible to observe an acoustical analog of superradiance in the spin-phonon interaction process in paramagnetic crystals.[27] In such a system the paramagnetic spins are coupled to the lattice vibrations (in a manner described by Jacobsen and Stevens[28]). As shown in Ref. 29, in the slowly varying envelope approximation the coupled spin-phonon equations become almost identical in form to Eqs. (1).[30] Acoustical gain can be suitably defined, and so in a high gain medium, it should be possible for an initially inverted ensemble of spins, perturbed by kT fluctuations, to rapidly transfer its stored energy to the lattice. The ensuing acoustic waves should have all the properties of the coherent emission observed in optical superradiance.

Recently, Hahn and Wilson[27] proposed a related experiment to observe superradiant emission in a spin-phonon system by preparing the spins in a phased array. The phonon avalanche experiment of Brya and Wagner,[31] although probably not a true coherent effect, was an interesting advance along these lines.

B. X-ray Lasers

The requirements for efficient superradiant emission should also be of interest to designers of x-ray laser systems. Due to the short lifetimes of the transitions and the lack of suitable mirrors in this regime, most proposed schemes use a single pass high gain swept-excitation system. Thus, x-ray lasers will superradiate. Consequently, the conventional rate equation analysis is not applicable, and the above considerations can be useful to estimate the output behavior. Some of the discussions in Section III are especially relevant to the x-ray regime; in particular, T_1 is usually so short that the inversion process will not be completed by the time superradiance occurs [see Eq. (10c)].

As mentioned above, the rate equation analysis gives incorrect results. For example, in the Na scheme of Duguay and Rentzepis,[32] rate equation analysis predicts (at the threshold value) I_p about 10 times smaller and T_w about 10 times larger than the semiclassical predictions.[9] In addition, the threshold inversion density is a factor of 10 smaller than the corresponding rate equation threshold.

Specific applications of these requirements to x-ray laser schemes are discussed further in Ref. 9.

C. Ultrashort Pulse Generation

Since superradiance is the optimum method for extracting coherent energy from an inverted medium, it is interesting to consider it as a method for generating ultrashort pulses. Although in the ideal case T_w decreases with increasing N, many of the conditions listed above restrict the shortness of output pulses one can hope to achieve. Combining Eqs. (7b), (7d), and (10c) shows that the inversion time places a particularly restrictive limit on the minimum T_w which can be generated superradiantly:

$$T_w \gtrsim 2\tau/\phi \quad . \tag{11}$$

Therefore, ultrashort pulse generation by this method requires swept excitation, small κ, and as short an inversion time as possible. Values of T_w/τ less than 1/10 appear possible.

Note that superradiance is a transient process, and so the generation of ultrashort pulses by this method is inherently different from the mode locking approach, where short pulses are generated by mixing a set of equally-spaced phase correlated modes to synthesize a Fourier spectrum.

References

(1) M. Gross, C. Fabre, P. Pillet, and S. Harouche, Phys. Rev. Lett. 36, 1035 (1976).

(2) Q. H. F. Vrehen, H. M. J. Hikspoors, and H. M. Gibbs, Phys. Rev. Lett., to be published; H. M. Gibbs, in this volume; and Q. H. F. Vrehen, in this volume.

(3) A. Flusberg, T. Mossberg, and S. R. Hartmann, Phys. Rev. Lett., to be published, and in this volume.

(4) T. A. DeTemple, in this volume.

(5) Derivations of these conditions and computer results which support them can be found in J. C. MacGillivray and M. S. Feld (unpublished).

(6) I. P. Herman, J. C. MacGillivray, N. Skribanowitz, and M. S. Feld, in Laser Spectroscopy, edited by R. G. Brewer and A. Mooradian, Plenum, 1974.

(7) J. C. MacGillivray and M. S. Feld, Phys., Rev. A 14, 1169 (1976).

(8) N. Skribanowitz, I. P. Herman, J. C. MacGillivray, and M. S. Feld, Phys. Rev. Lett. 30, 309 (1973).

(9) J. C. MacGillivray and M. S. Feld, Appl. Phys. Lett., to be published.

(10) D. C. Burnham and R. Y. Chiao, Phys. Rev. 188, 667 (1969).

(11) R. Friedberg and S. R. Hartmann, Phys. Lett. 38A, 227 (1972) and references contained therein.

(12) F. T. Arecchi and E. Courtens, Phys. Rev. A 2, 1730 (1970).

(13) R. Saunders, S. S. Hassan, and R. K. Bullough, J. Phys. A 9, 1725 (1976) and in this volume.

(14) R. H. Dicke, Phys. Rev. 93, 99 (1954) and in Proceedings of the Third International Conference on Quantum Electronics, Paris, 1963, edited by P. Grivet and N. Bloembergen, Columbia U. P., 1964.

(15) R. Bonifacio and L. A. Lugiato, Phys. Rev. A 11, 1507 (1975) and 12, 587 (1975).

(16) The notation is similar to that of A. Icsevgi and W. E. Lamb, Jr. [Phys. Rev. 185, 517 (1969)], extended to include level degeneracy and the polarization source term Λ_p.

(17) The delta-function assumption is needed so that a boundary condition can be written for Eq. (5). It does not affect the derivations of Eqs. (2)-(4).

(18) It follows from the assumption of a delta-function input E field that $E(x=0,t) = (h/\mu)\theta_0\delta(t)$.

(19) This interaction has been studied by R. Saunders, R. K. Bullough, and S. S. Hassan (Ref. 13).

(20) M. D. Crisp, Phys. Rev. A 1, 1604 (1970).

(21) Although population inversion created after this time can contribute to later radiation lobes, the contribution is small due to the filling of the lower state.

(22) F. T. Arecchi and R. Bonifacio, IEEE J. Quantum Electron. QE-1, 169 (1965); A. Icsevgi and W. E. Lamb, Jr. (Ref. 16); and R. Bonifacio, F. A. Hopf, P. Meystre, and M. O. Scully, Phys. Rev. A 12, 2568 (1975).

(23) In this sense the condition $\tau = T_R$ is the dividing line between self-induced transparency pulse evolution and superradiance.

(24) S. L. McCall and E. L. Hahn, Phys. Rev. 183, 457 (1969).

(25) P. T. Ho, J. C. MacGillivray, S. Liberman, and M. S. Feld (unpublished).

(26) N. E. Rehler and J. H. Eberly, Phys. Rev. A 3, 1735 (1971).

(27) See references contained in E. L. Hahn and R. Wilson, in the proceedings of the XIXth Congress Ampère, Heidelberg, 1976.

(28) E. H. Jacobsen and K. W. H. Stevens, Phys. Rev. 129, 2036 (1963).

(29) C. Leonardi, J. C. MacGillivray, S. Liberman, and M. S. Feld, Phys. Rev. B 11, 3298 (1975).

(30) The similarity between the Maxwell-Schrödinger and spin-phonon equations has also been exploited by N. Shiren [Phys. Rev. B 2, 2471 (1970)] to observe an acoustical analogue of self-induced transparency.

(31) W. J. Brya and P. E. Wagner, Phys. Rev. 157, 400 (1967).

(32) M. A. Duguay and P. M. Rentzepis, Appl. Phys. Lett. 10, 350 (1967).

EXPERIMENTS IN FIR SUPERRADIANCE*

A. T. Rosenberger**, S. J. Petuchowski**, and T. A. DeTemple†

**Department of Physics, University of Illinois

†Department of Electrical Engineering

University of Illinois

Abstract: The occurrence of superradiance in the far infrared has been extended to the homogeneously broadened regime with the observation of delayed single-pulse emission in methyl fluoride at 496 μm. The delay, width, intensity, and asymmetry of the observed pulses are compared with the predictions of several theoretical models. Reasonable quantitative agreement is found with a Maxwell-Bloch mean-field model and a Maxwell-Bloch model including unidirectional propagation. The relation of the observations to steady-state superradiance and the contraction of the emitted pulse is discussed.

I. INTRODUCTION

The cooperative spontaneous emission of radiation, or superradiance, has been a topic for much lively discussion in the past decade; that this is still true may be seen in the content of the articles and the spirit of the discussions included in this volume. Dicke[1] first showed that a gas of N excited molecules, mutually interacting via their common radiation field, can undergo cooperative radiative decay with the intensity of emission proportional to N^2. This implies a pulse width proportional to N^{-1}, and these two characteristics plus a delay of the peak of the emitted pulse

*Supported by Army Research Office: Durham.

which is also very nearly proportional to N^{-1}, are generally considered to reflect superradiant behavior. More detailed theoretical treatments[2-8] have shown that the following general conditions must be satisfied for superradiance to occur:

$$T_s \ll T_{sp} \quad , \tag{1}$$

$$\alpha_o L \gg 1 \quad , \tag{2}$$

$$L \lesssim L_c \quad . \tag{3}$$

The first follows from the N^2 dependence of the intensity; the superradiance time T_s is the approximate scale of the pulse width and must be much less than the spontaneous decay time T_{sp} for the intensity to be proportional to N^2 rather than to N, since $N \gg 1$. Condition (2) implies that the superradiance time T_s is small compared to collision or Doppler dephasing times T_2 or T_2^* (for more detail, see Ref. 8), and condition (3) requires that the sample be shorter than a cooperation length L_c, the maximum length over which cooperation can occur in the emission process.[2] This last condition assumes simultaneous excitation and may be relaxed when the excitation is swept along the length of the sample, increasing the effective cooperation length to infinity.[9]

In view of these conditions, we may see why superradiance can be investigated experimentally in the far infrared (FIR). First, the spontaneous lifetimes of pure rotational FIR transitions are on the order of seconds. This means that even in a reasonably long sample (∿ meters) at convenient pressures (10^{-3} - 1 torr), T_s will not be so small ($\ll 10^{-9}$ sec) so as to preclude temporal detection of the pulses. Second, in samples of such lengths and pressures, the gain $\alpha_o L$ is often $\gg 1$. Third, although the cooperation lengths in such systems are not always greater than the sample lengths, the molecules are excited by optical pumping on a coupled infrared (IR) transition, and this swept excitation should at least partially relax condition (3).

It is not surprising, then, that the first observation of Dicke superradiance took place on FIR rotational transitions in HF.[10] Early observations[11-12] of FIR emission in CH_3F (methyl fluoride) indicated the possibility of superradiance, and we have reported its observation.[14] Other molecules which are likely candidates for displaying cooperative behavior in spontaneous emission are D_2O[12,15] and NH_3.[12] A comparison of several interesting transitions is given in Table I. In Table I, λ is the wavelength of the FIR transition; T_2 is the collisional dephasing time, $T_2 = (\pi\Delta\nu_H)^{-1}$, where $\Delta\nu_H$ is the full homogeneous width at half-maximum; T_2^* is the Doppler dephasing time, $T_2^* = (2\sqrt{\pi}\,\delta\nu_D)^{-1}$, where $\delta\nu_D$ is the Doppler half-width at $1/e$; p_{hom} is the pressure above which the FIR transition can be considered to be homogeneously broadened; α_{max} $(p > p_{hom})$ is the line center gain which a π-pulse pump would produce; and Δ_p is the detuning of the pump line from the absorbing transition in units of infrared Doppler widths of the absorption, where we assume the pump is a CO_2 laser except in the case of HF, where an HF laser was used. From Table I it is evident that the gains on these FIR transitions can be very large. The more likely candidates for superradiant emission will be those for which the pump detuning is small: HF, $C^{12}H_2F$, and $C^{13}H_3F$. In fact, the 66 μm emission in D_2O has been investigated and found to be due to stimulated Raman emission.[15]

Table I. A comparison of several interesting FIR transitions. The notation used is defined in the text.

MOLECULE	D_2O	HF	NH_3	D_2O	$C^{12}H_3F$	$C^{13}H_3F$
λ (μm)	66	84	292	385	496	1222
pT_2 (nsec-torr)	8	6.9	12	8	8	8
T_2^* (nsec)	37	48	153	218	366	916
p_{hom} (torr)	0.77	0.51	0.27	0.13	0.08	0.03
α_{max} (cm^{-1})	4.7	49	5	2	1.6	0.45
Δ_p (Doppler widths)	30	0	7	6	1	0.3

The observation of superradiance in CH_3F is described in Section II, Section III is a comparison of our experimental results to several theories, and in Section IV some unresolved problems and possibilities for future experiments are discussed.

II. OBSERVATION OF SUPERRADIANCE IN CH_3F

Figure 1 shows the partial energy level diagram which is of interest in optically pumping CH_3F. The pump is the P(20) line of CO_2 at 9.55 μm; it lies midway in frequency between two absorptions in the ν_3 band, the Q(12,1) and Q(12,2) absorptions where the notation is $\Delta J(J_{lower},K)$. The pump is detuned by one infrared Doppler width (40 MHz) from either transition. There are therefore two possibilities for FIR emission, R(11,1) and R(11,2) which are separated in frequency by 40 MHz and have a wavelength of 496 μm; the K = 2 transition is expected to dominate, since its line center pump absorption coefficient is four times larger than that for K = 1. Also depicted in Fig. 1 are the various relaxation processes which take place, along with the population notation; the only one of these relaxation processes which is important at the short times involved in our experiment is the rotational (ΔJ),[16] which, since $T_1 = T_2$, yields a FIR homogeneous linewidth ($\Delta\nu_H$) of 40 MHz/torr. Notice that pumping via a coupled IR transition allows the creation of a full population inversion in the FIR and that because of the slowness of the ΔK relaxation, the equilibrium population difference will be greater than the thermal value.

The experimental apparatus is shown in Fig. 2. The pump pulse was produced by a CO_2 TEA laser operated on a single longitudinal and transverse mode by the inclusion of a low-pressure CW CO_2 cell in the laser resonator. The resulting output was a smooth 200 nsec pulse of about 150 mJ total energy (see Fig. 3), emitted at CO_2 line center ± 30 MHz with a single-pulse spectral purity of better than 10 MHz including chirp.[17] The pump pulse passed through an optical breakdown switch and was reflected down the CH_3F cell by a dielectric-coated Si Brewster window;[13] the pump and FIR output passed through a Si Brewster output window and the two were separately detected. The optical breakdown switch utilizes UV-triggered AC breakdown of clean N_2 to cut off the pump pulse in less than 100 psec.[18] The input coupler to the CH_3F cell permits passage of the counter-propagating FIR, polarized perpendicular to the pump, allowing it to be absorbed to minimize feedback.[19] The pump was monitored by a Ge photon drag detector and the FIR

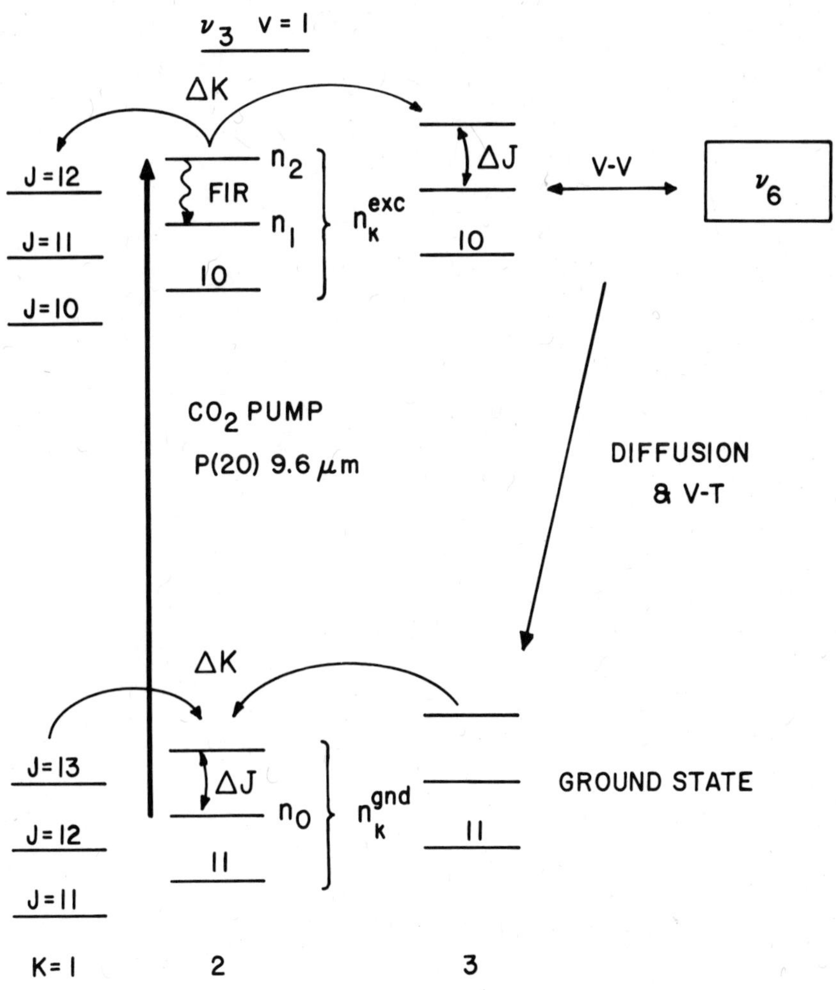

Figure 1. Partial energy level diagram for CH_3F pumped by a CO_2 laser. For the FIR, $T_1 = T_2 = 8$ nsec at 1 torr. The two fastest relaxation rates are $\Gamma_J = 1.3 \times 10^8$ sec^{-1} $torr^{-1}$ and $\Gamma_K = 1.2 \times 10^7$ sec^{-1} $torr^{-1}$. The fractional populations in the ground initial states of the two absorbing transitions, f(J,K), are: $f(12,1) = 6.89 \times 10^{-3}$, $f(12,2) = 6.49 \times 10^{-3}$.

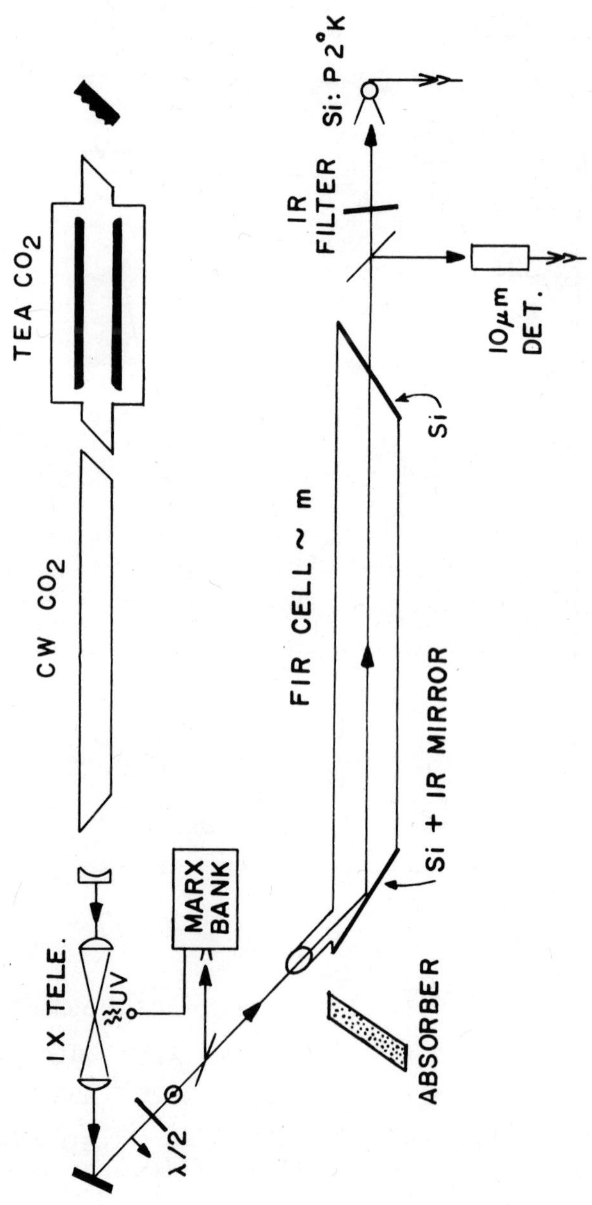

Figure 2. Experimental apparatus.

by a liquid helium-cooled phosphorus-doped Si photo-conductor having a 2 nsec response at 3°K.[20] The CH_3F pressure was measured by a calibrated capacitive manometer.

The optical breakdown switch, by cutting off the pump pulse, allows for the preparation of an excited state which may then decay in a cooperative manner without the further perturbation which would be caused by the presence of the pump during the superradiant emission process. Comparison of (a) and (b) in Fig. 3 shows the effect of this perturbation in the FIR emission; in (b) the FIR pulse is emitted after the cutoff of the pump. At higher pressures some pump-FIR overlap cannot be escaped, and FIR pulses which occur during the pump pulse exhibit varied time structure, of which Fig. 3(c) is an example;[12] FIR pulses of several lobes have been observed under these conditions, but those lobes often have random height and separation, unlike damped oscillations (ringing).

In the treatment of our data, we have excluded those FIR pulses which occurred simultaneously with the pump; Fig. 4 shows typical separated pulses and defines the terms, width (full width at half-maximum) and delay (time from pump cutoff to peak) of the FIR pulse. In Fig. 4, (a) and (b) show the effect of a variation in pressure on the superradiant pulse; one can see the delay and width decrease and the height increase as pressure (hence N) increases. It can also be noted that there is no ringing on the pulses and that they are not symmetric in time. Plotting the delay and pulse width as functions of inverse pressure, and the pulse height as a function of pressure squared, the data fall on straight lines as shown in Fig. 5. Note that there seem to be two classes of pulses (one indicated by circled points); our conjecture is that one class is due to emission for which $K = 1$ and the other due to $K = 2$ emission. The probable reason for this effect is a frequency instability of the pump laser.

In Fig. 5(a), the pulse delay at infinite pressure is seen to be a negative time. This simply indicates that the entire pump pulse is not necessary to prepare the fully inverted state; the pump cutoff serves as a convenient reference, but does not correspond to $t = 0$ for measurement of the delay. The significance of our data, then, as presented in Fig. 5, is that the measurements indicate the basic qualitative behavior expected of Dicke superradiance: (a) pulse delay proportional to p^{-1}, (b) pulse width proportional to p^{-1}, and (c) pulse height proportional to p^2. Furthermore, we have two distinct classes of pulses whose relative pressure dependence scales properly. We now wish to address the question of quantitative agreement with theory.

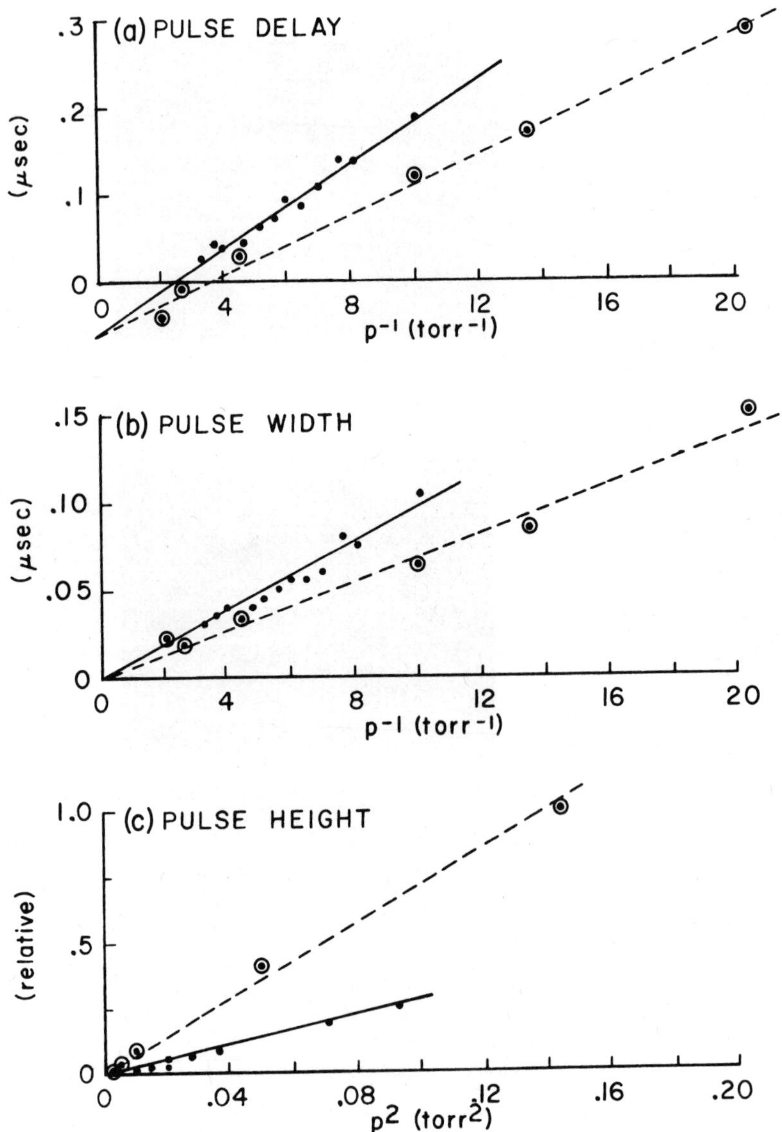

Figure 5. Experimental results. In (a) and (b) the homogeneously broadened region lies below $p^{-1} = 13$ torr^{-1}; in (c) it lies above $p^2 = 0.006$ torr^2. In (c), 1.0 on the relative height scale corresponds to a total emission of 22.8 W.

III. COMPARISON OF RESULTS WITH THEORY

The ideal model would give a full quantitative explanation of experimental results, including the effects of the pump, homogeneous and inhomogeneous broadening, propagation and spatially dependent chirping of the emission. In the comparisons we will make, however, the pump effects will be neglected and a suitably prepared initial state will be assumed; chirping will also be neglected. To make the comparisons, we have taken the slopes of the uncircled data (the larger set of points) and evaluated the delay, width, and height of a typical pulse which would occur at a pressure of 0.123 torr (in the homogeneous limit); these numbers are shown as the first row in Table II (the value of $I(t_o)$ may be a factor of 2 too large because of an uncertainty in the sensitivity of the detector). Also shown in Table II is a parameter δ which indicates the asymmetry in time of the FIR pulses: $\delta = \Delta t_F/\Delta t_R$, where Δt_R is the time for the pulse to rise from half-maximum to maximum, and Δt_F the time to fall back to half-maximum from the peak. We then put the experimental values of cell length (L = 350 cm), cross section of pump beam ($A = 2\ cm^2$), density of initially excited molecules ($\rho = 3.65 \times 10^{12}\ cm^{-3}$)[21] into various theories (with an initial tipping angle $\phi_o \simeq \sqrt{2/N}$, $N = \rho AL$, for all but the Rehler-Eberly case where full inversion is assumed) and calculated the predicted output for two limits: the disk and needle geometries.

A radiating system is characterized by its Fresnel number $F = 2A/\lambda L$ (0.23 in our experiment); for $F > 1$, the system is said to radiate as a disk, meaning the radiated field is essentially a plane wave and escapes primarily from the ends of the sample. In the needle configuration ($F < 1$), radiation escapes primarily due to diffraction and the field is more of a multimode spherical-like wave than a plane wave. The transition between the two cases is not sharp, but occurs over a range of F from 0.1 to 10 (see Ref. 3, Fig. 5). The importance of the Fresnel number in superradiance is that its value determines the scaling properties of the emission;[22] for example, in the disk limit, the intensity should increase as the square of the sample length, whereas in the needle limit, it should increase linearly with the length. This will be seen in the discussion of the first theory, that of Rehler and Eberly.[3]

A. Rehler and Eberly Model

The treatment of Rehler and Eberly for an extended system of two-level atoms, damped radiatively, was one of the earliest models

Table II. Agreement of several theories with experimental data. The conditions are: $p = 0.123$ torr, $A = 2\ cm^2$, $L = 350$ cm, $\rho = 3.65 \times 10^{12}\ cm^{-3}$, and $\phi_o \simeq \sqrt{(2/N)}$ (except for Rehler and Eberly, where full inversion is assumed); δ is a measure of the asymmetry of the pulse and is defined in the text.

		t_o (nsec)	Δt (nsec)	$I(t_o)$ (W)	δ
EXPERIMENT		187±9	73±8	1.1±0.3	1.3±.1
Rehler and Eberly Model	(Disk)	16.6	2.15	421	1.00
	(Needle)	43.4	5.98	151	1.00
Bonifacio and Lugiato Model	(Disk)	263	131	0.285	1.15
	(Needle)	195	117	2.29×10^{-6}	1.26
Pendulum Model	(Disk)	270	126	0.0245	1.56
	(Needle)	200	119	7.9×10^{-7}	1.26
Mean-Field Model	(Disk)	27	3.6	176	1.31
	(Needle)	162	71.5	0.525	1.20
Maxwell-Bloch Model	(Disk)	133	72	0.476	1.23
	(Needle)	144	86	0.223	1.31

of superradiance in a macroscopic system. It is a fully quantum treatment in which all the atoms experience the same field (mean-field model), so that propagation does not play a role in the development of the superradiant pulse. In comparing our results with the predictions of this model we can gain much insight. The existence of an analytical solution for the emission intensity permits the dependence of the pulse delay, width, and height on experimental parameters (ρ, L, A) to be made explicit. These dependences are expected to hold, at least qualitatively, even with the introduction of line broadening effects. However, because collision broadening is dominant in our experiment, quantitative agreement is not expected.

From a fully inverted two-level system, Rehler and Eberly predict a superradiant emission intensity of the form

$$I(t) = \frac{\hbar\omega_o}{4T_{sp}} \mu N^2 \,\mathrm{sech}^2\left(\frac{t - t_o}{2T_s}\right) \quad , \tag{4}$$

where $\hbar\omega_o$ is the energy separation of the levels and μ is a "shape factor" with the limiting values

$$\text{disk:} \quad \mu = 3\lambda^2/8\pi A \quad , \tag{5}$$

$$\text{needle:} \quad \mu = 3\lambda/8L \quad . \tag{6}$$

Also, $T_s = T_{sp}/\mu N$, so we have

$$\Delta t = 3.53\ T_s = 3.53\ T_{sp}/\mu N \quad , \tag{7}$$

$$t_o = T_s \ell n(\mu N) = \frac{T_{sp}}{\mu N} \ell n(\mu N) \quad , \tag{8}$$

$$I(t_o) = \frac{\hbar\omega_o}{4T_{sp}} \mu N^2 \quad . \tag{9}$$

The scaling behavior caused by μ is now evident; inserting our experimental conditions, the values in Table II result. A single symmetrical pulse is predicted, but it is far too intense and too short; as discussed above, this is to be expected since the only decay mechanism present is spontaneous emission.

B. Bonifacio and Lugiato Model

The next treatment is that of Bonifacio and Lugiato[7] who derived, from a general atom-field master equation, a semi-classical pendulum equation for the Bloch vector in the case of inhomogeneous broadening where the frequency distribution was assumed to be Lorentzian; writing their equation but replacing T_2^* with T_2 to correspond to our homogeneously broadened system, we have

$$\ddot{\phi} + \left[k + (2T_2)^{-1}\right] \dot{\phi} - \tau_c^{-2} \exp(-t/T_2) \sin\phi = 0 \quad , \tag{10}$$

where $\tau_c = [32\pi\ T_{sp}/3\lambda^2\rho c]^{1/2}$ and

$$\text{disk:} \quad k = \frac{c}{2L} \quad , \tag{11}$$

$$\text{needle:} \quad k = \frac{\lambda c}{2\pi A} \quad . \tag{12}$$

This was solved numerically for our experimental conditions with the initial conditions $\phi(o) = \phi_o = \sqrt{2/N}$ and $\dot{\phi}(o) = 0$ for both geometries. The intensity is found from

$$I(t) = \frac{\hbar\omega_o}{4} A\rho c \; \tau_c^2 \; \exp(t/T_2)\dot{\phi}^2(t) \quad . \tag{13}$$

Unlike the fits[23,24] to the data of the HF experiment of Ref. 10, no ringing is predicted; although the predicted pulse is asymmetric, the quantitative agreement is not good, as is seen in Table II. Also unlike Ref. 24, allowing k to be a free parameter does not give agreement; if, in fact, agreement could be found at this pressure for some value of k, another value of k would have to be used at a different pressure to give agreement. Although the two geometrical cases agree to within 35% in t_o and Δt, they differ by five orders of magnitude in their predictions for $I(t_o)$; this will be discussed later. At least the inclusion of homogeneous broadening in this fashion gives better agreement than the prototype treatment of Rehler and Eberly.

The balance of Table II can be understood from a simple semi-classical Maxwell-Bloch treatment of the problem. One arrives at resonant coupled equations of the form

$$\frac{\partial}{\partial t} P = -P/T_2 + ER \quad , \tag{14}$$

$$\frac{\partial}{\partial t} R = -(R - R_e)/T_1 - EP \quad , \tag{15}$$

$$\frac{1}{c}\frac{\partial}{\partial t} E + \frac{\partial}{\partial x} E + \kappa E = \alpha P \quad , \tag{16}$$

where $P = -\mathcal{P}/\rho\mu_{12}$ (where $\mathcal{P}$ is the out-of-phase component of the polarization and μ_{12} is the electric dipole matrix element between the two levels), $R = (N_2 - N_1)/N$ (the normalized population difference with an equilibrium value[25] R_e), T_1 = the phenomenological population decay time, $E = \mu_{12}\varepsilon/\hbar$ (where ε is the electric field envelope), κ is the radiation loss (= k/c), and $\alpha = \kappa/2T_s$.

C. Pendulum Model

One can recover a pendulum equation from (16) by setting $T_1 = T_2$, $R_e = 0$, $\partial E/\partial x \ll (1/c)\partial E/\partial t$; then $Pe^{t/T_2} = \sin\phi$, $Re^{t/T_2} = \cos\phi$, and $E = \dot{\phi}$. This pendulum equation is

$$\ddot{\phi} + k\dot{\phi} - \frac{k}{2T_s}\exp(-t/T_2)\sin\phi = 0 \quad , \tag{17}$$

where k has the same values as in the Bonifacio and Lugiato model in the limiting cases of disk and needle geometry. This was solved with the same initial conditions and parameters as the Bonifacio-Lugiato pendulum equation; energy balance gives the intensity as

$$I(t) = (\hbar\omega_o T_{sp} k^2/\mu)\dot{\phi}^2(t) \quad . \tag{18}$$

The only appreciable differences between the predictions of this model and the previous one are that the intensities are reduced by approximately an order of magnitude; this is due to the decay of population through T_1 which was not allowed previously. The intensity for the needle case is again much less than that for the disk limit; this will be discussed in the following section. Again, no ringing was predicted.

D. Mean-Field Model

If one assumes $\kappa E \gg \partial E/\partial x$, $\kappa E \gg (1/c)\partial E/\partial t$, $T_1 = T_2$, $R_e = 0$, then the intensity predicted by Eqs. (14)-(16) has the form

$$I(t) = (\hbar\omega_o \mu N^2/4T_{sp})z^2\mathrm{sech}^2[\beta(1-z) - \alpha] \quad , \tag{19}$$

where

$$z = \exp(-t/T_2) \quad , \tag{20}$$

$$\beta = T_2/2T_s \quad , \tag{21}$$

$$\alpha = \ell n(2/\phi_o) \quad . \tag{22}$$

The relationship between α and β strongly determines the character of the output pulse: if $\alpha > \beta$, the expression for I(t) can be cast in an asymptotic form in which the time scale depends only on T_2. The resulting peak intensity $I(t_o)$ is very small even though the delay and width of the pulses are not in bad agreement with the data. Increasing β such that $\beta \gtrsim \alpha$, the delay and width are not strongly affected although $I(t_o)$ becomes several orders of magnitude larger than the previous case; for $\beta > \alpha$, $((\beta-\alpha)/\alpha \lesssim 1)$ the delay and width become very short and the peak intensities very large. This can explain the wide difference in output power between the needle and disk configurations for the two pendulum equations, since spatial variations were also neglected in those cases. The existence of three regimes is not a new discovery; neglect of spatial variation is valid for a steady-state pulse, and the three regimes were predicted for steady-state swept-gain superradiant pulses.[9]

Equations (14)-(16) were solved numerically with the assumptions listed at the beginning of this subsection (R_e was assumed to have the value it would assume on a time scale short compared to the ΔK relaxation rate and long compared to the ΔJ relaxation rate: $R_e = -4.3 \times 10^{-3}$). This gave the results of Table II, where it may be seen that the needle case gives quite good agreement in all respects, whereas the disk case does not. Again, the seeming disagreement comes from the relation of β and α; α is the same for the two cases, but β is larger for the disk. This sensitivity can be seen in Fig. 1 of Ref. 9, where a slight change in the ratio of gain to loss can result in a large change in Δt or the pulse area; the numerical solution predicts an area of 0.66 π for the needle case, so the sensitivity to ϕ_o is great, even though ϕ_o appears only logarithmically. Using this value of R_e rather than zero increases t_o and Δt by about 5% and decreases $I(t_o)$ by about 15% implying a weak sensitivity to R_e.

E. Maxwell-Bloch Model

The agreement of the mean-field predictions with the measurements is surprising, because one would expect propagation to have an effect in the development of the superradiant pulse since L is not much shorter than L_c. For this reason, the Maxwell-Bloch equations (14)-(16) were solved treating propagation in one direction (for $R_e = 0$), and using a constant input field to stimulate an initial tipping angle, much as in Ref. 8. The results of this are also shown in Table II. Except for a shorter delay, the agreement is reasonable. The spatial variations in population, polarization, and field during the pulse evolution are evident in this model; at the time of emission, they vary appreciably only over the last one-third of the cell length.[26] This variation implies that the mean-field approximation is suspect, at least in comparison with this unidirectional treatment.

IV. DISCUSSION

The observation of superradiance in CH_3F may be summarized as follows: the experiments were performed in a pressure regime where the FIR transition is homogeneously broadened and delayed single pulses of width approximately equal to T_2 were observed to be emitted with equal intensity in both the forward and reverse directions. The presence of two waves implies that the phenomenon is not steady-state superradiance, in which the forward wave would be much larger than the reverse.[9] The Fresnel number of the system is calculated to be 0.23, implying a geometrical character more like that of a needle than a disk. This is supported by a measurement of the transverse intensity dependence of the FIR, which shows an angular divergence more consistent with the needle geometry.[3] Since $T_s \ll T_2$ and $\alpha_o L \gg |\ln \phi_o|$ in the experiment, we may call the effect strong, rather than limited, superradiance;[8] this is further supported by the observation of almost all the available energy being extracted by the pulse and by numerical calculations which give areas of nearly π for the emitted pulses. These considerations, then, make the CH_3F experiment very nearly the complement of the HF experiment in FIR superradiance.

The comparison of experimental results with theories, as summarized in Table II, shows the needle case of the mean-field model and the disk case of the Maxwell-Bloch model to have the most success in predicting experimental results. The mean-field model used is a one parameter fit - slight variations were allowed in N to get agreement, since the details of the action of the pump are uncertain. The applicability of the mean-field model to our experiment is not totally justified (for example, because the gain per pass is so large), but, similarly to the Rehler and Eberly model, it provides a convenient analytical expression for I(t), Eq. (19). The Maxwell-Bloch model used treats the one-wave case with a field loss made up of end emission and diffraction, and shows explicitly the spatial dependence of the population, polarization, and field.

It should be pointed out that the entries in Table II do not represent best fits of the various theories. Only N was allowed to vary to fit the mean-field model, and this value was used in calculating the predictions of the other models. A more exhaustive comparison would allow variations in field loss, in T_1, and perhaps in the cross-sectional area of the emitting cylinder of molecules. The Maxwell-Bloch model should also allow the possibility of propagation in both directions, coupling between the oppositely directed waves, and chirping.[26] All models neglect pump effects, and it should be noted that nutation on the IR pump transition will persist for the length of the pump pulse, and that infrared free induction decay due to the pump cutoff will have an effective T_2^* lifetime of approximately 10 nsec. These two effects may influence the superradiant behavior and may need to be included in a more complete model.

A more basic theoretical problem is the buildup of a macroscopic dipole in initially uncorrelated excited molecules, or, semiclassically, what is ϕ_o? This is a fully quantum problem; Rehler and Eberly's prescription, which amounts to $\phi_o = \sqrt{4/\mu N}$, does not seem to work for our situation indicating that relaxation and dephasing processes may have to be taken into account. Perhaps at this time, ϕ_o is best considered a free parameter.

The scaling behavior of the superradiant emission depends on geometry: if the system is radiating as a disk, the pulse intensity will increase as the square of the length but linearly with cross-sectional area, and vice versa if the needle configuration applies. In both cases, the intensity remains proportional to the square of the pressure. In addition, for the disk, t_o and Δt will vary little with area (only logarithmically through ϕ_o), but linearly with inverse length; again, the behavior should be opposite for the needle. A possibility of a dynamic lensing effect in the medium and subsequent FIR guided wave propagation could significantly affect the behavior in the needle limit, since the diffractive losses might be reduced.

In summary, the FIR seems to be a worthwhile spectral region in which to study superradiance. The CH_3F experiments have added the new regime of homogeneous broadening to the Doppler-broadened domain of the HF experiments. One possibility for future experiments involves scaling behavior, particularly in relation to swept-gain superradiance: a steady-state pulse which builds up in a longer cell will have a width small compared to T_2, a delay also short compared to T_2, and a correspondingly large intensity.[9,27] Another possibility is the measurement of gain in the lethargic regions.[28] We are anticipating a major contribution from experiments in FIR superradiance.

ACKNOWLEDGEMENTS

The authors wish to express their appreciation to C. M. Bowden, chairman of the Cooperative Effects Symposium, and to acknowledge many stimulating conversations with him, R. Bonifacio, R. K. Bullough, J. H. Eberly, M. S. Feld, F. A. Hopf, and P. Meystre.

References

(1) R. H. Dicke, Phys. Rev., 93, 99 (1954).

(2) F. T. Arecchi and E. Courtens, Phys. Rev. A., 2, 1730 (1970).

(3) N. E. Rehler and J. H. Eberly, Phys. Rev. A., 3, 1735 (1971).

(4) C. R. Stroud, Jr., J. H. Eberly, W. L. Lama, and L. Mandel, Phys. Rev. A., 5, 1094 (1972).

(5) R. Jodoin and L. Mandel, Phys. Rev., 9, 873 (1974).

(6) R. Friedberg and S. R. Hartmann, Phys. Rev. A., 10, 1728 (1974).

(7) R. Bonifacio and L. A. Lugiato, Phys. Rev. A., 11, 1507 (1975).

(8) J. C. MacGillivray and M. S. Feld, Phys. Rev. A., 14, 1169 (1976).

(9) R. Bonifacio, F. A. Hopf, P. Meystre, and M. O. Scully, Phys. Rev. A., 12, 2568 (1975).

(10) N. Skribanowitz, I. P. Herman, J. C. MacGillivray, and M. S. Feld, Phys. Rev. Lett., 30, 309 (1973).

(11) T. Y. Chang and T. J. Bridges, Opt. Commun., 1, 423 (1970).

(12) T. A. DeTemple, T. K. Plant, and P. D. Coleman, Appl. Phys. Lett., 22, 644 (1973); T. K. Plant, L. A. Newman, E. J. Danielewicz, T. A. DeTemple, and P. D. Coleman, IEEE Trans. Microwave Theory and Tech., MTT-22, 988 (1974).

(13) T. K. Plant and T. A. DeTemple, J. Appl. Phys., 47, 3042 (1976).

(14) A. T. Rosenberger, S. J. Petuchowski, and T. A. DeTemple, Conference Digest, Second International Conference and Winter School on Submillimeter Waves and Their Applications, San Juan, Dec. 6-11, 1976, and to be published.

(15) S. J. Petuchowski, A. T. Rosenberger, and T. A. DeTemple, to be published in IEEE J. Quantum Electron., June, 1977.

(16) T. A. DeTemple and E. J. Danielewicz, IEEE J. Quantum Electron., QE-12, 40 (1976); R. L. Shoemaker, S. Stenholm, and R. G. Brewer, Phys. Rev. A., 10, 2037 (1974).

(17) R. I. Rudko, IEEE J. Quantum Electron., QE-11, 540 (1975).

(18) E. Yablonovitch, Phys. Rev. A., 10, 1888 (1974); H. S. Kwok and E. Yablonovitch, Appl. Phys. Lett., 27, 583 (1975).

(19) Although the FIR is not completely polarized, the other component of polarization will be trapped by multiple reflection and absorption, and will not continue to feedback. The influence of introduced feedback has been observed, allowing us to claim that the feedback in our experiment was negligible.

(20) P. Norton, J. Appl. Phys., 47, 308 (1976); Phys. Rev. Lett., 37, 164 (1976).

(21) This density corresponds to an excitation of 14.2% of the available ground state population. This is only slightly less than the theoretical maximum of 16.7% for a linearly polarized pump.

(22) This has been discussed by E. Ressayre and A. Tallet, Phys. Rev. Lett., 37, 424 (1976).

(23) R. Friedberg and B. Coffey, Phys. Rev. A., 13, 1645 (1976).

(24) R. Bonifacio, L. A. Lugiato, and A. Airoldi Crescentini, Phys. Rev. A., 13, 1648 (1976).

(25) R_e will be a factor of 10 larger than its normal thermal value due to the population trapping effect discussed earlier.

(26) See Ref. 8 and R. Saunders, S. S. Hassan, and R. K. Bullough, J. Phys. A., 9, 1725 (1976); see also the articles by Bullough and Feld in this volume.

(27) In these experiments, only 5% of the initially excited molecules contribute to the pulse due to the fact that the pulse delays are $\sim 3\ T_1$. However, because of the geometric scaling behavior of T_s, under the appropriate conditions damping becomes sufficiently small to allow nearly full extraction of the initial excitation energy even in the non-steady state regime; see Eq. (19) and the discussion following.

(28) F. Hopf, P. Meystre, and D. W. McLaughlin, Phys. Rev. A., 13, 777 (1976).

COOPERATIVE EFFECTS IN ATOMIC METAL VAPORS*

A. Flusberg, T. Mossberg, and S. R. Hartmann

Columbia Radiation Laboratory, Department of Physics

Columbia University, New York, New York 10027

I. INTRODUCTION

In this article we present a summary of some of the preliminary work we have done at Columbia University on the observation of Dicke superradiance in atomic metal vapors. In addition we discuss optical three-wave mixing, a novel effect which is often observed to occur simultaneously with the Dicke superradiance.

This is primarily an experimental paper, and we make no attempt to define and delineate the various regimes of Dicke superradiance and/or superfluorescence. We leave this somewhat controversial topic to the other articles in this collection. Instead, we take a rather pragmatic approach: we consider Dicke superradiance (DS) to be the emission of a coherent burst of resonance radiation from an inverted system well after the external excitation of the inversion has passed. This is, of course, only one of the identifying features of DS. Our preliminary experimental goal has been to show that this "delayed burst" effect is fairly easy to obtain in extended media in the laboratory. That this is not obvious is clear from the small number of experimental studies of DS which have appeared in the literature until now.[1-5]

To see why only very weak pumping is necessary to obtain DS, consider a three-level atomic system consisting of a ground state

*This work is supported by the Joint Services Electronics Program (U. S. Army, U. S. Navy, and U. S. Air Force) under Contract DAAB07-74-C-0341.

$|g\rangle$, a "lower" state $|1\rangle$ and an upper state $|u\rangle$ of energies $W_g < W_1 < W_u$, respectively. DS is to be observed on the $|u\rangle \to |1\rangle$ electric-dipole-allowed transition, and for the sake of argument we suppose that the population n_u in state $|u\rangle$ is to be provided by $|g\rangle \to |u\rangle$ electric-quadrupole absorption, a weak pumping mechanism. In our experimental situation the thermal population n_1 of state $|1\rangle$ is negligible. For a Doppler-broadened gas, the gain αL at the $|u\rangle \to |1\rangle$ transition wavelength λ_1 is given by[6]

$$\alpha L \cong \frac{3}{4\pi} \lambda_1^2 \frac{T_2^*}{T_1} n_u L \quad , \tag{1}$$

where T_2^* is the Doppler dephasing time and T_1 the radiative lifetime of the $|u\rangle \to |1\rangle$ transition; L is the length of the excited part of the medium. By assumption, the atoms are in state $|u\rangle$ as a result of electric-quadrupole absorption of a short, broadband "pump" light pulse resonant with the $|g\rangle \to |u\rangle$ transition wavelength λ_2. The pump linewidth is assumed to be much larger than the atomic absorption linewidth. Thus the atomic density n_u in state $|u\rangle$ immediately after the passage of the pump pulse is given by[7]

$$n_u \cong n_g \lambda_2^2 \frac{\tau}{T_Q} \frac{\Phi}{A} \quad , \tag{2}$$

where T_Q is the radiative lifetime of the E2 (electric quadrupole) transition; τ, A and Φ are respectively the inverse linewidth, the cross-sectional area, and the number of photons in the pump pulse. Combining Eqs. (1) and (2) with the diffraction-limit condition $A/L \cong \lambda_2$ (i.e., assuming the confocal parameter of the pump beam is smaller than the length of the medium), we obtain

$$\alpha L \cong \frac{3}{4\pi} \lambda_1^2 \lambda_2 \frac{T_2^* \tau}{T_1 T_Q} \Phi \, n_g \quad . \tag{3}$$

In a typical situation, $\lambda_1 = 1.5 \times 10^{-4}$ cm, $\lambda_2 = 4 \times 10^{-5}$ cm, $T_2^* = 1$ nsec, $\tau = 0.1$ nsec, $T_1 = 100$ nsec and $T_Q = 0.1$ sec. Thus $\alpha L \cong 2 \times 10^{-24}\, \Phi\, n_g$. Now, a necessary condition for Dicke superradiance is $\alpha L \gg 1$.[3,6] This condition will be satisfied if

$\Phi n_g \gg 5 \times 10^{23}$ cm^{-3}. For $n_g = 10^{13}$ atoms/cm^3 the requisite number of photons in the pump pulse is $\Phi \gg 5 \times 10^{10}$, corresponding to a pulse energy of 0.02 μJ. A pulse of this energy (or as much as 3 orders of magnitude greater) is easily obtained from a dye laser pumped by a nitrogen laser. It is thus clear that optical electric-quadrupole pumping is strong enough to produce a sizeable value of αL. Similar considerations show that two-photon absorption far from resonance and stimulated Raman scattering are both adequate pumping mechanisms to obtain high values of αL in atomic media.

As the title indicates, we have restricted our study to atomic metal vapors. In particular, we have observed near-infrared DS in atomic thallium, rubidium, cesium, and sodium. These are, in fact, the only atomic media in which we have looked for DS. The wavelength range of the observed DS is 0.8 - 1.5 μm; the lower limit of 0.8 μm is particularly exciting, since it brings the observation of the DS phenomena into a wavelength range in which sensitive photomultiplier tubes are available for detection. The population inversion which initiates the DS has been achieved by several different optical-pumping schemes: directly from the ground state by electric-quadrupole absorption; by stimulated Raman scattering; by two-photon absorption; by electric-dipole absorption; and also (in thallium) by a more complicated process to be described below. We have not studied the DS phenomena in any great detail; we have chosen, instead, to focus our attention on the novel aspects of three-wave mixing, which we often found to be present at the same time as DS.

Three-wave mixing in an atomic vapor is forbidden in the electric-dipole approximation, and it was only recently that it was first observed (in Na vapor) by Bethune, Smith and Shen.[8] Here again we have found several instances of three-wave mixing (both sum and difference-frequency generation, SFG and DFG, respectively) in all the metal vapors we have studied. Of particular interest is the dramatic interference effect we have observed in some cases upon the application of a small magnetic field.

The unusual properties of three-wave mixing may be understood with the aid of the following argument.[9] Suppose two incident collinearly propagating photons are absorbed by an atom. (The photons may or may not have equal energy; alternatively, one may be absorbed and the other emitted). If we quantize along the direction of propagation, each photon carries ±1 units of angular momentum. Therefore the atom must absorb either 0 or ±2 units of angular momentum, and it follows that the atom cannot emit a third photon along the common direction of propagation while simultaneously returning to its original state. Thus conservation of the propagation-direction component of angular momentum rules out collinear three-wave mixing even when the electric-dipole approximation is no longer made, i.e., when higher multipole moments, such as magnetic dipole (M1) and electric quadrupole (E2) are taken into account. Three-wave mixing can be observed, however, by phase matching the three waves for noncollinear propagation,[8] or by applying an external magnetic field transverse to the direction of propagation.[10] Such a magnetic field transfers some angular momentum to the atom during the absorption-emission process and thus allows the three-wave mixing to occur. Similarly, noncollinear excitation provides a nonzero probability amplitude that one of the absorbed photons has a zero component of angular momentum along the direction of propagation of the generated third wave. As we will show below, the presence of both noncollinear excitation and an externally applied magnetic field produces an interesting interference effect, for which the intensity of the generated wave is sensitive to a reversal in direction of the magnetic field.

In the remainder of this paper we will discuss our experimental results (both DS and three-wave mixing) for each atomic element separately.

II. THALLIUM

This study began in the late spring of 1976, when we began to operate our home-made nitrogen laser and nitrogen-laser-pumped dye lasers. The N_2 laser produces 12 nsec-long 3371 Å pulses of peak power 100-200 kW at 5 Hz. Each of the two dye lasers (of the Hänsch design[11]) produces 7-nsec-long pulses with linewidths of $\cong 0.03\ cm^{-1}$ when an intra-cavity etalon is used and $\cong 0.2$-$0.4\ cm^{-1}$ without the etalon. The dye laser pulses have peak powers between 1-15 kW depending on whether the nitrogen laser energy is used to pump only one or both of the dye lasers and whether the intra-cavity etalons are used.

The first material with which we worked was atomic thallium, which was contained in a 15-cm-long fused-silica cell heated to 850°C (corresponding to a Tl number density n of about 8×10^{15} cm^{-3}). The rather unusual pumping scheme for the $7^2S_{1/2}$-$7^2P_{1/2}$ population inversion in thallium is shown in Fig. 1. A single dye laser pulse of wavelength 379.1 nm is focused by a 40-cm focal-length lens into the center of the thallium cell. It produces stimulated Raman scattering (Stokes) of wavelength 538.2 nm, and thereby creates a population in the $6^2P_{3/2}$ metastable level. (The stimulated Raman scattering is enhanced by the

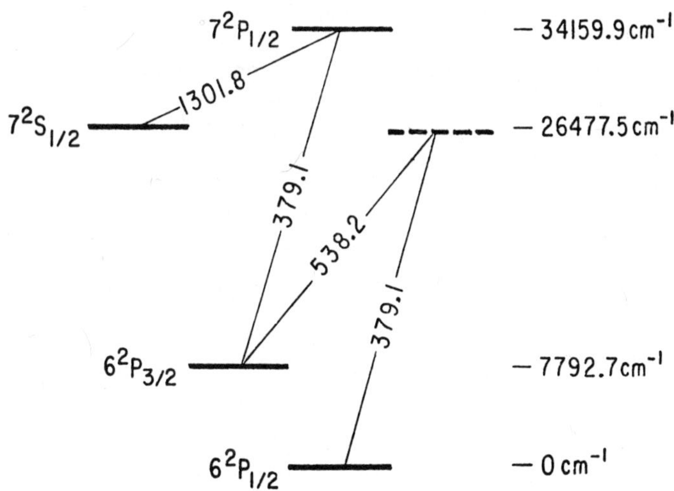

Figure 1. Energy level diagram of thallium relevant to Dicke-superradiance experiment. Indicated transition wavelengths are in nm. The laser at 379.1 nm is at exact resonance with the $7^2P_{1/2}$-$6^2P_{3/2}$ transition frequency. Dotted line indicates virtual level 99.2 cm^{-1} below $7^2S_{1/2}$ level. (Reprinted from Ref. 5).

near-resonance with the $6^2P_{1/2}$-$7^2S_{1/2}$ transition.) Electric-quadrupole absorption of the incident laser by the metastable $6^2P_{3/2}$ population produces a $7^2S_{1/2}$-$7^2P_{1/2}$ inversion. A DS burst is observed at 1301 nm. The excitation of the $7^2P_{1/2}$ population may, of course, be thought of as a single-step stimulated three-photon process. The DS pulse disappears if the exciting laser wavelength is detuned by as little as 0.01 nm.

Oscilloscope traces of the 1301-nm DS pulse are shown in Fig. 2. The DS pulse is detected in the forward direction (that of the incident laser beam) on a Ge avalanche photodiode (risetime < 1 nsec). Efficient detection is obtained by imaging the focal region in the cell onto the photodiode surface through a 1.2-μm cut-on filter (a silicon disc). Each trace of Fig. 2 represents a single sweep of the Tektronix 7904 oscilloscope using two 7A19 vertical-amplifiers (nominal risetime: 0.7 nsec) to simultaneously display the Ge detector signal as well as a cable-delayed signal from a Si photodiode (Monsanto MD-2) which detects the incident laser pulse. In each trace the laser pulse (the second peak) serves as a reference-time marker. For Fig. 2a the silicon filter has been replaced by several neutral-density filters and a 535-nm interference filter, so that the first pulse represents the Stokes radiation (at 538 nm) detected by the Ge photodiode. The jitter of the appearance of the Stokes pulse is found to be less than 1 nsec. In Figs. 2b, 2c, and 2d the infrared pulse appears delayed by 3, 4, and 9 nsec, respectively, with respect to the peak of the Stokes pulse. The long delays and short pulsewidths unmistakeably identify the process as Dicke superradiance.

At the same time as we observe DS, we also observe three-wave mixing in the same system. The formation of stimulated Stokes scattering during the laser pulse creates a coherent, macroscopic superposition of the $6^2P_{1/2}$ and $6^2P_{3/2}$ atomic states. In the presence of an applied magnetic field transverse to the direction of propagation, this superposition may radiate a difference frequency wave at the $6^2P_{1/2}$-$6^2P_{3/2}$ transition wavelength of 1283 nm via either the M1 or E2 moment of the $6^2P_{1/2}$-$6^2P_{3/2}$ superposition.[10] It may be shown that for small magnetic fields H, the difference-frequency signal should vary as H^2. The criterion of "smallness" is $\mu_B H \ll h\Gamma$, where μ_B is a Bohr magneton, $2\pi\hbar$ is Planck's constant, and Γ is the inverse dephasing time of the macroscopic superposition, which is given by the Doppler width in the low collision-rate limit. That the difference frequency generation (DFG) signal does indeed vary as H^2 is shown in Fig. 3. As expected for wave-mixing, which in a sense may be thought of as "weak superradiance",[3] the signal also varies as the square of the atomic number density n (Fig. 4). We note that unlike the DS, the DFG occurs even when the laser wavelength is tuned well away from 379.1 nm. The DFG may also be produced in the absence of stimulated Raman scattering if two independent laser beams, adjusted so that their frequency difference coincides with the $6^2P_{1/2}$-$6^2P_{3/2}$ splitting of 7793 cm^{-1}, are incident on the Tl vapor. It is clear that the DS and DFG are unrelated effects; however, the near coincidence of their wavelengths (1.30 versus 1.28 μm), and the fact that they may be observed simultaneously in the same system, makes for a strong experimental relationship between the two.

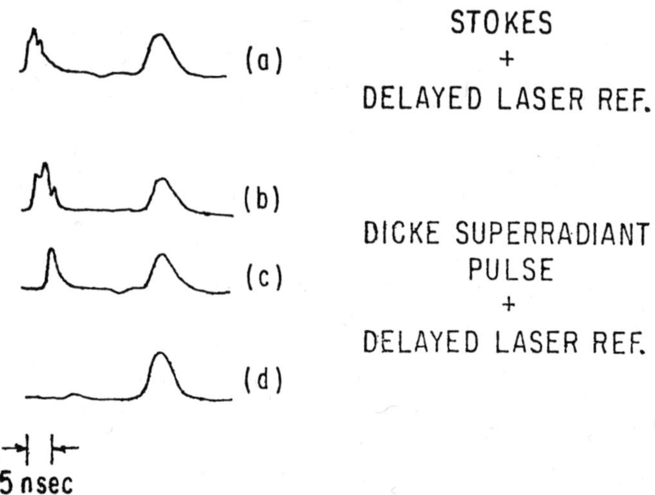

Figure 2. Tl $7^2P_{1/2} \rightarrow 7^2S_{1/2}$ superradiant pulse evolution and reference markers. Pulse on right side represents laser intensity and is electronically delayed by a constant amount ∿ 25 nsec. (a) Pulse on left represents Stokes emission at 538.2 nm. Tl cell temperature is T_c = 835°C, corresponding to a Tl number density n_c of 6×10^{15} cm^{-3}. (b) Pulse on left represents 1301-nm Dicke superradiant pulse under conditions of (a). (c) and (d) Pulse on left represents Dicke superradiant pulse for T_c = 770°C corresponding to $n_c = 2 \times 10^{15}$ cm^{-3}. Note correlation between superradiant delay and superradiant pulse intensity. We believe that the lack of correlation between superradiant delay and laser intensity is due to frequency jitter in the laser pulse. (Reprinted from Ref. 5.)

We have also observed DS when using laser-pump two-photon absorption from the ground state to create inversions on the $7^2S_{1/2}$-$7^2P_{1/2}$ and $7^2S_{1/2}$-$7^2P_{3/2}$ transitions (Fig. 5). Note in particular that in this experiment the laser frequency is

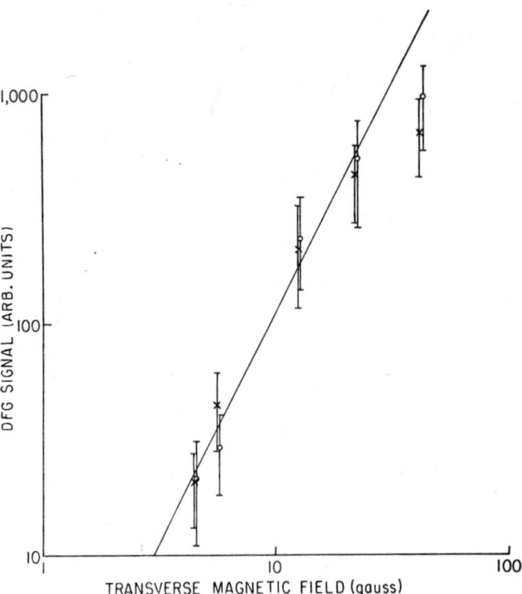

Figure 3. Thallium $6^2P_{1/2}$-$6^2P_{3/2}$: 1.28-μm DFG signal versus transverse magnetic field. The atomic number density is ≃ 5 × 10^{14} cm^{-3}, and the two collinear laser pulses incident on the cell have frequencies 26523 cm^{-1} and 18730 cm^{-1}, respectively. The data points denoted by circles correspond to one direction of the magnetic field, while those denoted by crosses correspond to the opposite direction. The straight line is drawn with a slope of 2 on the log-log plot. (Reprinted from Ref. 10.)

∿ 8000 cm^{-1} off resonance from the nearest ground-intermediate state splitting! This experiment truly demonstrates how easy it is to obtain a rapid inversion in an atomic system. The tie with three-wave mixing is again demonstrated in this experiment: we observe second-harmonic generation on both the $6^2P_{1/2}$-$7^2P_{3/2}$ and $6^2P_{1/2}$-$7^2P_{1/2}$ transitions. In the case of the former, the generation occurs as a result of the E2 moment of the superposition, and, as expected, the second-harmonic intensity is proportional to H^2. The $6^2P_{1/2}$-$7^2P_{1/2}$ second-harmonic generation is, however, independent of H. Indeed, this superposition has no E2 moment and a negligible M1 moment; the source of its associated second-harmonic generation remains to be explained.

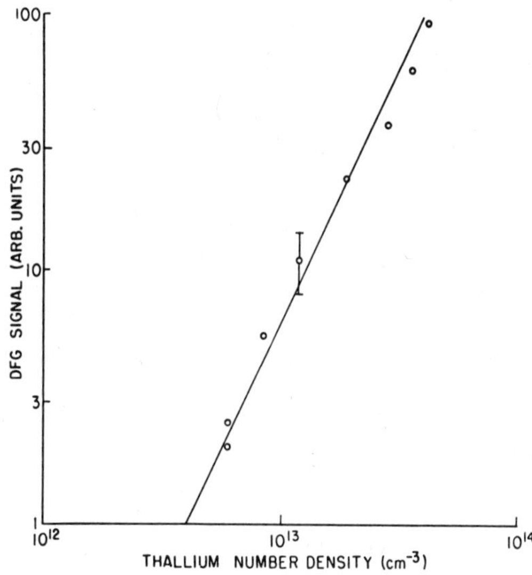

Figure 4. Thallium $6^2P_{1/2}$-$6^2P_{3/2}$: 1.28-μm difference-frequency generation intensity versus Tl atomic number density n. A transverse magnetic field of 90 G is applied throughout, and two collinear laser pulses, of frequencies 26484 cm^{-1} and 18691 cm^{-1}, respectively, are incident on the Tl cell. The straight line is drawn with a slope of 2 on the log-log plot (Reprinted from Ref. 10.)

III. CESIUM

We next turn our attention to DS in atomic cesium, for which we have used three optical-pumping schemes: E2 absorption (Fig. 6), stimulated Raman scattering (SRS), and resonant absorption (Fig. 7). For the first of these, pumping on the $6^2S_{1/2}$-$6^2D_{3/2}$ or $6^2S_{1/2}$-$6^2D_{5/2}$ transition produced pulses delayed by as much as 10 nsec at 1.36 μm ($6^2P_{1/2}$-$7^2S_{1/2}$) and 1.47 μm ($6^2P_{3/2}$-$7^2S_{1/2}$). Unfortunately, the absence of a suitable detector in our laboratory made it impossible for us to observe the upper legs of the cascade from the 6^2D states down to the $7^2S_{1/2}$ state (wavelengths: 16 μm and 3.1 μm). Thus, we do not know how much of the observed delay may be attributed to DS at these ir transitions.

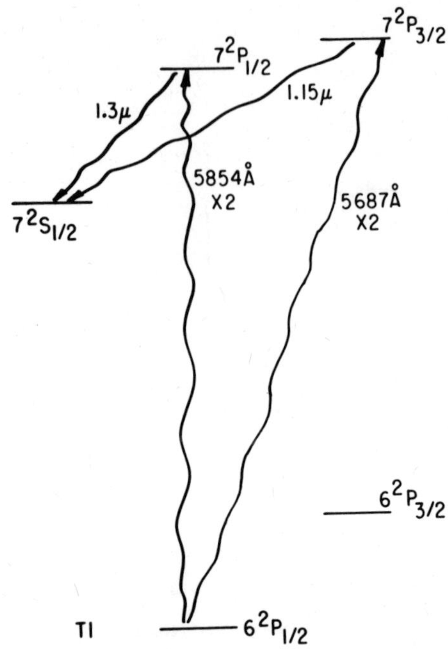

Figure 5. Thallium energy-level diagram for two-photon-absorption-pumped DS. When the incident laser frequency is half of the $6^2P_{1/2}-7^2P_{3/2}$ splitting (35161 cm^{-1}), we observe DS at 1.15 μm ($7^2S_{1/2}-7^2P_{3/2}$) and second-harmonic generation at 0.2843 μm. Similarly, when the laser frequency is half of the $6^2P_{1/2}-7^2P_{1/2}$ splitting (34160 cm^{-1}), we observe DS at 1.30 μm and second-harmonic generation at 0.2927 μm.

When we pumped with 455.5-nm-wavelength light (on resonance with the $6^2S_{1/2}-7^2P_{3/2}$ transition) we also observed delayed bursts at the various transition wavelengths between 1.34 and 1.47 μm, indicated in Fig. 7. The presence of a collimated, coherent burst of light at 5393.5 Å, corresponding to the forbidden $6^2S_{1/2}-7^2S_{1/2}$ transition, is another example of a three-wave mixing process. As in the case of the $6^2P_{1/2}-7^2P_{1/2}$ radiation observed in Tl, the 5393.5 Å light was emitted in the absence of a magnetic field. We have not yet clarified the mechanism of this radiation.

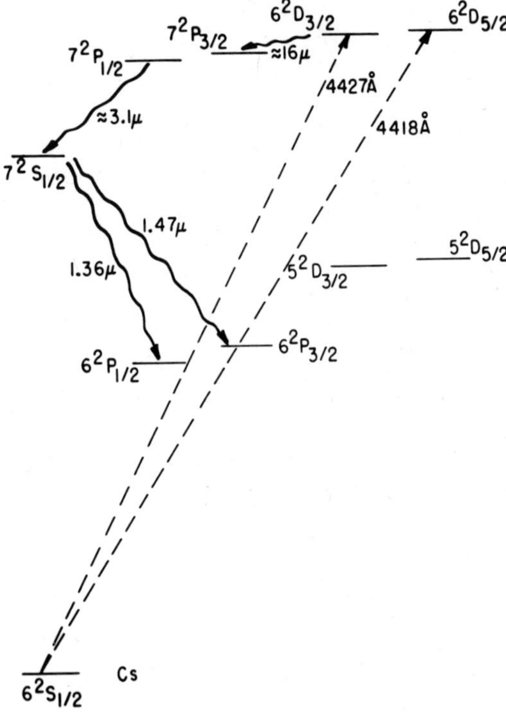

Figure 6. Cesium energy-level diagram for E2-absorption-pumped DS. Pump wavelengths (vacuum values) are denoted by dashed lines. Approximate signs (≈) denote wavelengths of pulses whose presence we deduce but do not detect (see text).

Figure 8 shows some examples of DS pulse-shapes in Cs at a Cs vapor pressure of about 10^{-2} torr and using a laser pump at 4555 Å. The 1.36-μm pulse associated with the $7^2P_{3/2}$-$5^2D_{5/2}$ transition is delayed by ≅ 10 nsec from the peak of the exciting laser pulse. Evidently there is some ringing in the DS pulse, but it would be premature to consider it true Dicke superradiant ringing in view of the many possible hyperfine-level beats in the $7^2P_{3/2}$-$5^2D_{5/2}$ transition (Fig. 7).

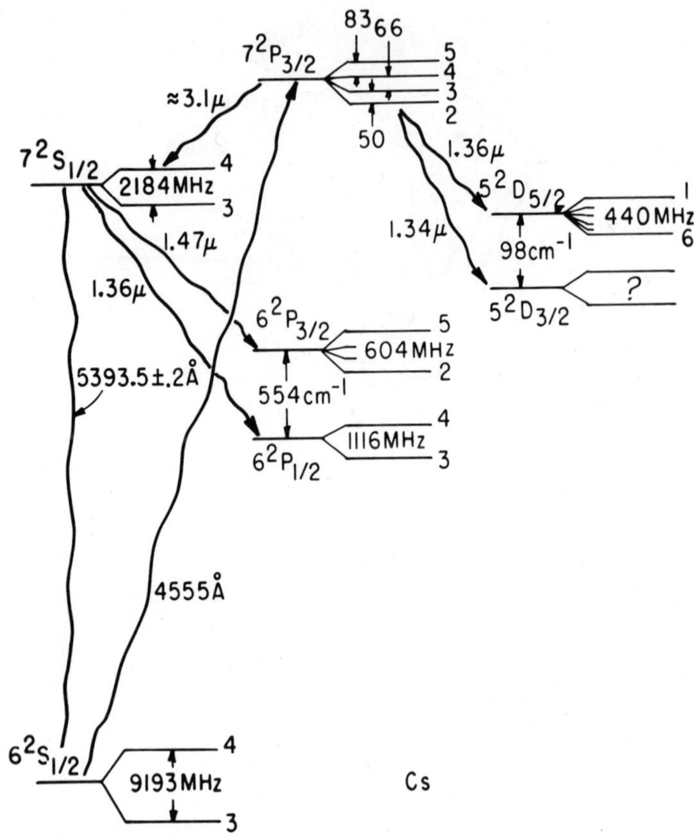

Figure 7. Cs energy-level diagram for DS in which the population inversion is brought about by either E1 absorption (pump at 4555 Å) or stimulated Raman scattering (pump near 4555 Å). The hyperfine splittings are all in MHz. Approximate signs (≈) denote wavelengths of pulses whose presence we deduce but are unable to detect (see text).

IV. RUBIDIUM

In rubidium (Rb^{87}, 125°C, ≈ 10^{-3} torr vapor pressure), we have used E2-absorption pumping (5166 Å vacuum wavelength) to produce an inversion between the 4^2D and $5^2P_{1/2,3/2}$ states. We observe DS on the 1.48 μm and 1.53 μm transitions. The relevant energy levels of this atomic system are depicted in Fig. 9. Again, we have explicitly labelled the hyperfine levels. The observed DS

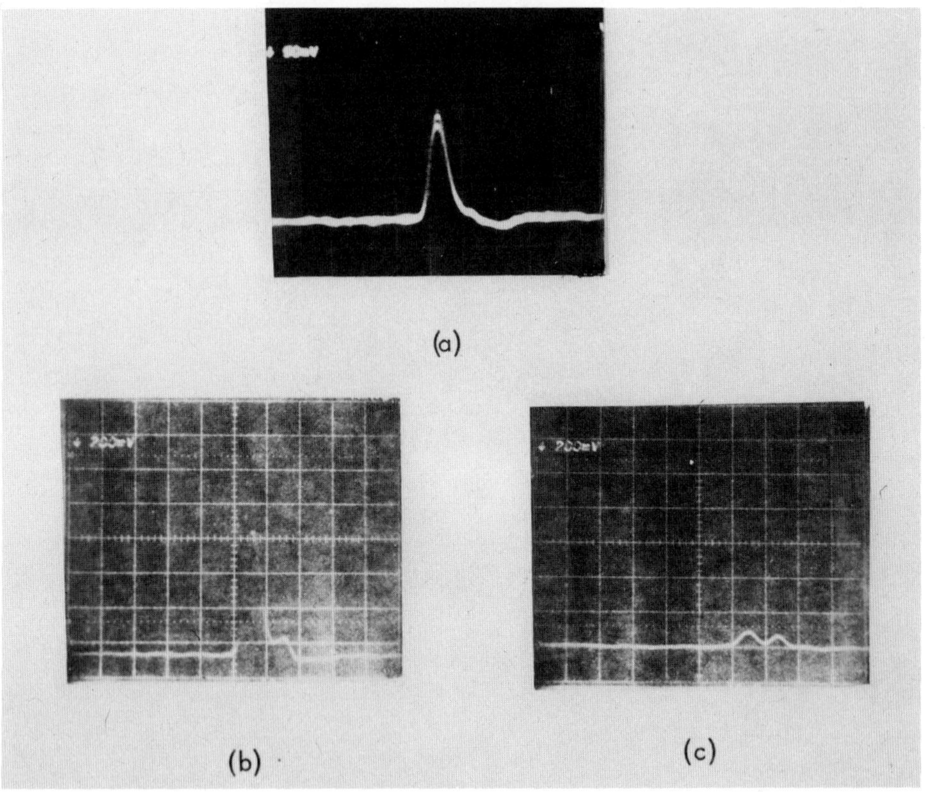

Figure 8. DS pulse shapes in Cs ($\simeq 10^{-2}$ torr vapor pressure). Time scale: 10 nsec/div. (a) Exciting laser pulse (several shots superimposed). Some detector overshoot is apparent. (b) Single DS pulse, $7^2S_{1/2} \rightarrow 6^2P_{1/2}$ transition. (c) Single DS pulse, $7^2P_{e/2} \rightarrow 5^2D_{5/2}$ transition.

pulses are delayed from the exciting pulses by as much as 12-15 nsec. Modulation is observed on either of the $4^2D \rightarrow 5^2P_{1/2}$ transitions. For this experiment an intra-cavity etalon narrowed the dye-laser linewidth to about 700 MHz; thus only one of the 4^2D states was populated by the 5166 Å absorption. The modulation frequency of about 900 MHz is suggestive of the hyperfine splitting of 818 MHz in the $5^2P_{1/2}$ state. It may thus be an example of a quantum-beat effect,[12] rather than superradiant ringing.

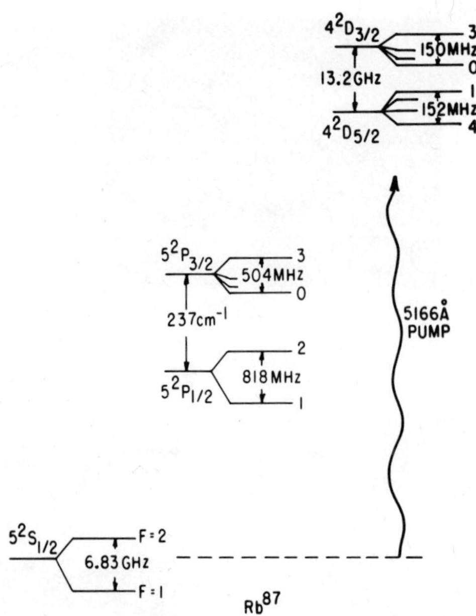

Figure 9. Rb^{87} energy-level diagram for E2-absorption-pumped DS (see text).

As shown in Fig. 10, the rubidium experiment was set up in a way that allowed the simultaneous observation of both forward and backward-emitted DS on the same detector. Our preliminary observations indicated that the essential features of both were the same. In an extension of this experiment, it should be possible to examine the correlation between the forward and backward DS.

SRS-induced three-wave mixing ($5^2S_{1/2}-6^2S_{1/2}$ and $5^2S_{1/2}-4^2D$) was observed in atomic Rb. The observation was made when the Rb cell temperature was 300°C (corresponding to $\cong 10^{16}$ atoms/cm^3) and the pump laser was about 1 Å off resonance from the $5^2S_{1/2}-6^2P_{3/2}$ absorption wavelength (Fig. 11). The emission wavelengths were measured to be those expected for the $5^2S_{1/2}-6^2S_{1/2}$ and $5^2S_{1/2}-4^2D$ atomic transitions[13] to within an uncertainty of about 0.5 Å. As in the case of Cs ($6^2S_{1/2}-7^2S_{1/2}$) and Tl ($6^2P_{1/2}-7^2P_{1/2}$), the collimated 4966-Å emission intensity was found to be independent of magnetic field. The 5166-Å radiation (E2 allowed) was not checked for magnetic field dependence.

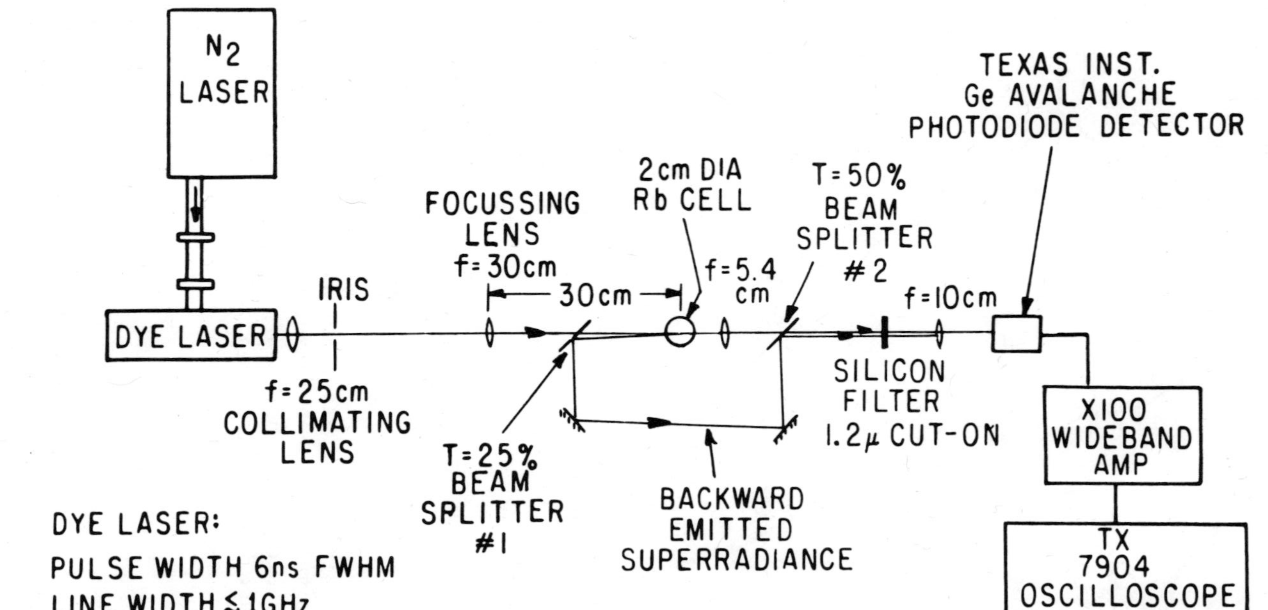

Figure 10. Experimental apparatus for Rb^{87} E2-absorption-pumped DS. The 250-MHz bandwidth, 100X amplifier was removed for the observation of 900 MHz modulation (see text).

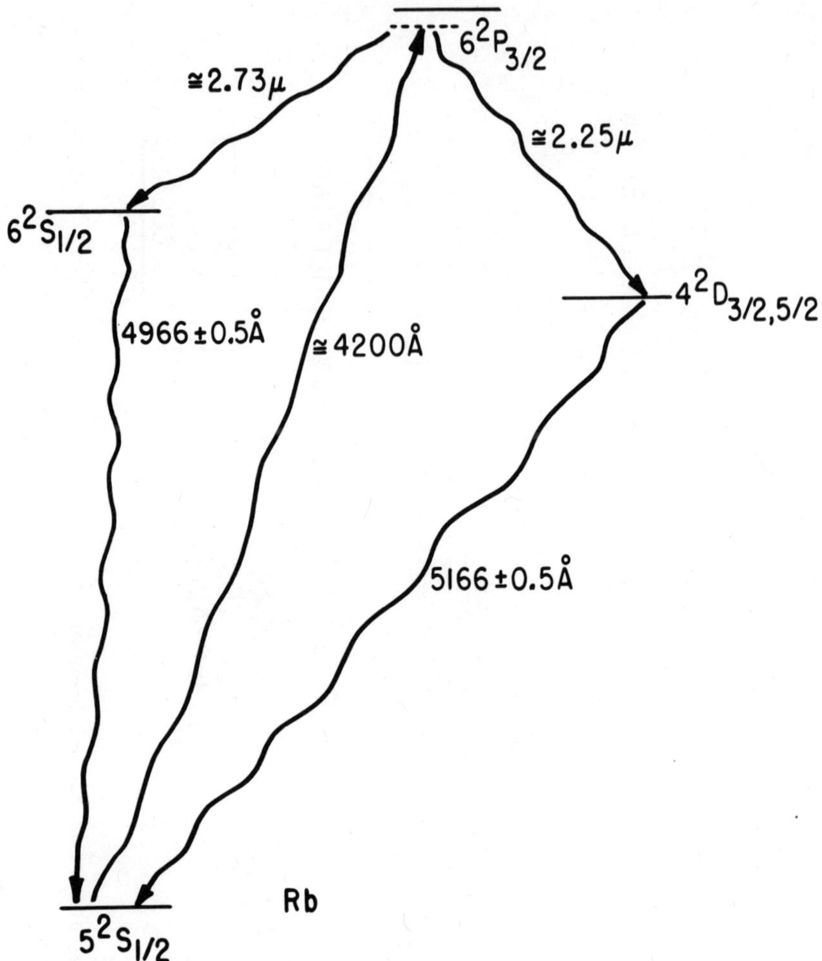

Figure 11. Energy-level diagram for three-wave mixing in atomic Rb. The vacuum value of each measured wavelength is indicated. Pulses whose wavelength values bear an approximate sign (≅) were presumed present, but not detected (see text).

V. SODIUM

The two-photon-absorption-pumped DS observed in atomic Na vapor (at a vapor pressure of $\simeq 10^{-4}$ torr) at 8183 Å ($3^2P_{1/2}$-3^2D)

and 8195 Å ($3^2P_{3/2}$-3^2D) (Fig. 12) is perhaps the most exciting, since it is "visible" to many photomultiplier tubes. Thus it should now be possible to study this emission from the fluorescence regime, through the weak-superradiance regime, all the way to the regime of strong superradiance. The short wavelength of this transition implies a proportionally shorter value of T_2^*, the Doppler dephasing time. Thus one expects a somewhat shorter delay until the onset of superradiance, and indeed the largest delay observed was about 5 nsec. A study of this transition would clearly be more valuable if the laser pump pulse were much shorter than about 5 nsec, or if the laser linewidth were sufficiently narrow to artificially increase T_2^*.[2,3]

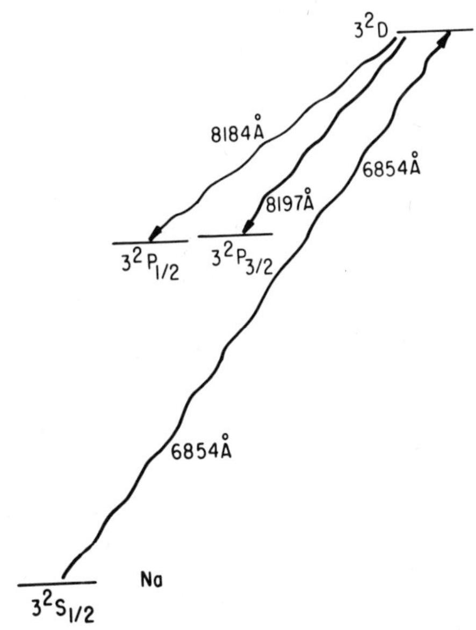

Figure 12. Energy-level diagram of 0.82-μm DS produced in Na by two-photon absorption from the ground state to the 3^2D states.

The last effect we discuss in this paper is SFG (three-wave mixing) on the $3^2S_{1/2}$-$4^2D_{5/2,3/2}$ transitions in Na vapor. We have investigated noncollinear electric-quadrupole SFG in the presence of a magnetic field.[14] What we have found seems rather astonishing at first glance: the intensity of the SFG wave changes by a factor as large as 10 under reversal of direction of a magnetic field (Fig. 13)!

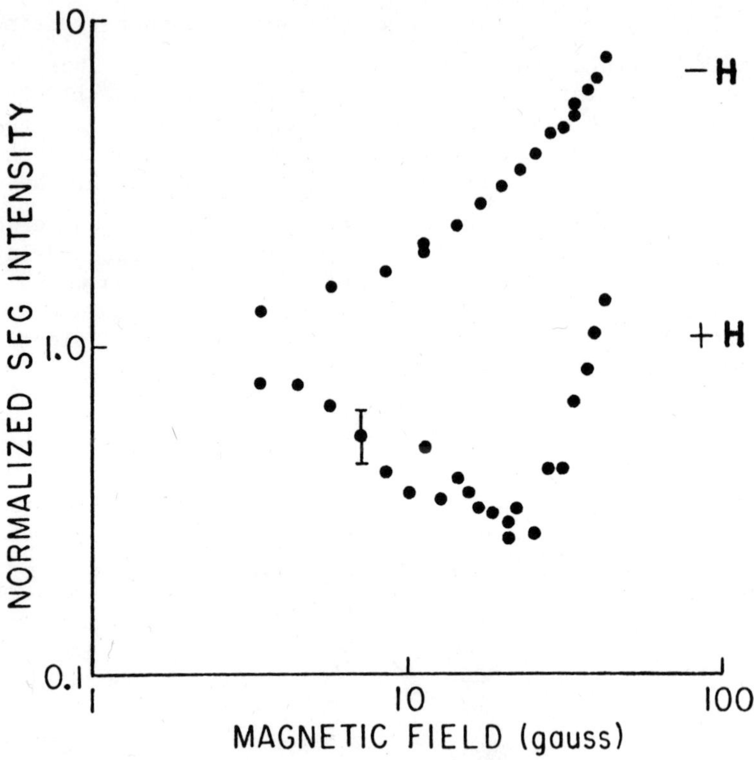

Figure 13. Intensity of noncollinear SFG in atomic sodium ($3^2S_{1/2}-4^2D_{3/2,5/2}$) as a function of magnetic field H. The magnetic field direction for the data labelled -H is reversed with respect to its direction for the data labelled +H. The vertical scale represents the H-dependent intensity I(H) normalized by $I_o = I(H = 0)$. For experimental details, see Ref. 14.

Our experimental procedure and apparatus are similar to that of Bethume, Smith, and Shen,[8] who first observed noncollinear SFG in atomic Na. Two dye lasers, pumped by the same N_2 lasers, are used. They are tuned near the 16956-cm^{-1} ($3^2S_{1/2}-3^2P_{1/2}$) and 17594-cm^{-1} ($3^2P_{1/2}-4^2D$) Na absorption lines, respectively. The laser frequencies are adjusted so that their sum $\omega_3 = \omega_1 + \omega_2$ overlaps the Doppler-broadened two-photon $3^2S_{1/2}$-4^2D absorption spectrum. The two noncollinear laser pulses are focused into the center of a cell containing Na vapor and argon buffer gas. The crossing angle θ ($\ll 1$) between the wave vectors (in the Na vapor) $\vec{k}_1$ and $\vec{k}_2$ of the incident laser beams is adjusted so that the phase-matching condition $\vec{k}_1 + \vec{k}_2 = \vec{k}_3$ is always satisfied, $\vec{k}_3$ being the wave vector of the SFG wave. This adjustment changes with Na density n, θ being proportional to $n^{1/2}$. The sum frequency wave at 289 nm is detected on a photomultiplier tube, and the signal is averaged over several hundred shots on a PDP8/e computer.

In the absence of a magnetic field only the noncollinearity of the two incident beams contributes to the macroscopic E2-moment density which radiates the sum-frequency wave. As shown in Fig. 14, the intensity of the SFG wave is then proportional to the square of the atomic Na density n. If the polarization of the incident laser beams are parallel, however, the noncollinear contribution nearly vanishes;[8] under these conditions, the application of a DC magnetic field H perpendicular to the laser polarization and to $\vec{k}_3$ produces an additional macroscopic E2-moment density, which for small H, is proportional to H. The SFG intensity radiated from it is proportional to $n^2H^2\ell^2$, where ℓ is the length of the pencil-shaped laser-excited volume of Na vapor. Under our experimental conditions $\ell \cong d/\theta$, where $d \cong 0.5$ mm is the diameter of the laser beam in the center of the cell. Since the crossing angle θ is proportional to $\sqrt{n}$, the magnetically induced SFG intensity is proportional to nH^2 (Figs. 14, 15).

If, on the other hand, the polarization of the incident lasers are perpendicular and the magnetic field is normal to the plane containing $\vec{k}_1$ and $\vec{k}_2$, both the noncollinear and the magnetically induced E2 moment density are present. The electric fields radiated from each interfere with one another; equivalently, we may say that the two E2-moment densities interfere. The resulting SFG intensity I may be written

$$I(H) = I_1 I_2 \times \{n^2 A + n^{3/2} HB + nH^2 C\} \qquad (4)$$

Here I_1 and I_2 are the intensities of the incident lasers. A, B, and C are constants which depend on the incident laser lineshapes,

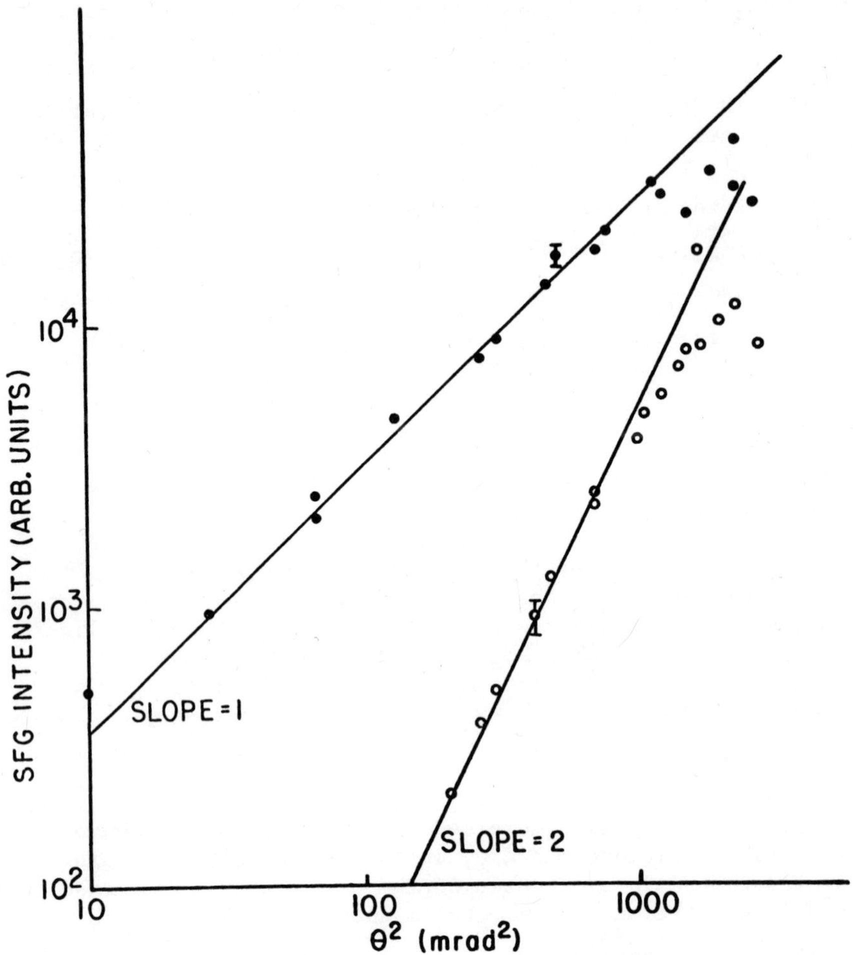

Figure 14. Na SFG intensity versus the square of the phase-matching angle θ. Open circles: magnetic field H = 0; laser polarizations normal to each other. Solid circles: H = 37 G, laser polarizations parallel to each other and normal to magnetic field direction.

the laser polarizations, and the atomic parameters (the two-photon-transition amplitude and its lineshape; the E2 moment; the M-1 moment matrix elements in the S and D states). The first term, which comes from the noncollinear effect only, has been studied by Shen et al.[8] The third term, which comes from the magnetic effect alone, is exemplified by the type of behavior shown in Fig. 15. Finally, the middle term, which comes from interference between the noncollinear and the magnetic effect, is present if the polarization of the noncollinearly induced SFG is not orthogonal to that of the magnetically induced SFG. It is just this middle term, which is sensitive to the sign of H, which is responsible for the unusual behavior shown in Fig. 13.

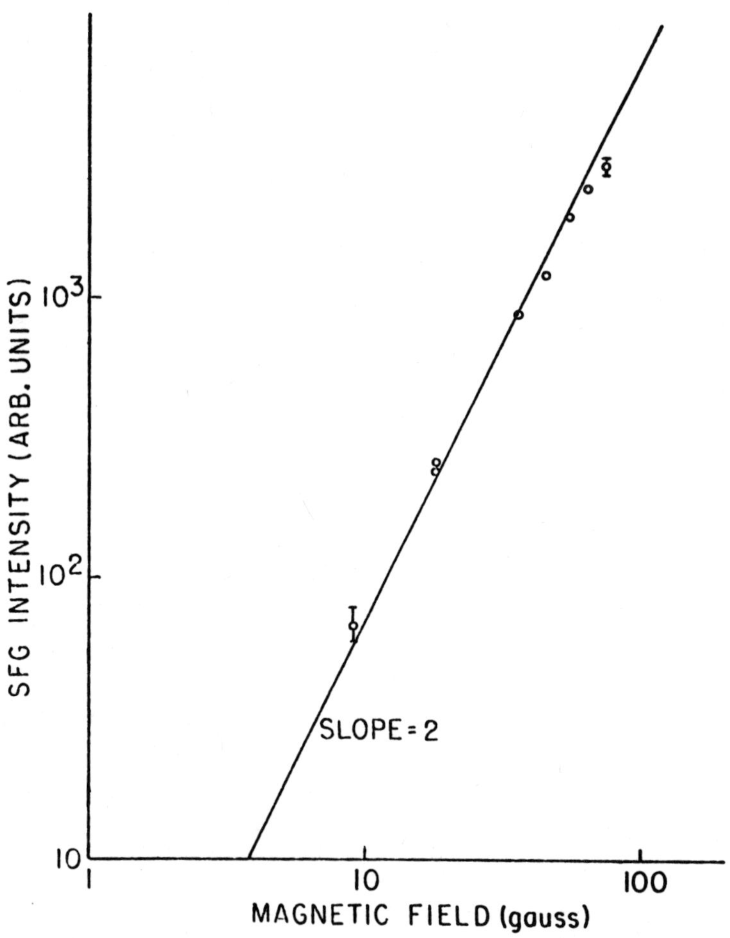

Figure 15. Na SFG intensity versus H. The laser polarizations are parallel and normal to the $\vec{k}_1$-$\vec{k}_2$ plane, while $\vec{H}$ is normal to the laser polarizations and lies in the $\vec{k}_1$-$\vec{k}_2$ plane.

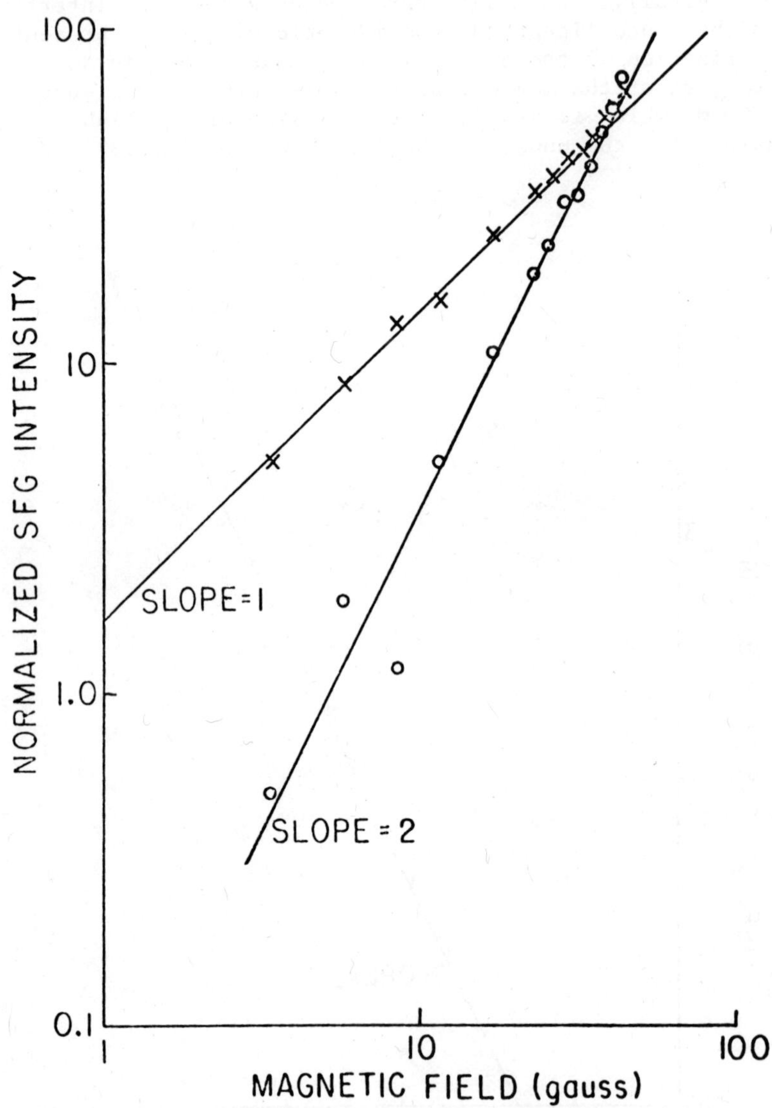

Figure 16. Difference and sum of plots of Fig. 13. Crosses; abscissa is $[I(+H) + I(-H) - 2I_o]/I_o$; circles; abscissa is $[I(+H) - I(-H)]/I_o$.

We note that Eq. (4) predicts that I(H) + I(-H) should be linear in H^2, whereas I(H) - I(-H) should be proportional to H. A plot of these quantities (taken from the data of Fig. 13) versus H is shown in Fig. 16. It can be seen that the data is in good agreement with Eq. (4).

VI. SUMMARY

To summarize: we have made preliminary observations of Dicke superradiance on many transitions in Tl, Cs, Rb, and Na in the wavelength range 0.8 - 1.5 μm. These observations confirm that Dicke superradiance is easily achieved in atomic vapors, even with very weak laser pumping schemes. At the same time, we have studied novel three-wave mixing effects in each of these atomic vapors. We have explained some of the three-wave mixing effects as resulting from either M1 or E2 coherent radiation in the presence of noncollinear excitation or an external magnetic field (or both). The mechanisms of others (e.g., Tl $7^2P_{1/2} \rightarrow 6^2P_{1/2}$) remain to be explained.

References

(1) N. Skribanowitz, I. P. Herman, J. C. MacGillivray, and M. S. Feld, Phys. Rev. Lett. 30, 309 (1973).

(2) Ping-Tong Ho, An Experimental Study of Superradiance, M. S. Thesis (M.I.T.), 1976 (unpublished).

(3) J. C. MacGillivray and M. S. Feld, Phys. Rev. A 14, 1169 (1976).

(4) M. Gross, C. Fabre, P. Pillet, and S. Haroche, Phys. Rev. Lett. 36, 1035 (1976).

(5) A. Flusberg, T. Mossberg, and S. R. Hartmann, Phys. Lett. 58A, 373 (1976).

(6) See, for example, R. Friedberg and S. R. Hartmann, Phys. Lett. 37A, 285 (1971) and Phys. Rev. A 13, 495 (1976).

(7) For an analogous expression for electric-dipole absorption, see E. U. Condon and G. H. Shortley, The Theory of Atomic Spectra, (Cambridge University Press), 1967.

(8) D. S. Bethune, R. W. Smith, and Y. R. Shen, Phys. Rev. Lett. 37, 431 (1976).

(9) T. Hänsch and P. Toschak, Z. Phys. 236, 373 (1970).

(10) A. Flusberg, T. Mossberg, and S. R. Hartmann, Phys. Rev. Lett. 38, 59 (1977).

(11) T. W. Hänsch, Applied Optics 11, 895 (1972).

(12) H. M. Gibbs, Q. H. F. Vrehen, and H. Hikspoor, Abstract, DEAP Meeting, (Lincoln, Nebraska) 6-8 Dec. 1976; also see the article by H. M. Gibbs and Q. H. F. Vrehen in this collection.

(13) Charlotte Moore, Atomic Energy Levels, NBS Circular 467 (Washington, D.C.), 1958.

(14) A. Flusberg, T. Mossberg, and S. R. Hartmann, Phys. Rev. Lett. 38, 694 (1977).

QUANTUM BEAT SUPERFLUORESCENCE IN Cs

Hyatt M. Gibbs*

Bell Laboratories
Murray Hill, NJ 07974

Abstract: Superfluorescence has been observed at 3 μm on the 7P to 7S transition in atomic cesium after 2 ns excitation of a 6S to 7P transition. It is shown that this transition is near ideal for studying superfluorescence. Under some conditions, however, the high degeneracy of Cs leads to excitation of several levels or superfluorescence of independent transitions. This gives rise to interference beats which have been observed with and without a magnetic field. The beat frequencies correspond not only to initial level splittings but also to combinations of initial and final level splittings. Superfluorescence beats are therefore basically different from single-atom quantum beats. Possibilities for spectroscopic applications of superfluorescence beats are also discussed.

I. INTRODUCTION

The superfluorescence experiments described here and in the next presentation were designed and performed with the collaboration of Dr. Quirin H. F. Vrehen at Philips Research Laboratories in Eindhoven, The Netherlands. The data were taken in his laboratory and mostly with his instrumentation. Many systematic studies were made and the final data were taken by him and Mr. H. M. J. Hikspoors after my return to Bell Laboratories. Let me take this occasion to express my sincere thanks to Dr. Vrehen for his kind hospitality and energetic and enthusiastic collaboration.

---

*Resident Visitor, 1975-76, Philips Research Laboratories, Eindhoven, The Netherlands.

The present study of superfluorescence (SF), i.e., the superradiance or cooperative emission of an initially inverted system, was motivated by the question of whether or not SF can occur with the emission of a single pulse without ringing. The original proposal by Dicke[1] and several subsequent treatments[2,3] had predicted single pulse $(\mathrm{sech})^2$ temporal profiles. The first clear observation[4] of SF was made in HF and indicated the presence of ringing, i.e., modulation of the output intensity, whenever Doppler dephasing was too slow to dampen the ringing. Computer solutions of the coupled Maxwell-Bloch equations permitted the inclusion of propagational effects, i.e., the polarization and field were not forced to have the same values at every position in the superfluorescing sample.[4] These solutions also indicated strong ringing for the conditions of the HF experiment. Bonifacio and Lugiato[5] then reemphasized their conviction of a regime of single-pulse emission as predicted by their mean-field approach. Our experiment was designed to satisfy the Bonifacio and Lugiato conditions for single-pulse emission to determine whether single-pulse SF actually occurs. In the first part of my presentation I will describe those conditions and how they are met almost ideally in cesium. Without a magnetic field, however, the Cs nuclear spin of 7/2 introduces degeneracies and splittings which result in simultaneous SF on more than one transition. In the second portion of my talk I will discuss the quantum beats which arise from these simultaneous SF emissions and the possibility of SF quantum beat spectroscopy. Dr. Vrehen will then describe careful measurements of single pulse SF on a single Cs transition in high magnetic field.

II. BONIFACIO AND LUGIATO CONDITIONS FOR PURE SUPERFLUORESCENCE

The Bonifacio and Lugiato[5] conditions for single pulse SF can be summarized by inequalities between important times in the experiment, i.e.,

$$\tau_E < \tau_c < \tau_R < \tau_D < T_1,\ T_2',\ T_2^* \quad . \tag{1}$$

The escape time τ_E is L/c. The Arecchi-Courtens[6] cooperation time τ_c is the maximum separation in time between atoms decaying cooperatively; for a system with Fresnel number of one or more, it depends on the density n and lifetime τ_0 but not on the sample length L: $\tau_c = (\tau_E\tau_R)^{1/2}$. The superfluorescent time τ_R is approximately the average time for the first photon to be emitted along the inversion cylinder: $\tau_R = \tau_0/\mu N$. Here τ_0 is the lifetime of the particular two-level transition undergoing SF, N is the total number of inverted atoms, and μ is a geometrical factor[2] depending upon the spatial distribution of inverted atoms. If the sample volume is smaller than λ^3, $\mu = 1$; but then near-field dipole-dipole interactions destroy SF except in very special spatial arrangements.[7] For a sample much larger than λ^3 and for a Fresnel number $A/\lambda L \geq 1$, $\mu \approx 3\lambda^2/8\pi A$, where A is the cross sectional area and L is the length of the cylinder of inverted atoms.[2,8] One then has

$$\tau_R = \frac{8\pi\tau_0}{(3n\lambda^2 L)} \quad ; \qquad (2)$$

note that this is τ^{RE} in the useful discussion of SF times by Friedberg and Hartmann.[9] It is the same as the τ of Rehler and Eberly[2], τ_N of Allen and Eberly[10], and τ_R of Burnham and Chiao[11] and of Skribanowitz et al.[4] if τ_0 is understood properly. These definitions are essential to a quantitative comparison of data with theories. Note that τ_c is the value of τ_R when L is replaced by the distance light travels in a time τ_c.

The next quantity in Eq. (1) is τ_D, the delay between initial inversion and the peak of the emitted pulse. This quantity is much more easily measured than τ_R. There is no agreement about its value theoretically except that it is the product of τ_R and a logarithmic factor, for example, τ_D might be $\tau_R \ln(N\mu)$ or $\tau_R \ln(N)$.[10,5] For our Cs experiment, τ_D/τ_R is about 20.

The decay time for energy is T_1 and for transverse polarization is T_2'; these are determined by radiative decay in Cs. The quantity T_2^* is the dephasing time; in Cs the Doppler shifts from the velocity distribution and the spatial inhomogeneities of any existing magnetic field contribute to T_2^*.

Before describing the details of the Cs 7P-7S 3 µm transition, it is instructive to list typical values of these times for our experiment to show that they satisfy well the inequalities of Eq. (1): $\tau_E \approx 0.067$ ns, $\tau_c \simeq 0.18$ ns, $\tau_R \approx 0.5$ ns, $\tau_D \approx 10$ ns, $T_1 \approx 70$ ns, $T_2' \approx 80$ ns, $T_2^* \approx 32$ ns. These values were obtained as follows: τ_E from a 2 cm sample length, τ_R from the 551 ns lifetime and the density measured for the observed τ_D of 10 ns, T_1 and T_2' from the radiative lifetimes of the upper and lower states, and T_2^* from the velocity and geometry of the atomic beam and the measured magnetic field inhomogeneities. The density was varied to yield delay times from a few ns to 35 ns.

A second requirement for simple interpretation is that the inversion preparation time or excitation pulse length, τ_P, be much less than the delay:

$$\tau_P \ll \tau_D \quad . \tag{3}$$

Our τ_P is slightly under 2 ns, so that the density could be adjusted to permit τ_D to satisfy both (1) and (3).

A third condition imposed by the single transverse mode theory of Bonifacio and Lugiato is a Fresnel number $\eta = A/\lambda L$ of about 1, so that diffraction losses will minimize the effects of other modes. A smaller η would result in poor communication between the two ends; larger η could result in many transverse modes.

These conditions for single pulse SF are reasonably intuitive. The atoms must be close enough to communicate well so that SF does not occur in different parts of the sample independently. If the delay time is less than all of the relaxation times, then neither the SF emission nor ringing will be destroyed by relaxation processes. Preparation of the system in a time short compared with its evolution time τ_D is important if the inversion process is not to be included in the dynamical equations. The reasons, why the 3 μm transition in Cs is well suited for a search for single pulse emission, are discussed after a description of the experimental apparatus.

III. CONCEPT AND APPARATUS OF THE EXPERIMENT

The basic elements of the experiment are indicated in Fig. 1. The experiments are performed simply by saturating one of the Cs 6S to 7P hyperfine transitions and observing SF to the 7S state (Fig. 2). This 3-level technique[4] results in complete population inversion. The 1 to 2 ns excitation pulses are produced by a N_2-laser-pumped oscillator and amplifier with a Fabry-Perot in between for good frequency stability. Peak powers are a few hundred watts; the bandwidth is 400-500 MHz. The Cs absorber is either contained in a 30 to 100°C cell of 1 to 10 cm length or sprayed from a 0.5 × 3 mm^2 slit in a 200 to 300°C oven containing a few grams of Cs. Both in the cells and in the beam thin cylinders are excited of Fresnel number equal to 1 within experimental error. In the atomic beam the length of the cylinder has

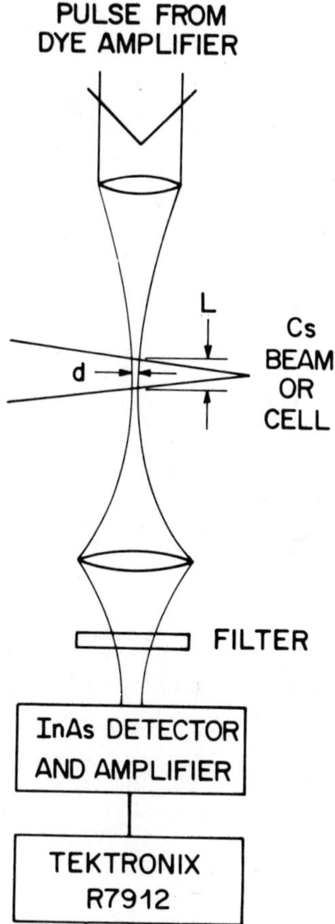

Figure 1. Simplified diagram of superfluorescence experimental apparatus.

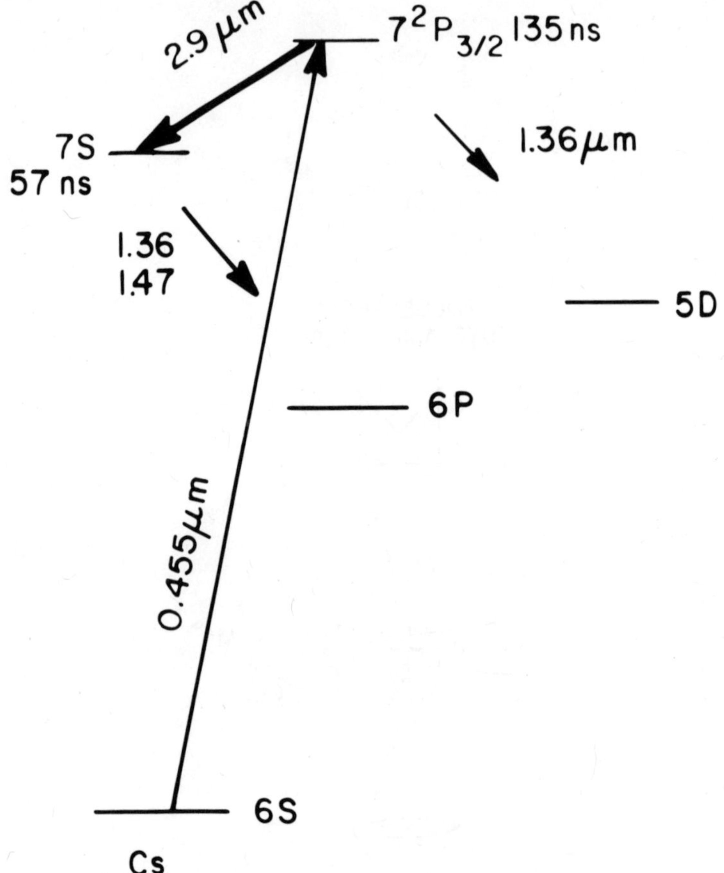

Figure 2. Simplified energy level diagram showing transitions coupled to the $7^2P_{3/2}$-to-7S transition of interest.

been varied from 1.3 to 3.5 cm. The SF signal is detected by an InAs photovoltaic detector with 2 mA bias current, resulting in a response time of about 1 ns (Judson J-12LD). The signals are amplified 22X with an Avantek 500 MHz amplifier and observed on a Tektronix R7912 transient digitizer. The pump radiation is removed with a germanium filter. Care is taken to avoid any feedback from windows, lenses or filters. SF signals are observed simultaneously in the forward and backward directions with equal delays.

IV. ADVANTAGES OF THE Cs 7P-7S TRANSITION FOR A SEARCH FOR SINGLE-PULSE SUPERFLUORESCENCE

Single-photon excitation at 455.5 nm permits the use of a N_2 pumped dye oscillator and amplifier. Short (< 2 ns) pulses are easily obtained with powers well above saturation for diameters less than 400 μm (as needed for unity Fresnel number and lengths less than 10 cm).

The narrow Doppler width of the 3 μm transition results in slow dephasing. The large mass (133) of Cs and the low (30 to 100°C) temperatures required for high vapor pressure result in a Doppler width $\Delta\nu_D$ of about 100 MHz, i.e., a dephasing time $T_2^* \simeq 0.47/\Delta\nu_D$ of about 5 ns in a cell. Only weak collimation is required in an atomic beam to lengthen T_2^* to 20 ns or more. In zero magnetic field with an oven-to-laser beam distance of 15 cm and transverse excitation lengths L of 1.3 to 3.5 cm, T_2^* varies from 80 to 30 ns. With the inhomogeneities of our 2.8 kOe field and L = 2 or 3.6 cm, $T_2^* \approx 32$ or 20 ns. This experiment is the first in which T_2^* could be made long enough to satisfy Eq. (1). Here Eq. (1) was satisfied even for $T_2^* > 3\ \tau_D$, so whether the effective dephasing time is T_2^*, or $\alpha L T_2^*$ (Ref. 4), dephasing should be insignificant.

Another advantage of Cs is its simple atomic structure. To be sure there is a high degeneracy since the nuclear spin is 7/2, but it was clear from self-induced transparency studies that a magnetic field could be applied and a two-level transition inverted.[12] It was less clear at first that it would be necessary or that high enough densities could be achieved. The observation of quantum beats in zero field made it necessary. The use of 5 gm oven loads and short running times made it possible.

Fig. 2 suggests that there may be strong competition for the 7P population to superfluoresce to the 5D state. From Eq. (2) one sees that for equal partial lifetimes, the longest wavelength transition has the shortest τ_R, and hence superfluoresces first and removes the upper level population. If one inserts the actual partial lifetimes[13], one finds τ_R for the 3 μm transition to be about 19 times more rapid than for the 1.36 μm 7P to 5D competitor and 64 times the 0.455 μm transition (if it were inverted). The strongest of the 7S to 6P transitions in high field has a τ_R about twice as long as that of the 7P to 7S. For most of the evolution time of the 7P to 7S SF, the density of the 7P state is much higher than that of the 7S density, making the buildup of SF on 7S to 6P very slow. At high densities, for which $\tau_D \lesssim \tau_P$, this is no longer true and simultaneous emission at several wavelengths is seen. But for $\tau_D \gg \tau_P$, as for studies of single-pulse SF, competition from coupled transitions is negligible.

The relaxation times and τ_0 are found from the radiative lifetimes: 135 ± 1 ns ($7^2P_{3/2}$, Ref. 14); 158 ± 5 ($7^2P_{1/2}$, Ref. 15); 57 ($7^2S_{1/2}$, Ref. 13); the last is a calculated value. If one studies SF between an upper level a and lower level b, one has[12]

$$\frac{1}{T_1} = \frac{1}{\tau_{ab}} + \frac{1}{2\tau_{ac}} + \frac{1}{2\tau_{bd}} \quad , \tag{4}$$

where τ_{ab} is the partial radiative lifetime from a to b and τ_{ac} and τ_{bd} are the radiative lifetimes of a and b to all other (lower) states. Since the branching ratio from $7^2P_{3/2}$ to 7S is 0.49 (Ref. 13), τ_{ab} = 135/0.49 = 275.5 ns and τ_{ac} = 135/0.51 = 264.7 ns; τ_{bd} = 57 ns. Then Eq. (5) gives T_1 = 70 ns. Similarly[12],

$$\frac{1}{T_2'} = \frac{1}{2}\left(\frac{1}{\tau_a} + \frac{1}{\tau_b}\right) \quad , \tag{5}$$

yielding T_2' = 80 ns.

The value of τ_0 needed for calculating τ_R from Eq. (2) can also be found. The $7^2P_{3/2}$ to $7^2S_{1/2}$ partial lifetime has been shown to be 275.5 ns. In a high magnetic field, one is able to selectively excite the $7^2P_{3/2}$ (M_J = -3/2) substate, and it must decay to $7^2S_{1/2}$ (M_J = -1/2). Thus, there is no further branching and the partial lifetime of the transition of interest is 275.5 ns. There is, however, a further factor arising from polarization considerations. The excitation cylinder determining the SF axis is perpendicular to the magnetic field. The ΔM = +1 transition undergoing SF has a partial lifetime of 275.5 ns, corresponding to a dipole moment $p = p_x + ip_y$, but only one linear component (say p_x) is able to participate in SF because of the geometry selected. Thus τ_0 is twice as large,

$$\tau_0 = 551 \text{ ns} \quad . \tag{6}$$

In other words, one is exciting a circular transition at right angles, for which its strength is one half.

In summary, the 7P to 7S 2931 nm transition in Cs has many advantages: small Doppler width, convenient wavelength for single-photon excitation, known relaxation times T_1 and T_2' (much longer than the easily achievable 2 ns excitation pulse duration), well understood atomic structure permitting the selection of a two level system, only weak competition from coupled transitions, and ease of handling in an atomic beam (high vapor pressure and single isotope). These advantages result in the opportunity to study SF under ideal conditions. One might have preferred Cs to have no nuclear spin rather than 7/2, but then we would have missed the fun with quantum beats described in the next section!

V. QUANTUM BEAT SUPERFLUORESCENCE

Single-atom quantum beat spectroscopy has expanded rapidly as tunable lasers have been developed.[16] Here we report the first observation of quantum beats in SF.

A. Upper State Beats

The complexities of the nuclear spin via the hyperfine interaction are shown in Fig. 3 for zero magnetic field. The laser bandwidth permits excitation from only one ground F state, as shown in Fig. 4. SF could develop on transitions from 7P to 7S. Any beats with a frequency of the 7S splitting of 2175 MHz are beyond our detector's response. But beats at the upper state 401 MHz splitting were seen; see Fig. 5. Pulses with delays longer than 10 ns appear to be single pulses modulated by 400 MHz beats. The inset shows the expected quantum beat fluorescence intensity I_F for weakly excited single atoms. The beats are less clear for τ_D < 10 ns, where ringing or cooperation length effects may reduce the beating. The importance of inhomogeneous broadening, i.e., T_2^*, was studied on this transition by varying the length of atomic beam traversed. The SF intensity I_{SF} was observed at a 20 ns delay. For fixed excitation diameter, the output intensity at fixed delay should not vary with excitation length except by T_2^* dephasing. No reduction was observed in changing L from 1.3 to 3.5 cm, i.e., T_2^* from 80 to 30 ns. Even in a cell where $T_2^* \approx 5$ ns the drop in I_{SF} with τ_D was about the same (τ_D^{-2}) in the cell and in the beam, for τ_D < 20 ns. For longer τ_D the cell I_{SF} vanished

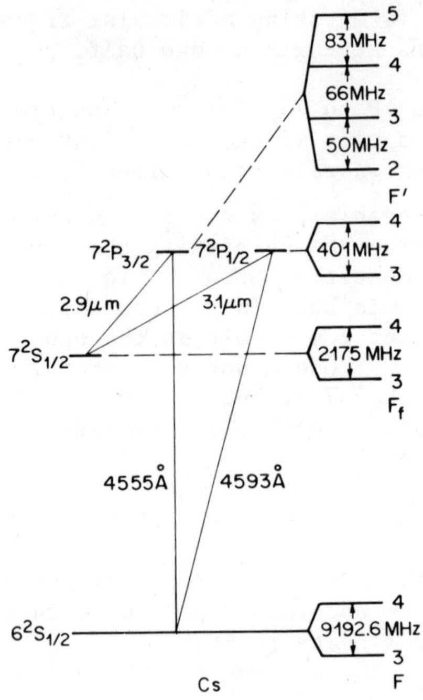

Figure 3. Hyperfine structures in zero magnetic field; see Ref. 20.

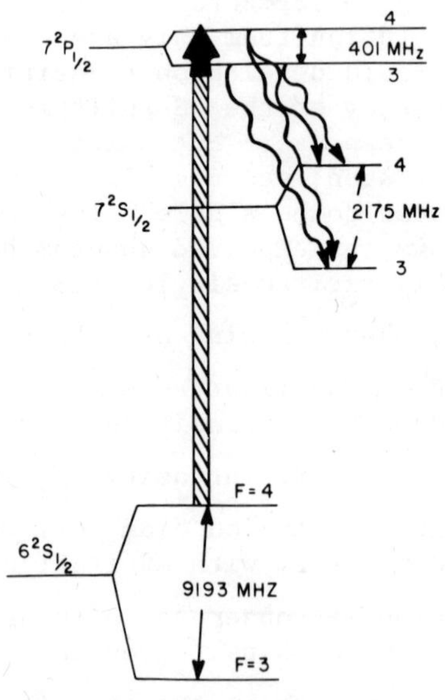

Figure 4. Selective excitation from one ground hyperfine state to the $7^2P_{1/2}$ states.

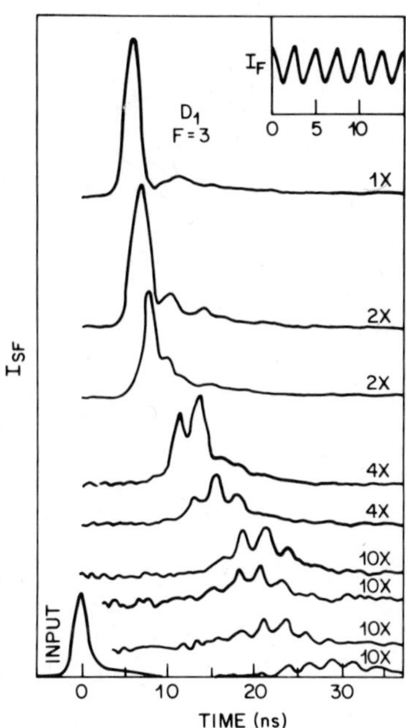

Figure 5. Quantum beats in the superfluorescence from an atomic beam. Circularly polarized 4593 Å excitation from 6S, F = 3 to $7^2P_{1/2}$, F = 3 and 4 is used; see pulse shape and time occurrence at bottom left. The inset shows I_F, the single-atom quantum beat fluorescence. 10X means the signal has been amplified by 10, etc.

rapidly, whereas I_{SF} in the beam continued as τ_D^{-2} until it was lost in the noise at 35 to 40 ns. Thus T_2^* was certainly unimportant in the beam data for $\tau_D < 20$ ns.

The SF from the $7^2P_{3/2}$ state is more complicated because 3 F states are simultaneously excited yielding 3 upper state beat frequencies[17]; see Fig. 6. Since these frequencies beat with each other, the interference can lead to complicated decay curves even in the single atom case.[17] The depth and frequency of modulation depend upon the details of the light polarization and the angular momenta, as well as the splittings. Consider the case of Fig. 7 with excitation from one ground hyperfine state J_gF_g to two upper states JF_1 and JF_2, which superfluoresce to final state

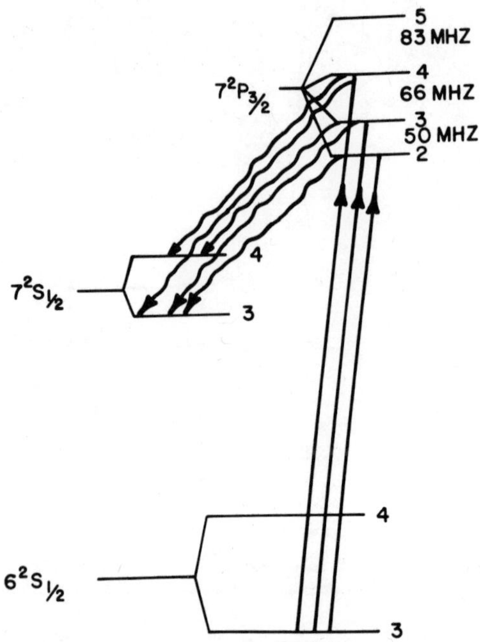

Figure 6. Selective excitation from 6S, F = 3 to $7^2P_{3/2}$, F = 2, 3, and 4 resulting in three beat frequencies.

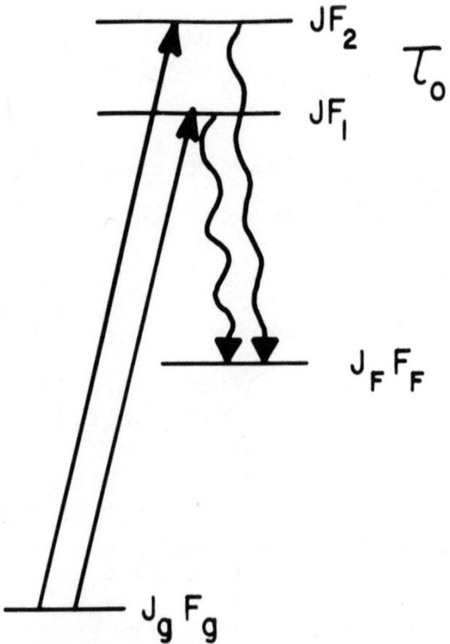

Figure 7. Notation for the single-atom quantum beat formula.

$J_F F_F$. Assuming weak broadband excitation, the single atom fluorescence intensity is given by[18]

$$I_F \propto e^{-t/\tau_0} \sum_{kq} (-1)^{k + 2J + J_F + I + F_g} (2 F_g + 1) \begin{Bmatrix} 1 & 1 & k \\ J & J & J_F \end{Bmatrix}$$

$$\times \phi^k_q \psi^k_q \sum_{F_1 F_2} (-1)^{F_1 + F_2} (2 F_1 + 1)(2 F_2 + 1)\, e^{-i\omega_{F_1 F_2} t}$$

$$\times \begin{Bmatrix} 1 & 1 & k \\ F_1 & F_2 & F_g \end{Bmatrix} \begin{Bmatrix} J & J & k \\ F_1 & F_2 & I \end{Bmatrix} \begin{Bmatrix} J_g & J & 1 \\ F_1 & F_g & I \end{Bmatrix} \begin{Bmatrix} J_g & J & I \\ F_2 & F_g & I \end{Bmatrix} ,$$

$$\phi^k_q = \sum_{\mu=-1}^{+1} e_\mu (e_{\mu-q})^* (-1)^{1+\mu} (2k + 1)^{1/2} \begin{pmatrix} 1 & 1 & k \\ \mu & q - \mu & -q \end{pmatrix} ,$$

where ϕ^k_q and ψ^k_q are the excitation and detection polarization tensors and e_μ are the spherical components of the polarization unit vector. For example, for linearly polarized excitation $e_o = 1$ and for circularly polarized excitation $e_{+1} = 1$. In calculating I_F for the SF cases, the detection polarization was assumed the same as the excitation, as was usually the case, although an analyzer was not always used. When these formulae are applied to the case of Fig. 6, one obtains the top curves of Fig. 8, showing the complicated interference pattern and polarization dependence. The SF intensity I_{SF} is, of course, further complicated by the SF process. However, note that SF emission ceases or is greatly diminished at delay times of destructive interference in I_F. The initial wave functions which determine the phases of the interfering frequencies are the same for the single atom and SF cases. Cooperation between atoms accelerates the emission in the SF case but does not remove the interference giving rise to beats. In fact, the modulation of the SF pulse by quantum beats is a dramatic illustration of the close coupling between the electromagnetic field and the atomic polarization in SF emission. When the polarization is reduced by destructive interference, I_{SF} is also reduced. But as soon as the polarization becomes strong again via constructive interference, the SF emission resumes where it left off; another τ_D evolution time is not required.

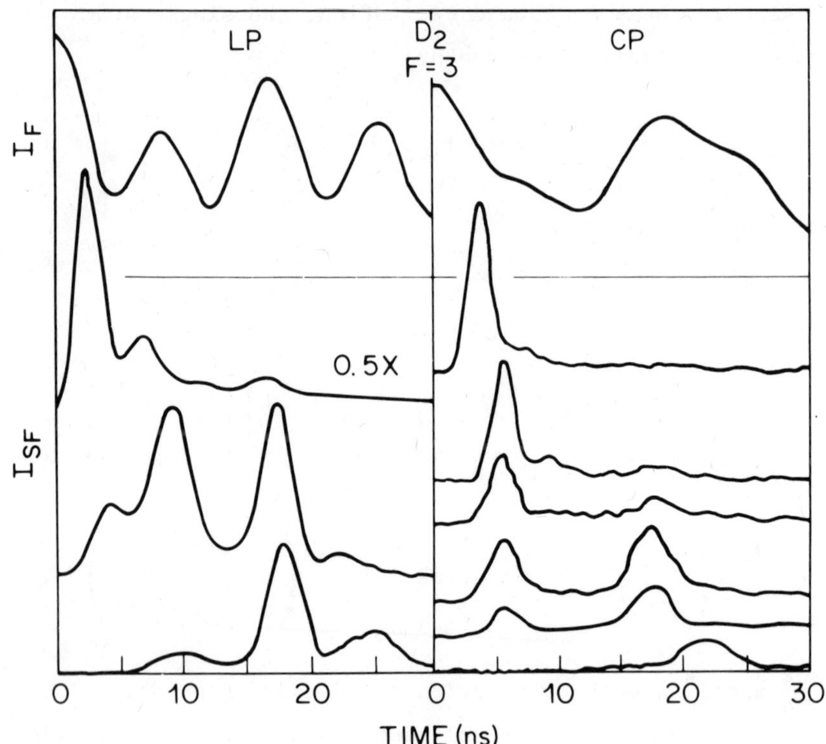

Figure 8. Quantum beats in superfluorescence with 6S, F = 3 to $7^2P_{3/2}$, F = 2, 3, 4 excitation, with linearly polarized (LP) light on a 10 cm cell and circularly polarized (CP) light on a 2.6 cm pathlength atomic beam. The top I_F curves show calculated single-atom beats. The lower curves are the observed SF beats. SF is strongly suppressed near the minima in I_F. In the CP case the delays as a function of density jump discontinuously, avoiding the minimum at 12 ns.

B. Lower State Beats

The SF beats from upper states discussed in the previous part have been shown to be very similar to single-atom beats. These beats arise from the fact that in calculating the emission intensity one must sum the amplitudes of two indistinguishable channels before performing the square. In emission from a single upper state to two lower states, the final states are in principle distinguishable. Hence, there is no interference or beats.[16]

Lower state beats are possible for many atom emission, where the indistinguishability of atoms introduces a sum of amplitudes once again, but care must be taken to avoid phase changes by motion or collisions.

The SF case is much simpler. Since SF is a coherent emission, two or more superfluorescing transitions can always beat with each other just as if they were separate laser sources. Consider the case of Fig. 9 showing (M_J = m, M_I = M) substates. At an appropriate excitation power only the 6S (-1/2, -3/2) to $7^2P_{3/2}$ (-3/2, -3/2) transition was excited, and SF was observed to 7S (-1/2, -3/2) without beats. Then by increasing the power, the M = -5/2 transition was simultaneously excited. SF then occurred simultaneously on two <u>independent</u> transitions <u>incoherently</u> excited. The observed beats are shown in Fig. 10, where the beat frequency crudely measured from the photographs is only slightly higher than the expected 295 MHz difference in transition frequencies. In this particular case this difference is dominated by the lower state splitting.

In intermediate fields of 350 to 700 Oe, beats were seen that tuned with magnetic field. They probably arise from 6S (F = 4, M_F = -2, -3) to $7^2P_{3/2}$ (m = -1/2, M = -5/2, -7/2) with SF to 7S (F = 4, M_F = -3, -4). The observation of Zeeman SF beats suggests another spectroscopic application of SF beats.

C. Superfluorescence Beat Spectroscopy

SF beats are fascinating, but they are a nuisance to careful studies of the SF process. It would seem that they are of no spectroscopic importance either, since the pulsed emission of the SF process only makes SF beats more complicated than single atom beats. However, there may well be situations where SF beats could be seen and single atom beats could not, either because the splitting is in the lower state or the single atom fluorescence is too weak. SF has been used to reduce the emission time of a weak transition from about 1 second to a few nanoseconds.[4] Also, the SF emission is not isotropic, but highly directional and easily focusable onto a single detector. This great reduction in emission time and collection losses greatly enhances the observed peak emission intensity. The limit of resolution is likely to be imposed by the Doppler dephasing time T_2^*, since the SF delay time and pulse width cannot be much longer than T_2^*.

Therefore, not only for its intrinsic interest, but also for possible spectroscopic applications, the theory of SF beats (including chirps) should be developed.

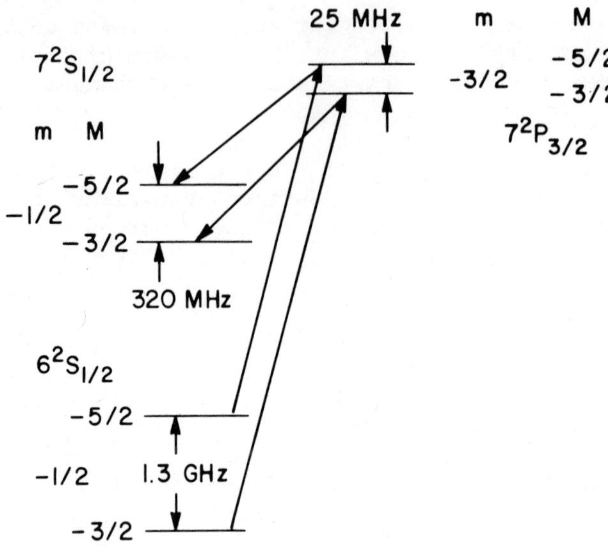

Figure 9. Relevant energy levels in a magnetic field of 2.8 kOe; m refers to the electronic and M to the nuclear magnetic quantum number. The $7^2P_{3/2}$ states are excited incoherently, namely from different ground states.

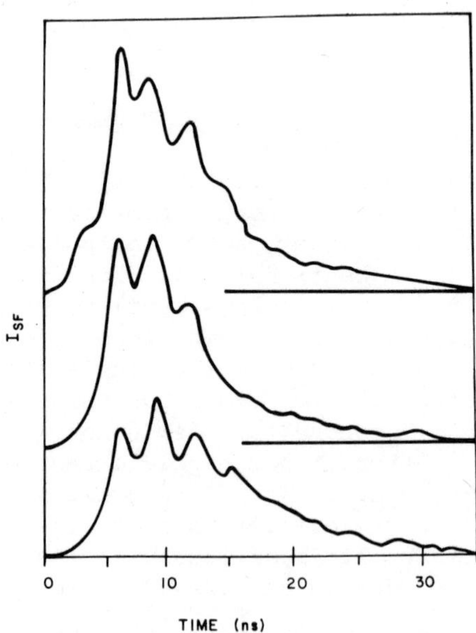

Figure 10. Quantum beats in a 2.0 cm pathlength atomic beam in 2.8 kOe.

VI. SUMMARY

The 2931 nm 7P to 7S transition in Cs is a convenient, well-understood, near-ideal system for studying superfluorescence. SF beats are easy to see at the difference frequencies of the SF emissions, whether they arise from upper or lower splittings. SF beats could be used to study SF chirps or measure splittings of nonfluorescing lower states or weakly decaying upper states. Beats complicate the study of SF; in particular, they could be mistaken for ringing. Dr. Vrehen describes, in the next presentation, a careful study of SF emission, free of beats.

References

(1) R. H. Dicke, Phys. Rev. 93, 99 (1954).

(2) N. E. Rehler and J. H. Eberly, Phys. Rev. A 3, 1735 (1971).

(3) R. Bonifacio, P. Schwendimann, and F. Haake, Phys. Rev. A 4, 302 and 854 (1971) and references therein.

(4) N. Skribanowitz, I. P. Herman, J. C. MacGillivray, and M. S. Feld, Phys. Rev. Lett. 30, 309 (1973). I. P. Herman, J. C. MacGillivray, N. Skribanowitz, and M. S. Feld in Laser Spectroscopy, R. G. Brewer and A. Mooradian, editors (Plenum, NY, 1974). J. C. MacGillivray and M. S. Feld, Phys. Rev. A 14, 1169 (1976).

(5) R. Bonifacio and L. A. Lugiato, Phys. Rev. A 11, 1507 (1975) and 12, 587 (1975).

(6) F. T. Arecchi and E. Courtens, Phys. Rev. A 2, 1730 (1970).

(7) R. Friedberg, S. R. Hartmann, and J. T. Manassah, Phys. Lett. 40A, 365 (1972). R. Friedberg and S. R. Hartmann, Opt. Commun. 10, 298 (1974).

(8) The integral given in Ref. 2 for μ for a circular cylinder has been numerically integrated yielding for a length of 2 cm, $\lambda = 2931$ nm and a Fresnel number $A/\lambda L$ of 1 the value $0.78\ (3\lambda^2/8\pi A)$.

(9) R. Friedberg and S. R. Hartmann, Phys. Rev. A 13, 495 (1976).

(10) L. Allen and J. H. Eberly, Optical Resonance and Two-Level Atoms (John Wiley, 1975). Chapter 8 is a good introduction to superradiance.

(11) D. C. Burnham and R. Y. Chiao, Phys. Rev. 188, 667 (1969).

(12) R. E. Slusher and H. M. Gibbs, Phys. Rev. A 5, 1634 (1972).

(13) O. S. Heavens, J. Opt. Soc. Am. 51, 1058 (1961).

(14) S. Svanberg and S. Rydberg, Z. Physik 227, 216 (1969).

(15) P. W. Pace and J. B. Atkinson, Can. J. Phys. 53, 937 (1975).

(16) Reviewed by S. Haroche in K. Shimoda, ed., High Resolution Laser Spectroscopy (Springer Verlag, 1976). For quantum beats in limited superradiance such as photon echoes see: A. Compaan, L. Q. Lambert, and I. D. Abella, Phys. Rev. A 8, 1641 (1973); P. F. Liao, P. Hu, R. Leigh, and S. R. Hartmann, Phys. Rev. A 9, 332 (1974); R. L. Shoemaker and F. A. Hopf, Phys. Rev. Lett. 33, 1527 (1974). Beats in SF or superradiance are discussed in J. H. Eberly, Lettere al Nuovo Cimento 1, 182 (1971). See also I. R. Senitzky, Phys. Rev. Lett. 35, 1755 (1975). For the absence of single-atom lower-state beats, see, in particular, W. W. Chow, M. O. Scully, J. O. Stoner, Jr., Phys. Rev. A 11, 1380 (1975).

(17) S. Haroche, J. A. Paisner, and A. L. Schawlow, Phys. Rev. Lett. 30, 948 (1973) and J. A. Paisner, Thesis, Stanford University, 1974.

(18) This agrees with a calculation by H. J. Andrä, private communication. See also Ref. 17.

(19) Beats were seen for linearly polarized D_1 excitation and detection even though they are not predicted for a single atom, under the usual assumptions of a nonsaturating excitation broad compared with the state separation. Our near-saturation and near-transform-limited (300 to 500 MHz width) pulses often violated these assumptions, invalidating the summation leading to this selection rule.

(20) Cs hyperfine structures: $6^2S_{1/2}$, M. Arditi and T. R. Carver, Phys. Rev. 109, 1012 (1958); $7^2S_{1/2}$, R. Gupta, W. Happer, L. K. Lam, and S. Svanberg, Phys. Rev. A 8, 2792 (1973); $7^2P_{1/2}$, H. Bucka, Z. Physik 151, 328 (1958); $7^2P_{3/2}$, H. Bucka, H. Kopfermann, and E. W. Otten, Ann. Physik (7) 4, 39 (1959).

SINGLE-PULSE SUPERFLUORESCENCE IN CESIUM

Q. H. F. Vrehen

Philips Research Laboratories
Eindhoven, The Netherlands

Abstract: Superfluorescence has been observed at 2931 nm from the $7P_{3/2}$ to $7S_{1/2}$ transition in cesium. It is shown that a completely inverted two-level system can be prepared in a magnetic field. Experimental results on the superfluorescent output are reported for a range of sample lengths (1.0 cm to 5.0 cm) and densities (8×10^9 cm^{-3} to 2×10^{11} cm^{-3}) and for several values of the inhomogeneous dephasing time (5 ns, 18 ns, and 32 ns). For sufficiently long delay times the emission from the pencil-shaped volume of Fresnel number one consists of a single pulse for all sample lengths, both in an atomic beam and in a cell. Neither homogeneous relaxation nor inhomogeneous dephasing can explain the absence of ringing. For increased densities and reduced delay times, depending on the length of the sample, multiple-pulse output occurs. The occurrence of single and multiple pulses, the delay times, and the pulsewidths are discussed.

I. INTRODUCTION

The experiments on single-pulse superfluorescence described in this presentation were undertaken in collaboration with Dr. Hyatt M. Gibbs of Bell Laboratories during his stay as a Resident Visitor, 1975-1976, at the Philips Research Laboratories in Eindhoven, The Netherlands. He clearly recognized the possible significance of the experiment and he identified the cesium system as nearly ideal for our purposes. His competence and good fellowship has made our cooperative effort a most enjoyable experience.

The motivation for the present study is discussed in detail by Dr. Gibbs in the preceding paper, henceforth to be referred to as [I], and will be restated here only briefly. While early theoretical work predicted single-pulse emission[1-3], the first experiment[4] clearly showed the presence of ringing, which has been attributed to propagational effects[5]. The question remained, however, whether single-pulse outputs would be observed in an experiment satisfying the conditions specified by Bonifacio and Lugiato[6]. Recent experiments[7,8], results of which were published after the present study had been initiated, did not provide an answer to this question. A discussion of the Bonifacio and Lugiato conditions for "pure superfluorescence" and of how these conditions can be met in cesium is to be found in sections II and IV of [I].

The present paper is devoted to a description of the experiment that has led to the observation of single-pulse superfluorescence in cesium under conditions that eliminate inhomogeneous dephasing or homogeneous relaxation as possible causes for the absence of ringing. Particulars of the experiment are presented in section II, with special attention given to the preparation of a completely inverted two-level system in the $7P_{3/2}$ to $7S_{1/2}$ manifold of Cs and to the determination of the initial excited state density. In section III the experimental results are summarized and in section IV these results are discussed and compared with the theoretical work of MacGillivray and Feld[5] and of Bonifacio and Lugiato[6].

Throughout this paper the same nomenclature and symbols are used as in [I]. For the convenience of the reader the definitions of the important times are repeated here:

τ_R superfluorescence time $\tau_R \equiv \dfrac{8\pi\tau_o}{3\lambda^2 Ln}$,

τ_o spontaneous emission decay time,

τ_D delay time of the superfluorescence pulse with respect to the pump pulse,

τ_E escape time ($\tau_E \equiv L/c$),

τ_C cooperation time ($\tau_C \equiv \sqrt{\tau_E \tau_R}$).

In these definitions n is the initial excited state density, L the sample length, λ the wavelength, and c the velocity of light in vacuum.

II. EXPERIMENT

The basic elements of the experimental apparatus have been described in section III of [I]. Some of those elements that are of particular importance for the controlled excitation of a two-level system in the 7P to 7S transition of atomic cesium will be considered here in more detail.

The N_2-laser-pumped dye-laser oscillator, which incorporates a telescope and grating, delivers a pulse of 6 ns duration and 20 GHz bandwidth. Spectral filtering is performed with a pressure-tuned Fabry-Perot of 15 GHz free spectral range (FSR) in combination with an angle-tuned etalon of 33 GHz FSR. Pulse shaping is then carried out in an amplifier by means of gain quenching[9]. The resulting pulse shape is presented in Fig. 1. The FWHM pulse length is less than 2 ns; a weak tail extends to about 4 ns after the peak. The jitter is less than 0.5 ns. The bandwidth is 400-500 MHz, as measured from the fringe width of Fabry-Perot rings. Peak powers amount to a few hundred watts; only 10% or less of this power is needed to saturate a volume of Fresnel number one for a length of 5 cm or smaller.

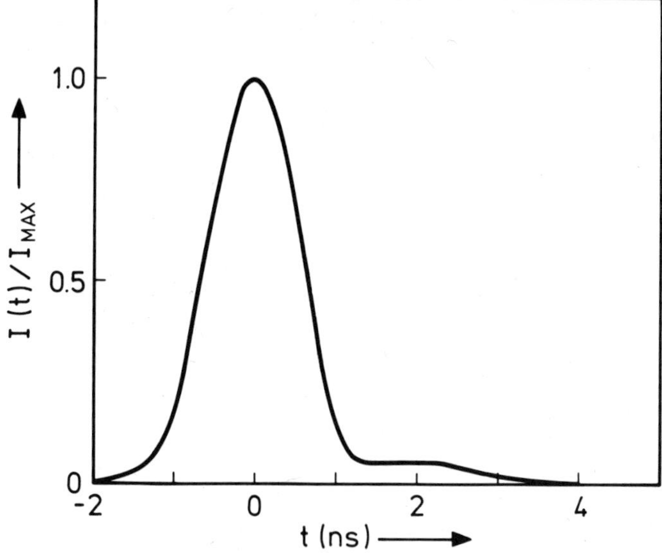

Figure 1. Shape of the excitation pulse.

In Fig. 2 the transition frequencies are indicated as a function of magnetic field for the D2 line ($6S_{1/2}$ to $7P_{3/2}$) of Cs. Only the transitions that are allowed in the high-field limit

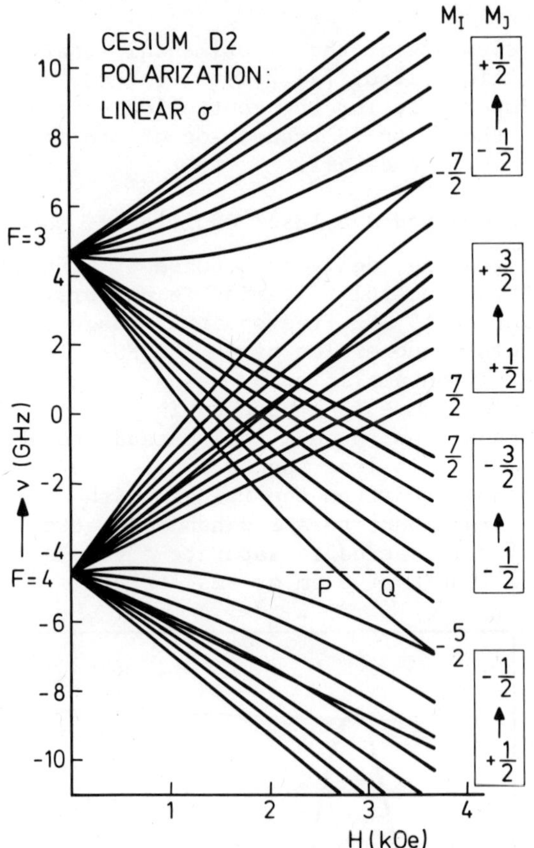

Figure 2. Transition frequencies of atomic cesium in a magnetic field. Only transitions that are allowed in the high-field limit are indicated. Excited state hyperfine splittings have been neglected. The points P and Q correspond to the operating points in the experiment.

are shown; the (small) hyperfine splittings of the excited state are not exhibited. Most measurements have been made with excitation at the points P or Q, which correspond to population of the levels ($m_J = 3/2$, $m_I = -5/2$) or ($m_J = -3/2$, $m_I = -3/2$) of $7P_{3/2}$ respectively. Each of these levels decays to only one level of the $7S_{1/2}$ manifold, ($m_J = -1/2$, $m_I = -5/2$) or ($m_J = -1/2$, $m_I = -3/2$). With the help of an auxiliary cesium cell in zero magnetic field the laser frequency is adjusted for optimum superfluorescence (SF) output at the F = 4 transition ($6S_{1/2}$, F = 4 to $7P_{3/2}$). This frequency setting is accurate to one hundred MHz. Next the magnetic field at the sample is tuned for maximum SF output at the point P or Q. The average SF output power as a function of magnetic field is shown in Fig. 3. The SF output has clear maxima at the field values corresponding to P and Q and a near zero in between. The half-width of the maxima is 250 Oersted or 600 MHz (roughly half the frequency separation between successive transitions). The pronounced minimum between P and Q disappears when the excitation intensity is increased enough to cause appreciable power-broadening of the levels. The presence or absence of the minimum is a convenient indicator of the level of excitation (see below).

A pencil-shaped volume of length L and cross-sectional area A is excited in either a cell, or, for increased T_2^*, an atomic beam. In all experiments the Fresnel number $A/\lambda L$ is made close to one by adjusting the laser beam diameter, which is essentially constant along the length of the sample. The diameter is measured by moving a slit transversely through the laser beam and measuring the transmitted power as a function of slit position, see Fig. 4. Evidently, the diameter of the pumped volume is not sharply defined and depends on the pump intensity; the diameter at 1/e of maximum power is used to calculate the Fresnel number, and from the possible variations in pump intensity it is estimated that the actual Fresnel number lies between 0.5 and 2.0.

The superfluorescent pulses are detected with an InAs detector of 1 ns response time. Each individual pulse is recorded by the Tektronix Transient Digitizer R 7912 and stored on video-tape for later inspection. For the highest densities in the atomic beam the running times are very short, less than five minutes. Nevertheless, with a repetition rate of 10 Hz, thousands of shots can be collected.

As already noted in [I], care was always taken to avoid feedback from windows, lenses, and filters. The apparatus has been disassembled and reassembled many times and pulse shapes and delays proved to be reproducible within the limits indicated. It is believed therefore that the influence of feedback on the experimental results is negligible.

Much attention has been given to a careful determination of the density n of the excited Cs atoms in the sample volume. With S = 1/2 and I = 7/2 the ground state splits up into 16 levels. If atoms are pumped to saturation from just one of these levels, then n is related to the ground state density prior to the pump pulse n_{at} by the expression $n = n_{at}/32$. The determination of n

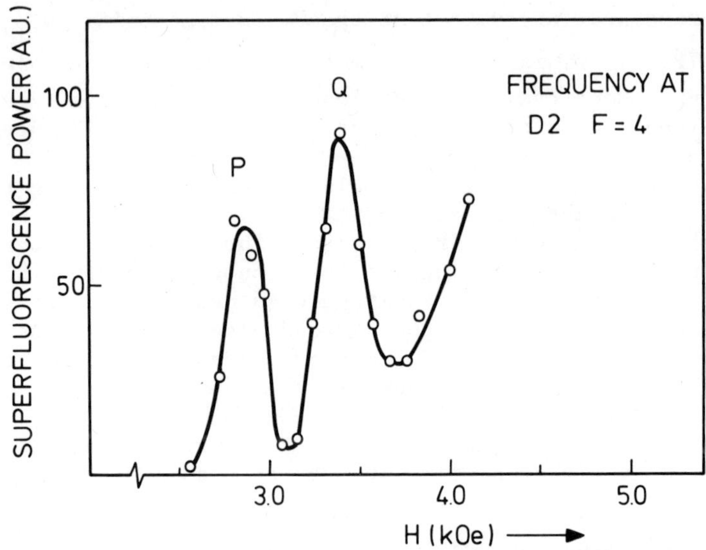

Figure 3. Average intensity of superfluorescence emission as a function of magnetic field at the frequency corresponding to the points P and Q in Fig. 2. Each data point is an average for 20 shots. The minimum between P and Q disappears at high excitation levels because of power broadening.

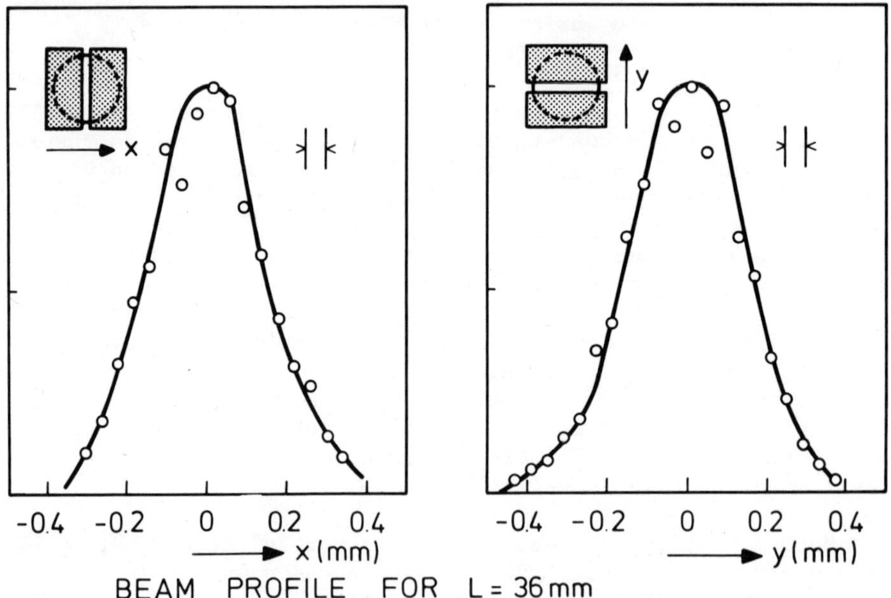

Figure 4. Transverse profile of the laser beam for a sample length of 3.6 cm as measured by moving a slit through the beam. The diameter at 1/e of maximum intensity is used to define the Fresnel number.

requires the complete saturation of one and only one transition and the measurement of n_{at}. Complete saturation of just one level is realized as follows. At a low pump power level the laser frequency and the magnetic field are adjusted to a particular transition, e.g., the point P in Fig. 2. It follows from the discussion of Fig. 3 that just one transition is then pumped. By changing neutral density filters the intensity of the pump beam is increased, and this is accompanied by a decrease in the delay time of the SF output pulse, until the selected transition is fully saturated. When the pump intensity is further increased, adjacent levels will eventually be excited, which results in the appearance of beats in the output (see [I]) and in the disappearance of the minimum between P and Q in Fig. 3. The range of pump powers, for which complete saturation of just one transition takes place, is limited to about a factor of three. This experimental observation is easily understood from the fact that the pump pulse duration is less than 2 ns while the separation of adjacent transitions is only slightly more than 1 GHz. For the measurement of n_{at} in the cell, the temperature of the cold spot is measured with a chromel-alumel thermocouple. Vapor pressure data[10] are used to determine the atomic density in the cell. The determination of the density in the atomic beam will be discussed with the help of Fig. 5. At a distance R from the aperture in the oven, n_{at} is given by[11,13]

$$n_{at} = \frac{\eta F}{\pi R^2 \bar{v}} , \tag{1}$$

where F is the number of atoms that leave the oven per second, $\bar{v}$ is the average velocity of the atoms in the beam, and η is a channeling factor. For an effusive beam emerging from a small hole in a thin wall, η equals 1. In the experiment the beam is not effusive; at the operating pressures in the oven of 1 to 10 torr the mean free path of the atoms[12] is much smaller than the dimension of the aperture which has a height of 3 mm, a width of 0.6 mm, and a depth along the beam axis of 2 mm. The flow from the oven probably resembles that of a free jet. For an atomic free jet a value 1.93 for η is quoted by Anderson[13]. Experimentally η was determined as follows. A copper plate of known area was placed in the beam with its normal along the beam direction at a given distance from the oven. A known amount of Cs was evaporated from the oven and the mass condensed on the copper plate was measured with various techniques of analytical chemistry (atomic absorption, atomic emission and titration). The measurement was repeated for several evaporation rates. In this way a value of $\eta = 2.1 \pm 0.2$ was found independent of the evaporation rate, in reasonable agreement with the value for a free jet.

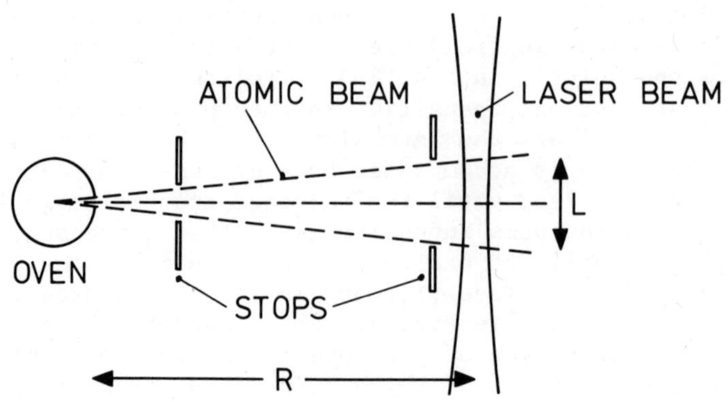

Figure 5. Geometry of the atomic beam.

The average velocity in the beam is not precisely known. The value of the average velocity in an effusive beam[11] has been used:

$$\bar{v} = \sqrt{\frac{9 \pi k T}{8 m}} \quad . \tag{2}$$

It is about 18% higher than the average velocity in a cell[11], and approximately 19% lower than the velocity in a supersonic flow of high Mach number[13,14]. The uncertainty in the velocity may thus introduce a possible systematic error of at most ±20%.

Finally the evaporation rate F has to be established. In principle this is done by setting F equal to the ratio of M_o, the number of atoms originally present in the oven, to t_{eff}, the effective running time during which the atoms evaporate from the oven. Before a run the oven is loaded with about 5 gram of Cs. In roughly five minutes the oven is brought to its final temperature, which is kept constant thereafter. Since F(t) is proportional to the density and the average velocity of the atoms in the oven[11,13], or equivalently, to the density and the square root of the absolute temperature in the oven, t_{eff} is calculated from the relation

$$t_{eff} = \frac{\int_{t_o}^{t_1} \rho(t) \sqrt{T(t)} \, dt}{\rho_f \sqrt{T_f}} , \qquad (3)$$

where $\rho(t)$ and $T(t)$ are respectively the density and the temperature of the oven at time t and ρ_f and T_f are those at the time of the measurements (the run begins at t_o and ends at t_1). The end of a run is always marked by a sudden increase of the oven temperature and the abrupt disappearance of the SF signal.

The random error in the measurement of n is estimated to be ±40%; it arises mainly from the random error in the measurement of t_{eff}, which is estimated at ±30%. As mentioned above a possible systematic error of ±20% must be reckoned with. To improve the accuracy of the density measurements, a Langmuir-Taylor probe is presently being installed in the beam apparatus.

III. EXPERIMENTAL RESULTS

Measurements have been made for sample lengths of 1.0, 2.0, 3.6 and 5.0 cm, for large (beam) and small (cell) values of T_2^* and for a range of densities. A survey is given in Table I. Only a limited number of experiments was made for the 1.0 cm cell, but extensive studies were made for the 2.0 cm and 3.6 cm beam and for the 5.0 cm cell. Figures 6, 7, and 8 present pulse shapes for decreasing densities, i.e., increasing values of τ_R, from top to bottom. For all sample lengths single pulses occur for sufficiently large delay times. The typical shape of the single pulses is somewhat asymmetric, the front being steeper than the tail. Occasionally, however, very nearly symmetric pulses were found, with shapes close to $(\text{sech})^2$. An example is shown in Fig. 9. On increasing the density a transition from single pulses to multiple pulses is found. The transition as a function of density is rather abrupt for each length. In the multiple-pulse regime the pulse shapes fluctuate strongly from shot to shot even for fixed delay; for this reason several pulses are reproduced in Figs. 6, 7, and 8.

	CELL	BEAM	BEAM	CELL
Length (cm)	5.0	3.6	2.0	1.0
T_2^* (ns)	5	18	32	5
τ_E (ns)	0.17	0.12	0.07	0.035
τ_R (ns)	0.15-1	0.1-1.8	0.15-1.3	0.12-0.5
τ_D (ns)	6-20	5-35	6-25	5-12
αL	35-5	90-10	215-25	45-8

Table I. Survey of experimental conditions.

T_2^* inhomogeneous dephasing time; τ_E escape time = L/c; τ_R superfluorescence time; τ_D delay time; αL amplitude gain at the center of the atomic line.

In the atomic beam the pulse amplitude is a linear function of the inverse delay time squared, as illustrated in Fig. 10 for the 3.6 cm beam and for delays from 8 ns to 35 ns. The signal disappears in the detector noise for delays of 40 to 45 ns, without any evidence for an anomalous decrease of signal strength due to dephasing. In contrast, for the 5.0 cm cell, the signal vanishes abruptly for delay times beyond 20 ns, apparently because of the finite dephasing time $T_2^* = 5$ ns in the cell. It may be concluded that T_2^* is certainly larger than 10 ns in the 3.6 cm beam; a value of 18 ns is calculated from the velocity spread of the atoms and from magnetic field inhomogeneities. Moreover, it may be concluded that in the 3.6 cm beam, and *a fortiori* in the 2.0 cm beam, the emission process is not affected significantly by atomic dephasing.

An experimental estimate has been made of the total energy emitted in the pulse, using the observed pulse amplitudes, the nominal sensitivity of the detector and the transmission coefficients of filters, windows and lenses. In an atomic beam, at least 20% of the energy initially stored in the sample is emitted in the forward direction alone.

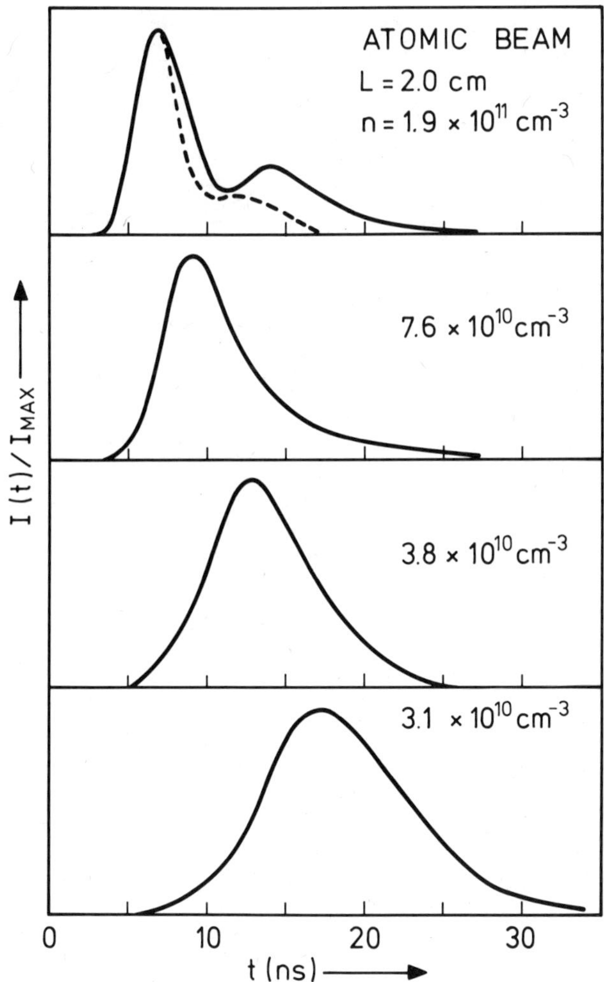

Figure 6. Superfluorescence pulse shapes for several densities n in an atomic beam of 2.0 cm length.

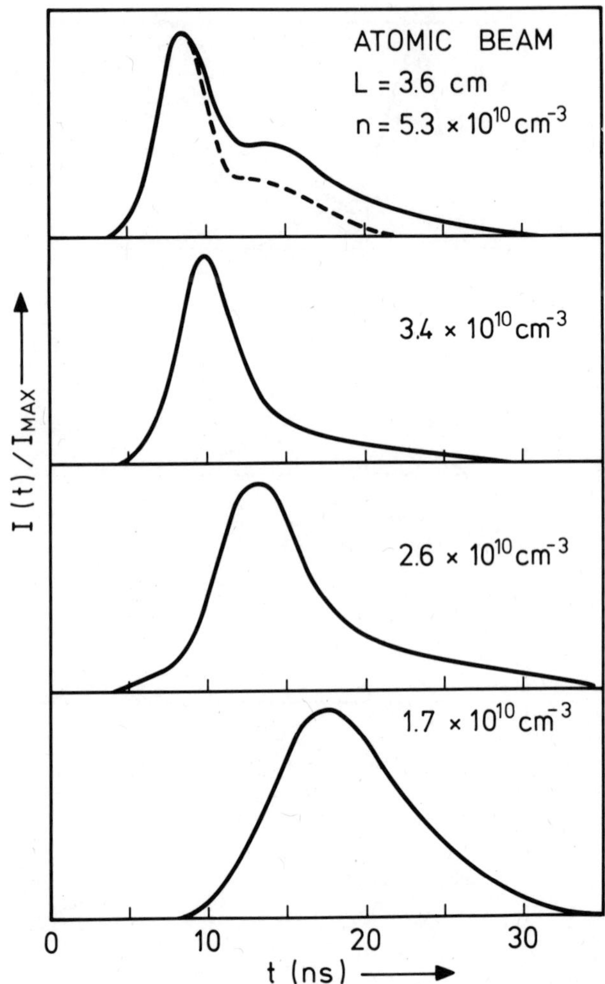

Figure 7. Superfluorescence pulse shapes for several densities n in an atomic beam of 3.6 cm length.

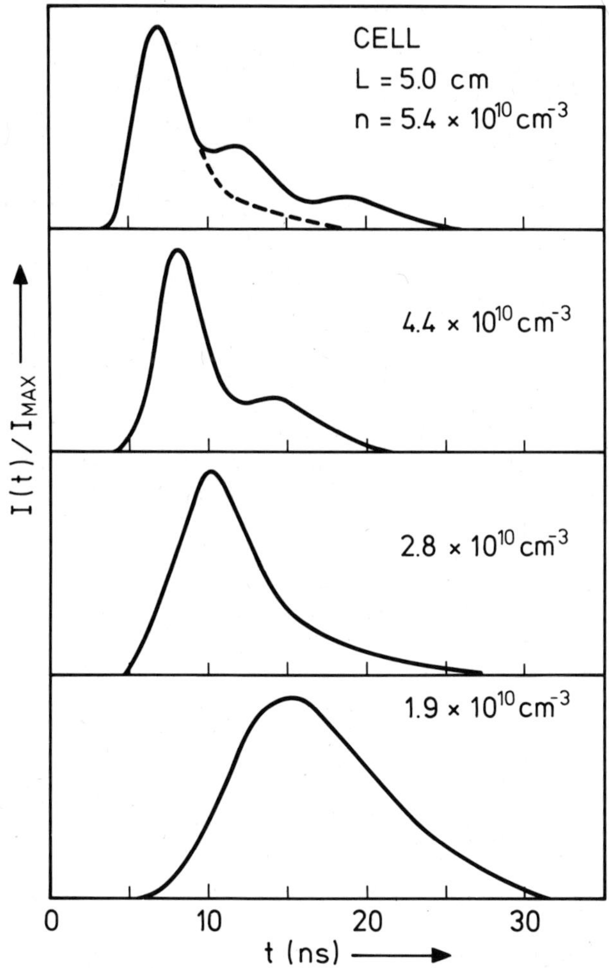

Figure 8. Superfluorescence pulse shapes for several densities n in a cell of 5.0 cm length.

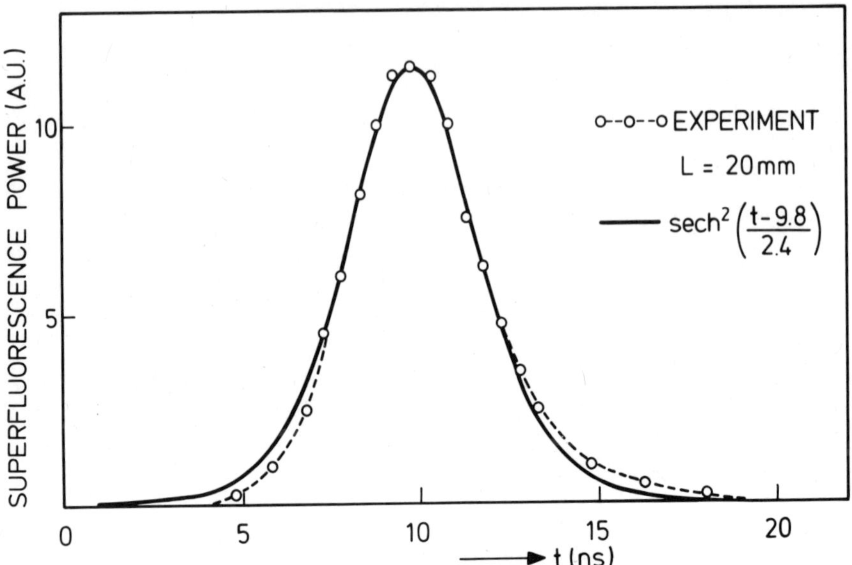

Figure 9. Example of the very symmetrical pulses that have often been observed. Open circles show the experimental pulse shape. For the fit to a $(\mathrm{sech})^2$ the delay time, the pulse width and the peak intensity are treated as free parameters.

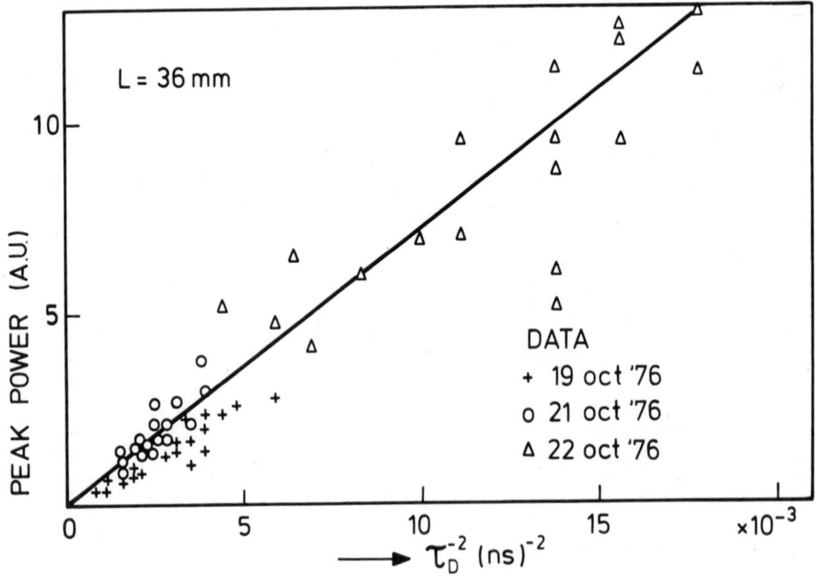

Figure 10. Peak intensity of superfluorescence pulse as a function of inverse delay time squared. Beam 3.6 cm.

The pulse width τ_W has been plotted as a function of delay time in Fig. 11 for the 3.6 cm beam. The width is proportional to the delay time, $\tau_W \approx 0.5\ \tau_D$. The same holds true for other sample lengths. The delay time as a function of τ_R is shown in Fig. 12. The values of the delay are in between those expected from the work of MacGillivray and Feld[5], and those predicted by the theory of Bonifacio and Lugiato[6], curves I and II respectively.

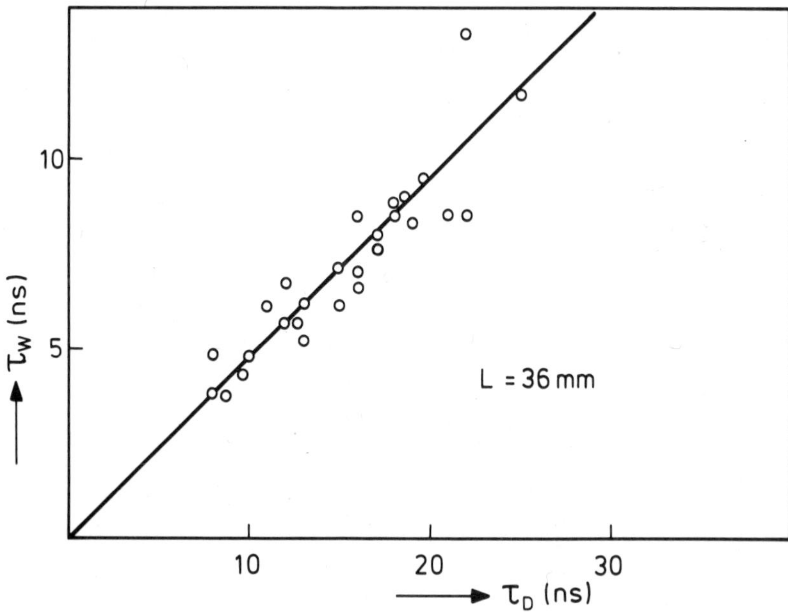

Figure 11. Pulse width as a function of delay time. Beam 3.6 cm.

The possibility of superfluorescence on competing transitions, either in parallel or in cascade with the $7P_{3/2}$ to $7S_{1/2}$ transition has been studied in a 10 cm cell in zero magnetic field. At sufficiently high densities superfluorescent output was observed at 1469 nm ($7S_{1/2}$ to $6P_{3/2}$), at 1359 nm ($7S_{1/2}$ to $6P_{1/2}$) and at 1360 nm ($7P_{3/2}$ to $5D_{5/2}$), while parametric emission occurred at 852 nm ($6P_{3/2}$ to $6S_{1/2}$) and at 894 nm ($6P_{1/2}$ to $6S_{1/2}$). For a given delay of 15 ns the cascade output at 1469 nm required a four times higher density than the emission at 2932 nm ($7P_{3/2}$ to $7S_{1/2}$). All other transitions required at least a ten times higher density. A similar experiment with a 5.0 cm cell in a magnetic field again confirmed that the 1469 nm transition requires a four times higher density than the 2932 nm transition

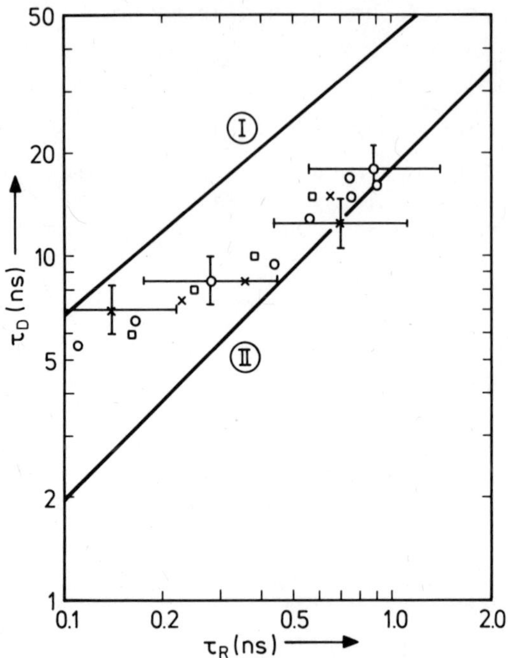

Figure 12. Delay time τ_D versus superfluorescence time τ_R. Crosses (2.0 cm beam); open circles (3.6 cm beam); open squares (5.0 cm cell). Curves I and II show the delay times predicted by MacGillivray and Feld (ref. 5) and by Bonifacio and Lugiato for single-pulse output (Ref. 6), respectively. Horizontal error bars indicate the random errors in τ_R caused by uncertainties in the density measurements.

for the same delay time. It follows from these results that competition plays only a minor part and can be neglected.

IV. DISCUSSION OF RESULTS

The experimental results will now be considered in the light of recent theoretical work. The data will be compared with the mean-field theory of Bonifacio and Lugiato (BL)[6], and with the numerical analysis of MacGillivray and Feld (MF)[5]. In the BL theory the inhomogeneous electromagnetic field generated by the oscillating dipoles is replaced by a mean field, common to all dipoles. This approximation enables the authors to obtain analytical results. In the MF treatment the coupled Maxwell-Bloch equations are solved numerically. A strict comparison of experiment and MF theory requires numerical calculations for the particular experimental conditions. Such calculations are not yet available for the Cs experiment. Therefore, use will be made of the approximate expressions that MF have given for a number of quantities.

The experimental quantities that lend themselves most easily for a comparison with theory are: (1) the occurrence of a regime of single pulses without ringing or oscillations and the transition of such a regime of single-pulse output into a regime of ringing, (2) the pulse delay time τ_D as a function of τ_R, and (3) the pulse width τ_W as a function of τ_R.

The most significant result of the measurements is the observation of single-pulse outputs. For all lengths of the sample volume single pulses occur for sufficiently large values of τ_R. On increasing the density a transition from single pulses to multiple pulses is found. A survey of the results is shown in Table II. Here, $\tau_{R_{osc}}$ is the largest value of τ_R for which multiple pulses have been observed. For $\tau_R > \tau_{R_{osc}}$ the output always consisted of a single pulse. From the MF calculations it appears that ringing should be present in the superfluorescence output. Such ringing is similar to that described by Burnham and Chiao[15]; it depends logarithmically on the atomic density. MF find the number of pulses or lobes K to be given by

$$K \approx \frac{|\ell n(\theta_o/2\pi)|}{4} \quad , \tag{4}$$

where θ_o is given by

$$\theta_o \approx \{ \sqrt{2\pi}\ N\ (\sqrt{\alpha L})^3 \}^{-1/2} \quad . \tag{5}$$

Here N is the total number of initially excited atoms and αL is the initial amplitude gain at line center.

In the present experiment K ≈ 3 to 4, essentially independent of the density over the range of densities involved. A decrease of the ringing or even a complete suppression of it for long delay times, however, could conceivably take place through inhomogeneous dephasing or homogeneous relaxation processes. It is difficult, however, to explain the results of Table II in that way. For an

	L(cm)	τ_E	$\tau_{R_{osc}}$	$\tau_{D_{osc}}$	Comment
Cell	1.0	0.035	< 0.13	< 5	No ringing observed
Beam	2.0	0.07	0.15	6-7	Weak
Beam	3.6	0.12	0.25	8-9	
Cell	5.0	0.17	0.4	10	Strong ringing occasionally

Table II. Survey of results with respect to the occurrence of multiple pulses. Here L is the length of sample; τ_E, the escape time = L/c; $\tau_{R_{osc}}$ the largest value of τ_R for which multiple pulses have been observed and $\tau_{D_{osc}}$ is the delay time corresponding to $\tau_{R_{soc}}$. All times are in ns.

atomic beam of 2 cm length, single pulses have been recorded for delay times as short as 8 ns, i.e., about 4 times shorter than the dephasing time T_2^* and nearly ten times shorter than the relaxation times T_1 and T_2'. On the other hand, multiple pulses have been found from a cell of 5 cm length for delay times as long as 10 ns, i.e., nearly twice as long as the dephasing time T_2^* in the cell. Thus, the conclusion seems justified that the single pulses and the multiple pulses in the cesium experiment are basically different from the ringing studied by MF. It can be seen from Table II that $\tau_{R_{osc}}$ is related to the length of the sample. In fact, it is found that $\tau_{R_{osc}} \approx 2\ \tau_E$. Even though the range of lengths is limited, and even though the random errors in the values of τ_R are sizeable, a correlation between the sample length and $\tau_{R_{osc}}$ seems well established. Such a correlation has been anticipated by BL. In fact, from their work single pulses are expected when the cooperation length is much longer than the sample length, or $\tau_R \gg \tau_E$, while oscillations in the output are predicted when this condition is no longer satisfied.

In the BL theory the oscillatory output is described as that of an underdamped pendulum. Experimentally the situation seems more complex. In the regime of multiple pulses large fluctuations exist in the pulse shapes, i.e., for a given length of the sample and for a fixed delay time of the first pulse, large fluctuations occur from shot to shot in the peak intensity of the second pulse relative to that of the first pulse and in the delay of the second pulse with respect to the first pulse. These fluctuations are so pronounced that it seems more appropriate to speak of multiple pulses than of either ringing or oscillations.

Next will be discussed the delay time as a function of τ_R. According to MF the delay time is related to τ_R by the following expression:

$$\tau_D \approx (\tau_R/4)\ [\ell n\ (\theta_o/2\pi)]^2 \quad . \tag{6}$$

Equation 6 predicts τ_D to be nearly proportional to τ_R. According to BL, one may expect[16]

$$\tau_D = \tau_R\ \ell n\ N \quad , \tag{7}$$

provided $\tau_R \gg \tau_E$; when the latter condition is not satisfied an increase of the delay over that predicted by Eq. 7 will occur. Again, a linear relation between τ_D and τ_R can be anticipated from expression (7). As can be seen from Fig. 12, the experimental results lie in between the curves I and II, which correspond to Eqs. 6 and 7 respectively. For decreasing values of τ_R, the delay time decreases more slowly than for either of the theoretical curves. As already mentioned above, a relative increase of τ_D has been foreseen by BL when τ_R approaches τ_E. From their argument, a dependence of τ_D on the sample length for a given τ_R might be expected; the experiment has not yielded evidence for such a dependence.

Finally the pulse widths will be discussed. According to BL the output intensity as a function of time is given by

$$I(t) = I_{max}\ \mathrm{sech}^2\left(\frac{t - \tau_D}{2\ \tau_R}\right) \quad , \tag{8}$$

in the regime of single pulses. From expression (8) it follows that $\tau_W = 3.52\ \tau_R$. According to MF one has

$$\tau_W \approx \tau_R \ |\ell n(\theta_o/2\pi)| \quad . \tag{9}$$

For the present experiment, expression (9) predicts $\tau_W \approx 14\ \tau_R$. Combining the data presented in Figs. 11 and 12, one finds that experimentally τ_W equals about 10 to 15 times τ_R for $\tau_R > 0.3$ ns. The experimental result is thus close to the MF value.

In conclusion, single pulse SF has been observed. The absence of ringing cannot be explained by relaxation processes. With increasing density the single pulses give way to multiple pulses, which seem to be different both from the MF-type ringing and from the BL-type oscillations. The width of the pulses is larger than foreseen by BL and close to the estimate of MF. The delay time varies more slowly with τ_R than predicted by either theory. All experimental results discussed in this section have been obtained with a Fresnel number of one. Further experiments, with different Fresnel numbers, are being planned to establish the possible influence of the transverse sample dimensions on the superfluorescence output.

V. SUMMARY

It has been shown that a completely inverted two-level system can be prepared in the $7P_{3/2}$ to $7S_{1/2}$ manifold of atomic cesium in a transverse magnetic field. Both in a cell and in an atomic beam the excited state density can be determined with fair accuracy. The use of an atomic beam allows one to satisfy the Bonifacio and Lugiato conditions for the observation of "pure superfluorescence". For long delay times single-pulse outputs are observed; for shorter delay times, depending on the length of the sample, the single pulses break up into two or more pulses. The multiple pulses seem to be different, both from the ringing predicted by MacGillivray and Feld and from the oscillations predicted by Bonifacio and Lugiato. Pulse widths and delay times take on values in between those predicted by the above mentioned authors.

ACKNOWLEDGEMENTS

The author would like to thank Prof. D. Polder for illuminating discussions and Mr. H. M. J. Hikspoors for performing many of the measurements.

References

(1) R. H. Dicke, Phys. Rev. 93, 99 (1954).

(2) N. E. Rehler and J. H. Eberly, Phys. Rev. A 3, 1735 (1971).

(3) R. Bonifacio, P. Schwendimann, and F. Haake, Phys. Rev. A 4, 302 and 854 (1971) and references therein.

(4) N. Skribanowitz, I. P. Herman, J. C. MacGillivray, and M. S. Feld, Phys. Rev. Lett. 30, 309 (1973); I. P. Herman, J. C. MacGillivray, N. Skribanowitz, and M. S. Feld in Laser Spectroscopy, R. G. Brewer and A. Mooradian, editors (Plenum, N.Y. 1974).

(5) J. C. MacGillivray and M. S. Feld, Phys. Rev. A 14, 1169 (1976).

(6) R. Bonifacio and L. A. Lugiato, Phys. Rev. A 11, 1507 (1975) and 12, 587 (1975).

(7) M. Gross, C. Fabre, P. Pillet, and S. Haroche, Phys. Rev. Lett. 36, 1035 (1976).

(8) A. Flusberg, T. Mossberg, and S. R. Hartmann, Phys. Lett. 58A, 373 (1976).

(9) A. Eranian, P. Dezauzier and O. De Witte, Optics Commun. 7, 150 (1973); A. Andreoni, P. Benetti, and C. A. Sacchi, Appl. Phys. 7, 61 (1975); B. Bölger, L. Baede and H. M. Gibbs, Optics. Commun. 19, 346 (1976).

(10) R. E. Honig, RCA Review 23, 567 (1962); D. R. Stull in American Institute of Physics Handbook, D. E. Gray, editor (McGraw-Hill, N.Y., 1972), p. 4-298.

(11) N. F. Ramsey, Molecular Beams (Oxford University Press, Oxford, 1956).

(12) The collision cross-section for Cs-Ar collisions is as large as 572×10^{-16} cm^2 (ref. 11, p. 34). In view of the larger polarizability of Cs as compared to Ar, the cross-section for Cs-Cs collisions can be estimated at 1×10^{-13} cm^2. A pressure of 10 torr, corresponding to a density of 1.3×10^{17} cm^{-3}, then yields a mean free path of 0.6×10^{-4} cm.

(13) J. B. Anderson in Molecular Beams and Low Density Gas Dynamics, P. P. Wegner, editor (Marcel Dekker, N.Y., 1974).

(14) A. Kantrowitz and J. Grey, Rev. Sci. Instr. 22, 328 (1951).

(15) D. C. Burnham and R. Y. Chiao, Phys. Rev. 188, 667 (1969).

(16) Following Rehler and Eberly, ref. 2, Banfi and Bonifacio have used the relation $\tau_D = \tau_R \ln \mu N$, where μ is the geometrical shape factor of ref. 2. Inclusion of the factor μ in the logarithm reduces the delay time by roughly a factor of 2 in the present experiment. G. Banfi and R. Bonifacio, Phys. Rev. A. 12, 2068 (1975).

VISIBLE COOPERATIVE EMISSION IN INCOHERENTLY EXCITED COPPER VAPOR

T. W. Karras, R. S. Anderson, B. G. Bricks, and C. E. Anderson

General Electric Company, Space Sciences Laboratory

King of Prussia, Pennsylvania

Abstract: It is suggested that optical emissions from copper vapor excited by a pulsed electrical discharge are caused by cooperative emission. Pulse shape evolution and several quantitative comparisons support this conclusion.

I. INTRODUCTION

The few experiments that have been used to corroborate theoretical predictions regarding cooperative emission (superradiance or superfluorescence) have generally used coherent preparation of the medium in the 1-100 nsec time scales[1-4]. This has allowed analytical treatment of the data and has led to limited quantitative agreement with theory.[1,2] However, there is as yet no description of a single set of experiments displaying the full range of expected pulse behavior (e.g., pure to oscillatory superfluorescence[5,6]).

This paper will describe a set of experiments that shows the evolution of single hyperbolic-secant shaped pulses into strong oscillatory structures. The interpretation of this data as cooperative emission will also be supported by limited quantitative evidence.

Two factors both complicate the analysis and add to the interest of the data. The wavelength of the emission is in the visible, and the medium is incoherently excited by electrical discharges lasting tens of nanoseconds. Consequently, the time scale of the optical pulse delay generally causes the emission to take place while discharge current is still flowing.

Further complicating the analysis, the characteristic dephasing time, as measured by the Doppler line width, is significantly shorter than that of the observed optical pulse. However, as indicated in earlier work[1], this does not preclude an interpretation of cooperative emission since the gain of the devices used is very high. That is, if the dipole coupling time is shorter than the dephasing time, a net dipole moment can be built up.

A complete quantitative analysis will not be attempted. Only a presentation of the data and support for our interpretation of first identification of cooperative emission both in the visible and with incoherent excitation will be given.

It is important to note that the devices used in these experiments are of potentially great practical significance. They are currently contemplated for use in several ERDA and DOD applications.[7,8] Thus, a thorough understanding of their behavior would be very useful.

II. THE PULSED COPPER VAPOR LASER

Pulsed metal vapor lasers are a class of device first described by Walter et al.[9] Operation follows a cyclic pattern (see Fig. 1 which shows the energy level diagram of the copper atom as an example) in which electrical discharge excitation of a resonance level is followed by transition to a lower metastable level and consequent optical emission. Subsequent relaxation of the metastable to the ground state allows a repeat of the process with another discharge pulse. Trapping of the resonance radiation between the upper laser state and the ground state with adequate metal atom density insures that the transitions to the metastable lower laser level are favored.

While many metals have been used, the copper vapor laser has been the most attractive. The cross-section for excitation by electron collision of the upper laser state is very large,[10] and it shows great promise as an efficient, moderate power device.[9,11,12] As a result, information regarding experiments using this medium is readily available and will be emphasized here.

Data will be presented that was obtained with two different kinds of devices (i.e., flowing and discharge heated). In the flowing system[13] a stream of copper vapor passed through a chamber filled with neon and was excited by a fast pulsed transverse discharge operating at rates varying from a few hundred hertz to over 20 kilohertz. An excited copper medium with distorted elliptical

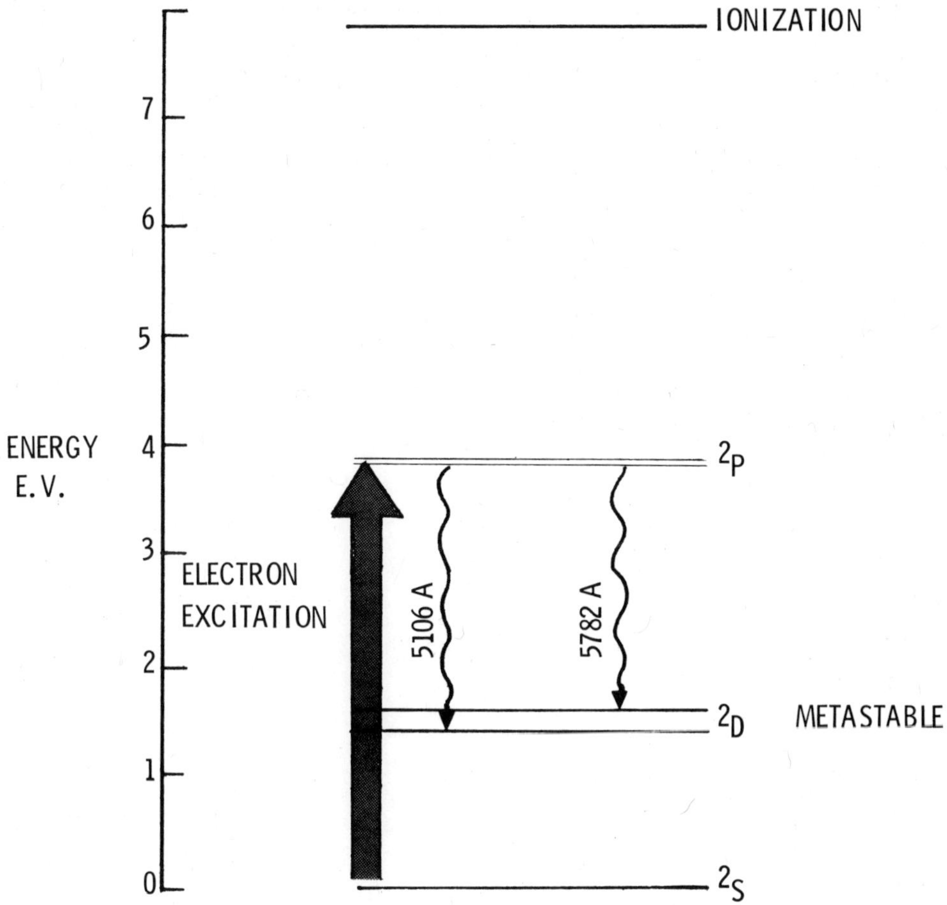

Figure 1. Copper Atom Energy Level Diagram.

cross-section resulted. The emitted power was measured with an EG&G radiometer system or a 1P28 photomultiplier tube and displayed on a Tektronix 7904 oscilloscope. A discharge-heated system[12,14] was used for most experiments. A ceramic discharge tube with electrodes sealed at each end was surrounded by a radiation shield and was filled with helium or neon. A fast pulsed discharge operating at 1-20 kHz ultimately heated the tube and small pieces of copper within it, producing the desired copper vapor pressure. At this point the discharge produced excitation in the cylindrical volume of copper vapor. Optical power and waveforms were measured with the EG&G radiometer, a Hamamatsu photodiode and a Scientec calorimeter.

In most experiments with either system, a 99.9% reflectivity mirror and a glass flat were used to form an optical resonator. The results of this laser configuration will be used in most of the discussion that follows with the output pulse energy (the number of photons emitted) being the measure of the number of cooperating atoms, N. However, a limited number of experiments with no mirrors were performed producing similar pulse shape evolution.

Table 1 gives a list of some relevant characteristics. The 510.5 nm emission overwhelmingly dominated that at 578.2 nm except in strongly oscillating pulses when it fell to 70% of the total.

Table 1. Relevant Copper Vapor Laser Characteristics

Wavelength (nm)	510.5	578.2
Single Atom Decay Constant[15] γ_o (sec^{-1})	0.195×10^7	0.206×10^7
Measured Laser Line Width (Hz)	$2.5\text{–}5.0 \times 10^9$	

III. QUALITATIVE PULSE SHAPE BEHAVIOR

In previous experiments involving cooperative emission, the most convenient method of displaying optical pulse shape behavior was to vary the number of cooperating atoms. Since our experiments were generally conducted so that this parameter increased (i.e., as the system heated up), the following description of optical pulse shape evolution will follow that pattern.

Just above threshold the temporal pulse width and the delay with respect to the discharge current were largest. As N increased, the pulse narrowed, the delay decreased, and the peak power rose rapidly. At some point the rate of pulse width decrease slowed, and the peak power became proportional to N^2 (see Section IV). The delay was no longer large, but it continued to decrease. Figure 2 shows examples of these regimes.

With continued increase in N, the pulse narrowed further, and the delay decreased to the point that emission and discharge current began to overlap strongly. Ultimately the smooth pulse shape was observed to lengthen and then develop some structure on its trailing edge. This evolution can be seen on Figure 3.

The structure of the trailing edge increased in amplitude and could be resolved into a series of oscillations when N was increased further. In addition, and in contrast to the pure single pulse regime discussed above, this increase of the number of cooper-

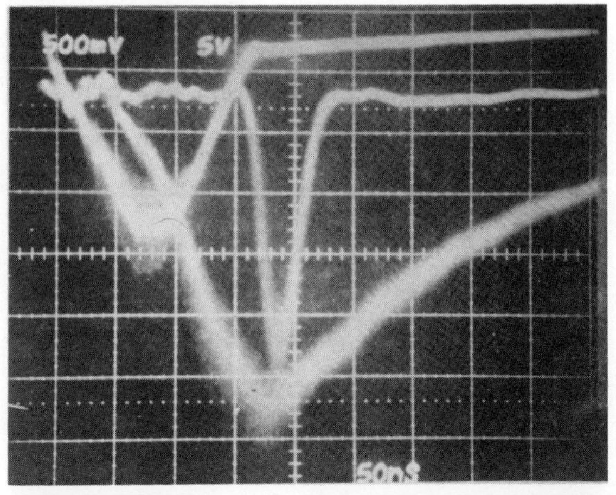

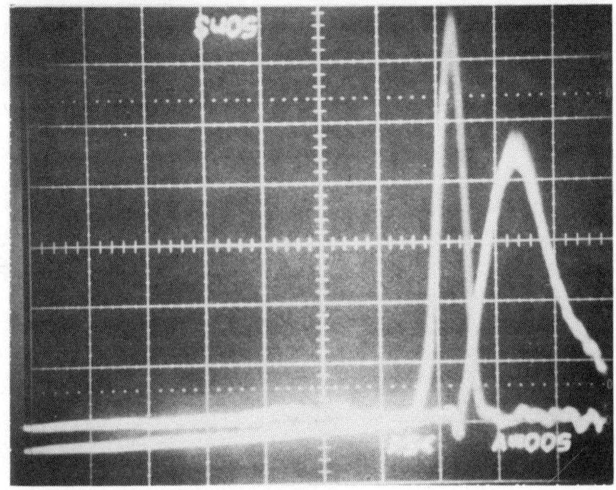

Figure 2. Emission from Flowing Copper Vapor Excited in Transverse Discharge. Sweep rate 50 nsec/div. for both. Amplitudes uncalibrated. Emission was largely 510.5 nm. Emitting region had a rectangular-elliptic cross-section 1.3 cm × 0.3 cm and was about 4 cm long. (a) $N \sim 0.3 \times 10^{14}$. Traces from top to bottom on the right are discharge current, optical emission pulse with mirrors and emission pulse without mirrors. (b) $N = 1.8 \times 10^{14}$. The early pulse is discharge current and the late pulse optical emission with mirrors.

ating atoms caused comparatively little decrease in the width of each of these oscillations. Furthermore, the emission now began

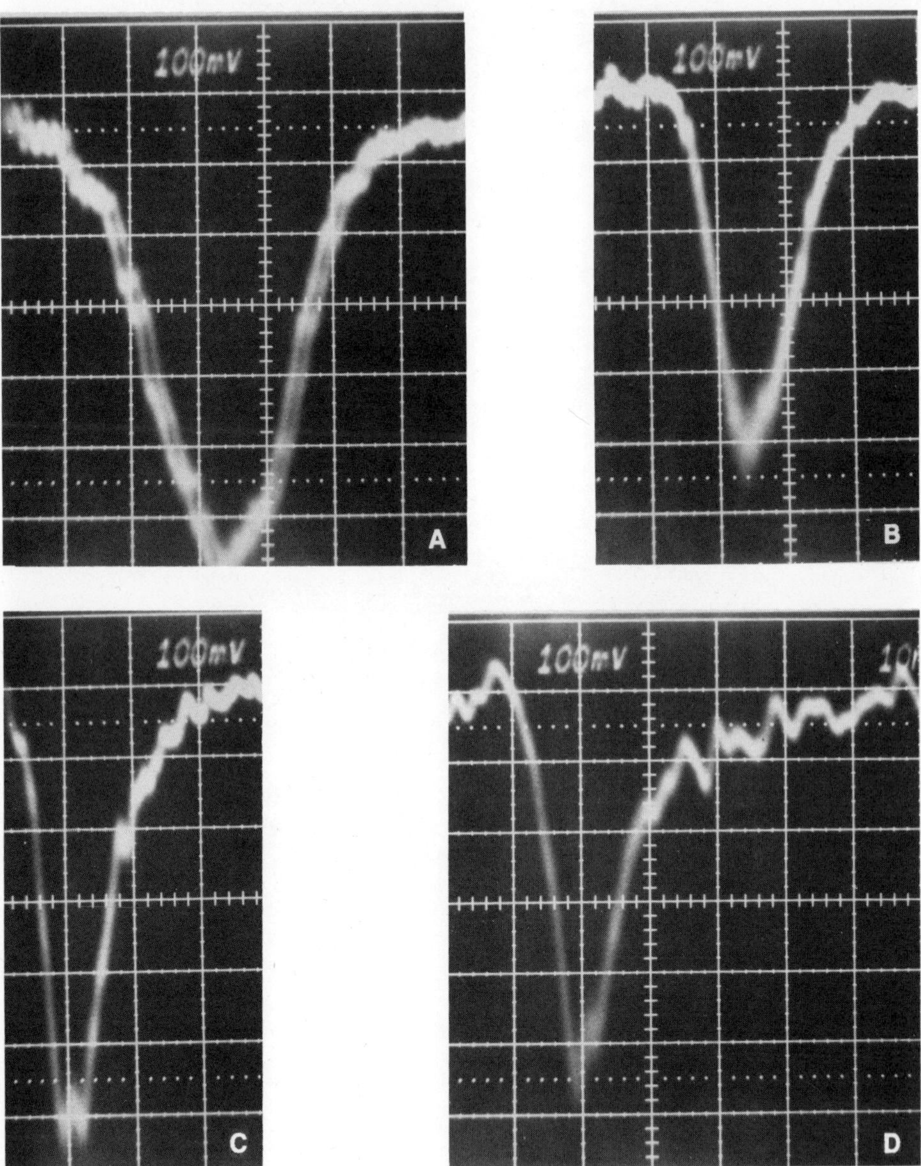

Figure 3. Emission from Discharge-Heated Copper Vapor: Nominal 1-1/2" Tube. These photographs show the complete evolution of a single pulse from threshold to the development of incipient oscillation. Tube clear inside diameter was 3.7 cm and hot zone about 45 cm long. (a) Threshold; 20 nsec/cm. (b) $N = 0.65 \times 10^{14}$; 20 nsec/cm. (c) $N = 6 \times 10^{14}$; 20 nsec/cm. (d) $N = 11.6 \times 10^{14}$; 10 nsec/cm.

early in the discharge and further decrease in the delay was not measurable (see Ref. 14). Figure 4 summarizes this development.

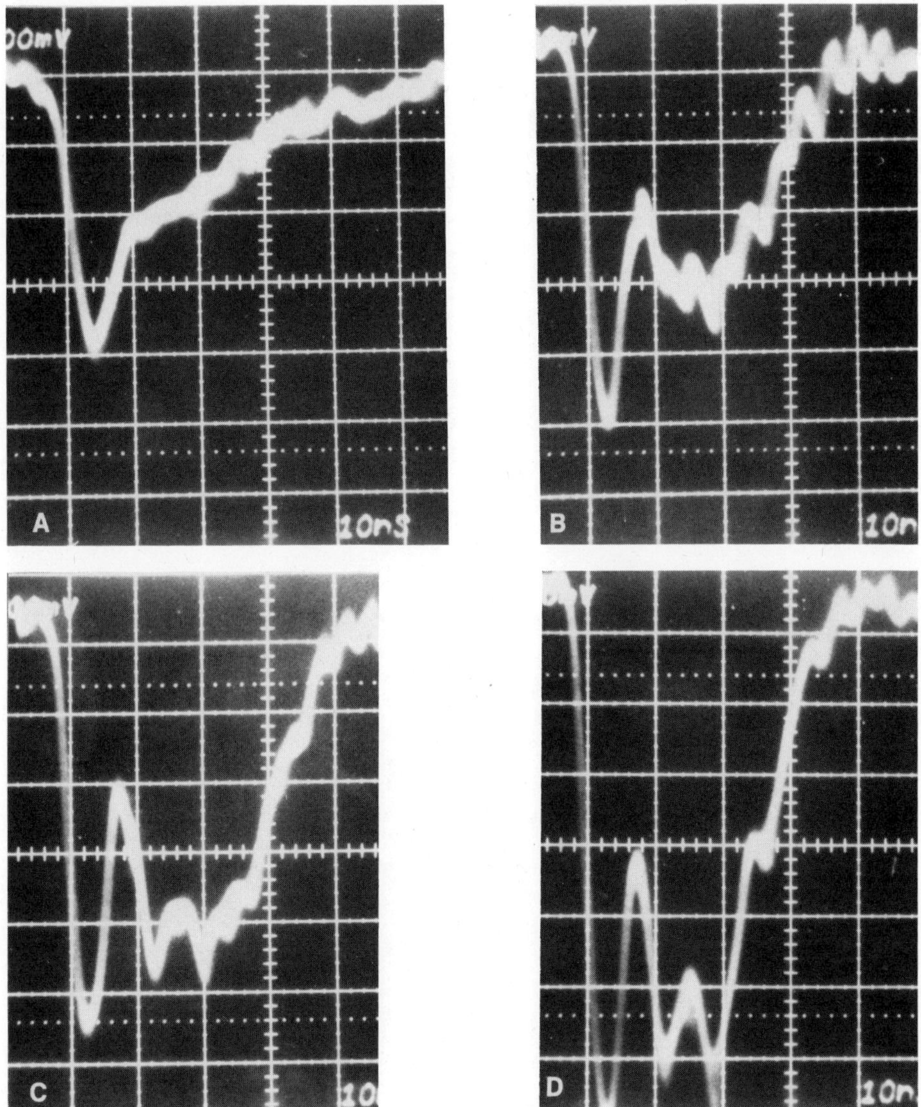

Figure 4. Emission from Discharge-Heated Copper Vapor: Nominal 7/8" Tube. These photographs show the evolution of emission into an oscillatory form. Sweep was 10 nsec/cm. Tube clear inside diameter was 2.1 cm and hot zone length about 20 cm.

(a) $N = 6.6 \times 10^{14}$. (b) $N = 13.2 \times 10^{14}$. (c) $N = 21.0 \times 10^{14}$.

(d) $N = 24.4 \times 10^{14}$.

The effect of the removal of mirrors from the system when it was operating in this regime was to produce a temporal pulse shape similar to that obtained at moderately lower N. That is, an experiment producing an oscillating pulse such as those in Figure 4 would, without mirrors, produce a single pulse of about ten nanoseconds FWHM.

The qualitative behavior with variations of discharge intensity was the same without mirrors as that described for measurements with mirrors except that the strong oscillatory regime was not observed. The spatial divergence of the emission was considerably greater than that obtained with mirrors, and consistent data are consequently not available to establish the value of N for such experiments.

Pulse shape evolution has been observed to continue the pattern indicated in Figure 5 with increased N leading to an increasing number of oscillation lobes and greater overall pulse width.

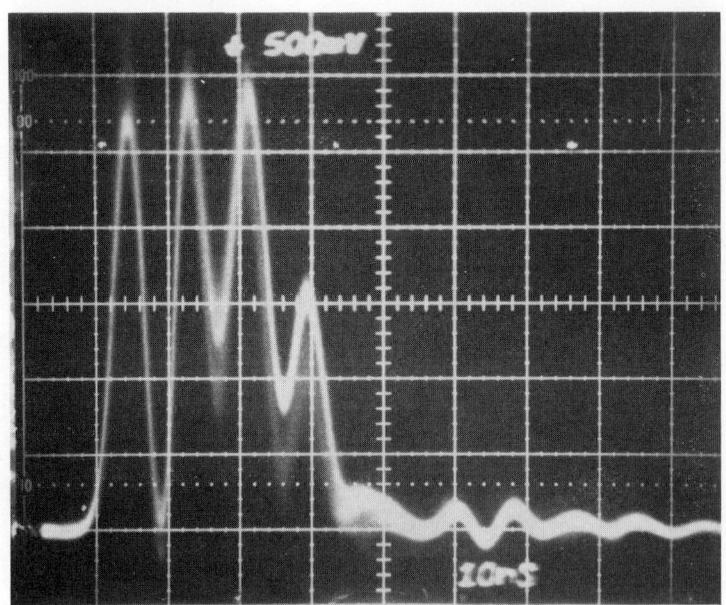

Figure 5. Emission from Discharge-Heated Copper Vapor: Nominal 3/8" Tube. Tube clear inside diameter was 0.85 cm and hot zone length about 8.5 cm. $N = 5.7 \times 10^{14}$.

It should be noted that as smaller diameter discharge tubes were used, smaller N was needed to produce a pulse with oscillation. An interesting contrast can be made between Figure 5, Figure 4(a), and Figure 3(c). All three have about the same value of N but their pulse shape varies dramatically with the smallest tube producing the narrowest pulses and having the strongest oscillation.

IV. ANALYSIS

The basic pattern just described follows that expected for cooperative emission.[1-6] Quantitative comparisons can also be made. Using data obtained with a nominal 1-1/2" inside diameter discharge tube [and only the N values between those corresponding to Figures 3(b) and 3(c)], the peak power has been plotted as a function of N^2 (see Figure 6). As expected[1-6] there is a linear fit. Additionally, optical pulse shape fits a hyperbolic secant form (see Figure 7) as predicted,[1,5,6] except during the rise where it occurs simultaneously with the discharge current.

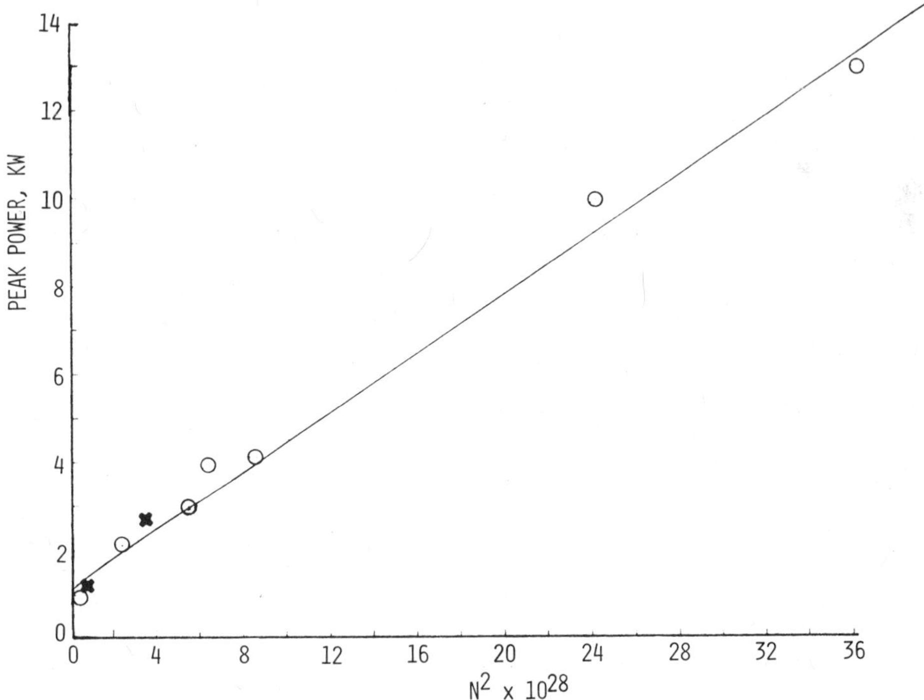

Figure 6. Peak Power Dependence Supporting Interpretation of Cooperative Emission. Data represented by spots obtained with nominal 1-1/2" tube (see Figure 3) and represented by crosses with flowing system (see Figure 2).

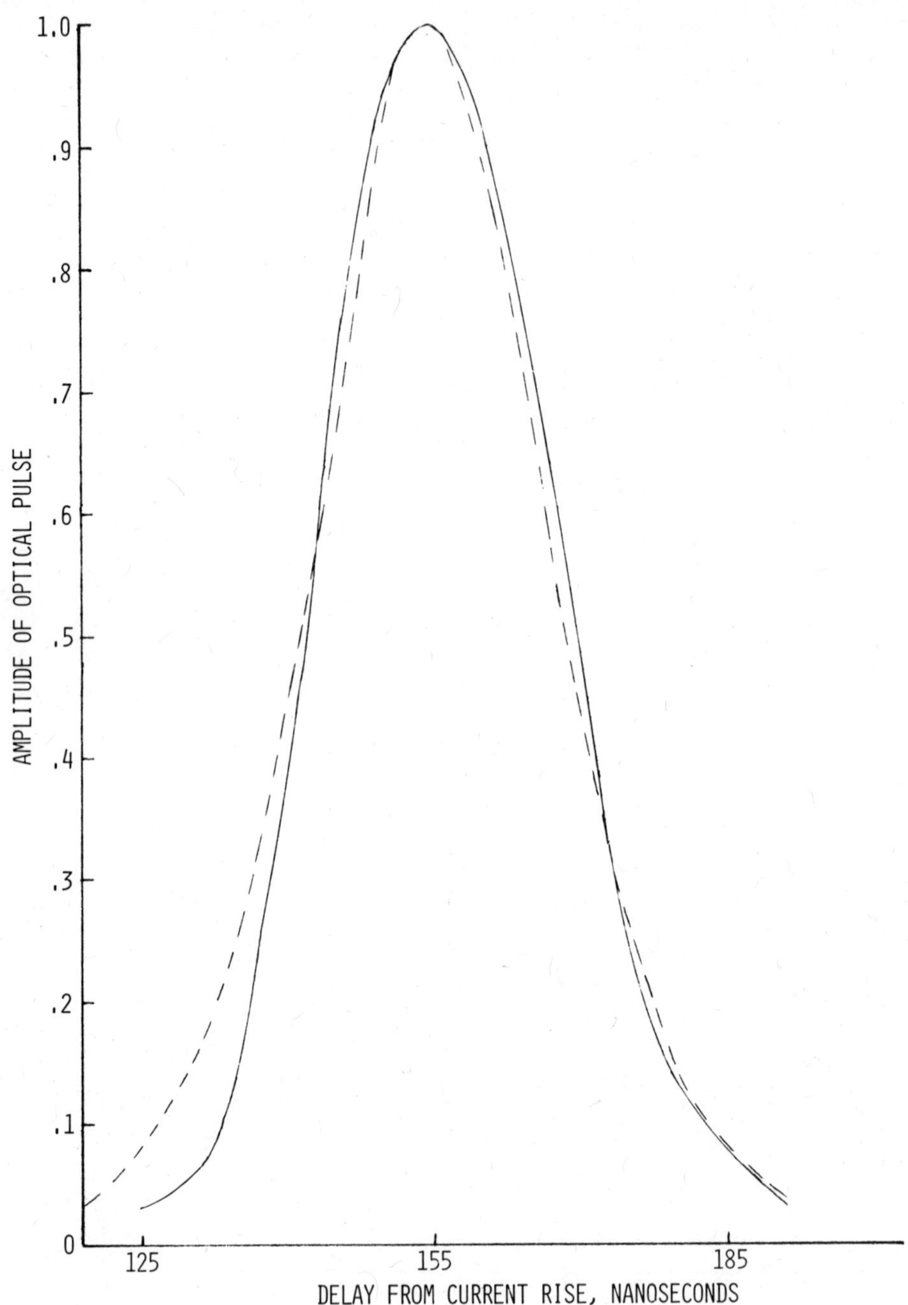

Figure 7. Optical Pulse Shape. Experimental pulse shape (solid line) is fit with sech^2 (t-155/15.3) (dashed line) where t is in nanoseconds.

The minimum value for N required to produce peak power proportional to N^2 can be determined from Equation 45′ of Reference 6 to be

$$N_{min} = \frac{\gamma_t \; 4\pi \; V}{\gamma_o \; \lambda^2 \; L} \quad .$$

This can be applied to the data of Figures 2 and 6. Using the inverse line Doppler width at half maximum for γ_t[16] and 90 cm for L, since a high reflectivity mirror is used at one end,[5] the minimum number of atoms necessary for cooperative emission is 0.1×10^{14}. If the data supporting Figure 6 is extended to values of N^2 below this minimum number, indeed the peak power is no longer proportional to N^2.

The treatment of Reference 4 can be used with the concept of effective dephasing time[1] $[(T_2^*)_{eff} \; T_2^* \; \alpha L]$ to fit a pulse shape with N just beyond the end of the curve in Figure 6 [e.g., Figure 3(d)]. The dephasing time, T_2^*, is 1.4×10^{-9} sec.[16] Since measurements indicate that for conditions similar to those of Figure 3(d) $\alpha L \sim 5$, then $(T_2^*)_{eff} \sim 7$ nsec. Reference 5 indicates that exp $(-t/T_2^*)$ will begin to dominate optical pulse decay when $T_2^* \sim t_R \; \ell n(N)/2$ (determined from Reference 5), as is roughly the case here. An examination of Figure 3(d) indicates an exponential decay with constant of about 8 nsec, close to that of the corresponding $(T_2^*)_{eff}$. The experiment producing Figure 5 gives similar agreement. The measured gain was $\alpha L = 18$. Consequently, now $(T_2^*)_{eff} \approx 25$ nsec, in adequate agreement with the nominal 30 nsec decay of the oscillation amplitude in Figure 5.

Finally, in comparing the pulse shapes in Figure 5, Figure 4(a) and Figure 3(c), all different size tubes but with the same N, it should be noted that L/L_c (see Reference 6) decreases monatonically as would be expected.

V. CONCLUSION

We suggest that high gain optical emissions from copper vapor excited by an electrical discharge are, under some conditions, caused by cooperative emission processes. The qualitative evolution of pulse shapes and several quantitative comparisons with theory support this conclusion.

We believe that this is the first such demonstration with incoherent excitation and emission in the visible part of the spectrum.

ACKNOWLEDGMENTS

The authors would like to acknowledge the experimental assistance of L. W. Springer and B. P. Fox.

References

(1) N. Skribanowitz, I. P. Herman, J. C. MacGillivray and M. S. Feld, Phys. Rev. Lett. 30, 309 (1973); and I. P. Herman, J. C. MacGillivray, N. Skribanowitz and M. S. Feld, Laser Spectroscopy, edited by R. G. Brewer and A. Mooradian, Plenum Press, 1975.

(2) M. Gross, C. Fabre, P. Pillet and S. Haroche, Phys. Rev. Lett. 36, 1035 (1976).

(3) Papers by H. M. Gibbs, Q. H. F. Vrehen, S. R. Hartman and T. A. DeTemple. Presented at the First Cooperative Effects Meeting, Redstone Arsenal, Alabama, December 1, 1976.

(4) Q. H. F. Vrehen, H. M. J. Hikspoors and H. M. Gibbs. Submitted for publication to Phys. Rev. Lett.

(5) R. Bonifacio and L. Lugiato, Phys. Rev. 11, 1507 (1975) and 12, 5871 (1975).

(6) G. Banfi and R. Bonifacio, Phys. Rev. 12, 2068 (1975).

(7) A. A. Pease and W. M. Pearson, Appl. Optics 16, 57 (1977).

(8) G. D. Ferguson, SPIE Ocean Optics 64, 15D (1975).

(9) W. T. Walter, N. Solimene, M. Piltch and G. Gould, IEEE JQE QE-2, 474 (1966).

(10) W. Williams and P. S. Trajmar, Phys. Rev. Lett. 33, 187 (1974).

(11) G. G. Petrash, Sov. Phys. Uspekhii 14, 747 (1972).

(12) A. A. Isaev, M. A. Kazaryan and G. G. Petrash, JETP Lett. 16, 27 (1972).

(13) B. G. Bricks, T. W. Karras, T. E. Buczacki, L. W. Springer and R. S. Anderson, IEEE JQE QE-11, 57 (1975).

(14) R. S. Anderson, L. W. Springer, B. G. Bricks and T. W. Karras, IEEE JQE QE-11, 172 (1975).

(15) L. Krause, Review of copper data in AFAL-TR-73-439 (1974); also see M. Rieman, Z. Physik 179, 38 (1964).

(16) Determined from 3 times the inverse Doppler width at half maximum (2.09×10^9) as in Reference 2.

SOME EFFECTS OF RADIATION TRAPPING ON STIMULATED VUV EMISSION IN Ar XIII

K. G. Whitney*, J. Davis†, and J. P. Apruzese*

*Science Applications, Inc., Arlington, Va. 22202

†Naval Research Laboratory, Washington, D.C. 20375

Abstract: The transient emission of VUV laser radiation from the carbon-like ionization stage of a laser-heated argon plasma is calculated. The plasma is taken to be spherical in shape, of uniform ion density, and to be uniformly heated by short pulses of laser energy, which are absorbed by the plasma electrons. Excitation of the carbon-like ionization stage is controlled by the trapping of resonance line radiation within the ionization stage. Two resonance lines, along with the VUV laser line, are self-consistently coupled to the rate equations describing the excitation dynamics of the argon plasma. Some effects of this coupling on the output of amplified spontaneous VUV emission are determined by comparing population densities, gain coefficients, and VUV laser power output as functions of time for a number of different cases in which the amount of radiation coupling to the plasma is varied.

I. INTRODUCTION

Two important properties of high density plasmas, which are generated and heated by nanosecond or subnanosecond laser pulses, are their lack of ionization equilibrium and their potential for large line optical depths.[1-4] The temperatures and ion densities within these plasmas can be, also, in general, highly non-uniform since the laser radiation that heats the plasma reacts strongly with it, leading to filamentation and/or focusing of the light pulse,[5,6] to plasma density profile modifications,[7,8] and, in the case of laser-solid target interactions, to resonant absorption and scattering of the pulse in strong density gradients.[9,10] Non-equilibrium ionization and radiation transport processes that occur in the plasma will acquire these same gradients. Thus, even though the line and continuum radiation emitted from a laser heated plasma offers one of the best means for diagnosing the plasmas' rapidly evolving density and temperature structure, this information must be carefully and self-consistently deconvolved from a wide variety of measurements having both spatial and spectral resolution.

It has been argued for a number of years that laser produced plasmas will have a third important property that derives from or is enhanced by the first. It is thought that the nonequilibrium ionization dyanmics can, under favorable conditions, produce large level inversions and significant outputs of coherent VUV radiation.[11-16] Population inversions in lower density, electron discharge heated plasmas have already been obtained in some of the low-lying ionization stages of low Z elements.[17-20] These ion laser plasmas are a current source of coherent UV radiation. By scaling to plasmas with larger atomic numbers at higher ion densities and by using shorter, more energetic, heating pulses, one expects to find similar behavior further up these same isoelectronic sequences.[21,22] However, as the frequency of the laser transition and the plasma density increase, the size of the gain medium needed for wave amplification will decrease. The concentration of pump energy into smaller plasma volumes may add to the difficulties of obtaining large and uniform population inversions in cylindrically shaped laser media. If the plasma is poorly shaped, radiation feedback may quench the inversion. In addition, the laser intensity pattern will reflect the fact that many modes will be amplified when the plasma does not have a long axis of maximum gain.

To examine some of these effects we will present results of gain calculations in a spherically shaped argon plasma that are based on a five level laser scheme in Ar XIII,[21,22] the twelve-times ionized, carbon-like ionization stage. Time independent and time dependent calculations are compared. Radiation from both the lasing transition and from two resonance lines, which terminate on the Ar XIII ground state, is transported throughout the plasma and self-consistently coupled to the excitation and ionization dynamics of the Ar XIII ionization stage. Trapping of one of the resonance lines indirectly pumps the upper laser level increasing the population inversion, while absorption of the other counteracts this effect by populating the lower laser level. In these calculations, the laser emission is isotropic; i.e., all free space modes receive the same amplification. Also, two argon plasmas are studied, each at uniform ion densities for which heating by inverse bremsstrahlung absorption of laser radiation is assumed to take place.

II. MODEL

In order to clearly distinguish radiation trapping from hydrodynamic effects, the calculations described in this paper were carried out in spherical stationary plasmas whose total ion densities were constant and uniform. Pump energy was also uniformly deposited throughout the sphere. Thus, the only temperature, ion density, electron density, or pressure gradients that were established in the plasma were set up by gradients in the trapped radiation field energy densities. Different versions of this model have been described in our earlier papers.[2,23] The ions are heated primarily through electron collisions:

$$\frac{3}{2} N_i \, k \, \partial_t \, T_i(r,t) = \frac{3 \, m_e}{m_i} \frac{N_e k}{\tau_e} (T_e - T_i) \quad , \tag{1}$$

where T_e and T_i are the electron and ion temperatures respectively and N_e and N_i, the electron and total ion densities (N_i = constant). τ_e is the electron-ion $\pi/2$ collision time, which itself depends on T_e, N_e, and N_i[24]:

$$\tau_e = 3.5 \times 10^5 \times T_e \, [\mathrm{eV}]^{3/2} \, \frac{N_i}{N_e^2 \log \Lambda \, [T_e, N_e]} \quad . \tag{2}$$

The electrons also undergo a wide variety of inelastic collisions with the ions. In our modeling, we monitor a subset of these interactions by solving a set of rate equations for the ground states of all of the ionization stages and for a select set of excited states in those ionization stages which are being studied in detail:

$$\partial_t \, N_\mu \, (r,t) = \sum_{\nu=1}^{n_s} W_{\mu\nu} \, N_\nu \quad , \quad \mu = 1, \ldots, n_s \quad , \tag{3}$$

where N_μ are population densities of the n_s ion states in the calculation. The rate coefficients $W_{\mu\nu}$ depend on N_e, T_e, and on trapped or stimulated radiation fields within the plasma.

Our argon model contains 19 ground states of the different ionization stages and 12 excited states in Ar XII, XIII, and XIV. Four excited states are added to nitrogen-like Ar XII, and five excited states to boron-like Ar XIV, to build structure around the carbon-like Ar XIII ionization stage, which contains the three excited states, shown in Figure 1, whose behavior we wish to study*. Steady state population inversions between the 3s and 3p states can be collisionally excited over a range of electron densities and temperatures because of the rapid radiative decay of the 3s state and the metastability of the 3p state.[22] In addition, collisional excitation of the Ar XIII ground state favors the pumping of 3p and 3d states[21], while cascade from the Ar XIV ground state will also preferentially populate the states of higher angular momentum. Experimentally, inversions between the $3s\,^3P$ and $3p\,^3D$ states in carbon-like ions have been observed in oxygen and fluorine plasmas.[17] By modeling the Ar XIII dynamics using the five level scheme in Figure 1, the effects on the time evolution of the inversion density of collisional mixing between the 3s, 3p, and 3d states, of couplings of these excited states to the Ar XIII and XIV ground states, and of radiation trapping of the $2p^2$–$2p3s$ and $2p^2$–$2p3d$ resonance lines can be investigated with a minimum of complication in the theoretical model.

Rate coefficients for only four different processes involving ground ionization states are calculated for eqs. (3): collisional ionization and recombination rates between ground states and between ground and excited states,

$$e + N_z(p) \underset{W^{CR}}{\overset{W^{CI}}{\rightleftarrows}} e + N_{z+1}(q) + e \quad ,$$

collisional excitation and deexcitation of excited states,

$$e + N_z(p) \underset{W^{CD}}{\overset{W^{CE}}{\rightleftarrows}} e + N_z(q) \quad ,$$

radiative recombination between ground states and from ground to excited states,

$$e + N_{z+1}(p) \overset{W^{RR}}{\rightarrow} N_z(q) + h\nu \quad ,$$

*NOTE, the $2p^2$–$2p3s$ and $2p^2$–$2p3d$ spontaneous emission decay rates in Figure 1 are corrected values from those originally estimated and used by us in Ref. 21. The statement in that paper, which claimed that the gain could be sustained in time by the inclusion of level structure in Ar XIV, was regretably also in error.

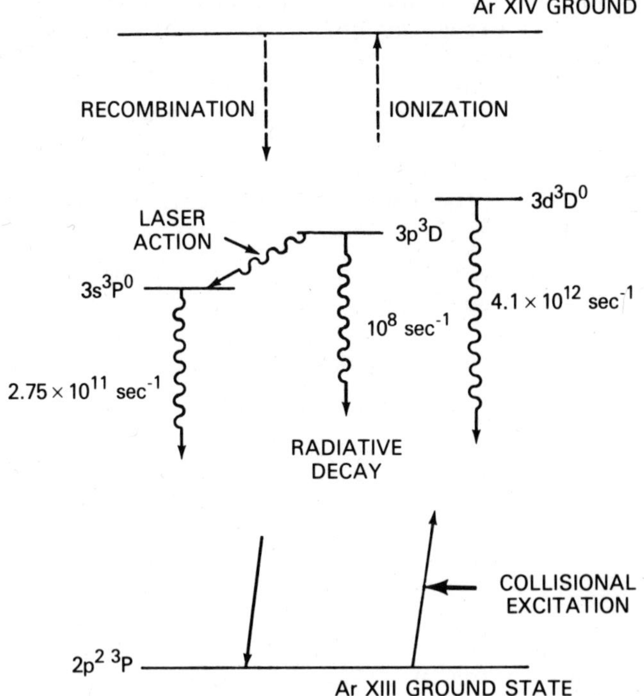

Figure 1. Energy level diagram of the Ar XIII ionization stage. The radiative decay channels are indicated in the diagram along with the calculated decay rates in sec^{-1} of the n = 3 states to the ground state. Not shown on the diagram are the spontaneous radiative decay rates of the 3s-3p and 3p-3d transitions, which are calculated to be 1.2×10^9 and 5.6×10^8 sec^{-1} respectively.

spontaneous decay of excited states,

$$N_z(p) \overset{A}{\rightarrow} N_z(q) + h\nu$$

and stimulated emission and self-absorption of the $2p^2$-2p3s and $2p^2$-2p3d resonance line radiation in Ar XIII:

$$N_z + (n + 1)\ h\nu \underset{\leftarrow}{\overset{BU}{\rightarrow}} N_z^* + nh\nu \Big|_{z\,=\,13} \quad .$$

The calculation of the collisional rate coefficients as functions of temperature is done either by use of analytic formulas for ionization or recombination processes, or by more detailed, distorted wave calculations to obtain the excitation rates. In the latter case, excited state ortho-normalized wave functions are first generated and then excitation cross sections are calculated and averaged over a Maxwellian velocity distribution function yielding excitation rate coefficients. Smooth excitation and deexcitation rates, as functions of T_e, are obtained by detailed balancing and by interpolation. More complete discussions of these procedures can be found in our earlier publications.[2,21,23]

In our model, only the excited states in Ar XIII are collisionally coupled. The calculated rate coefficients for these transitions are listed in Reference 21. The 3s and 3p states are also coupled by stimulated emission and absorption of the VUV laser radiation:

$$N_{3s} + (n + 1)\, h\nu \overset{BU}{\rightleftarrows} N_{3p} + nh\nu \quad .$$

The assumption of complete frequency redistribution is made; thus, the emission and absorption profiles are taken to be identical. Since the 3s-3p line profile has a Doppler core, this coupling of the 3s-3p radiation to the argon plasma excludes the possibility of hole burning within the gain profile.

The radiation field energy densities U_ν to which the 3s, 3p, and 3d Ar XIII states couple are found by solving an equation of radiation transport for an ensemble of specific intensities $I_\nu(r,\mu,t)$[25]:

$$\left\{\mu\, \partial_r + \frac{1-\mu^2}{r^2}\, \partial_\mu\right\} I_\nu\, (r,\mu,t) = -k_\nu(r,t) \left(I_\nu(r,\mu,t) - S_\nu(r,t)\right) . \tag{4}$$

Here, μ is the cosine of the angle between the ray and radius vectors, and k_ν and S_ν are self-absorption and source functions respectively at the frequency ν. For example, for the 3s-3p transition:

$$k_\nu = \frac{c^2}{8\pi\nu^2_{3s-3p}} \frac{A_{3s-3p}}{g_{3s}} \left(g_{3p}N_{3s}(r,t) - g_{3s}N_{3p}(r,t)\right) \phi_{3s-3p}(\nu) \tag{5}$$

and

$$S_\nu = \frac{2h\nu^3_{3s-3p}/c^2}{\left(\frac{N_{3s}(r,t)/g_{3s}}{N_{3p}(r,t)/g_{3p}} - 1\right)} \quad , \tag{6}$$

where $\phi_{3s-3p}(\nu)$ is the line profile function:

$$\int d\nu \ \phi_{3s-3p}(\nu) = 1 \quad . \tag{7}$$

Both Doppler and Voigt functions with time dependent Doppler widths (T_i dependent) are used in our calculations. The energy densities and Poynting vectors $\vec{\Pi}$, which are directed radially, are found from two moments of I_ν:

$$U_\nu(r,t) = \frac{2\pi}{c} \int_{-1}^{1} d\mu \ I_\nu(r,\mu,t) \tag{8}$$

and

$$\Pi_\nu(r,t) = 2\pi \int_{-1}^{1} d\mu \ \mu \ I_\nu(r,\mu,t) \quad . \tag{9}$$

The rate coefficient for stimulated emission of, for example, 3s-3p radiation is then given by

$$W_{3s-3p}\,[U_\nu] = \frac{c^3 A_{3s-3p}}{8\pi h\nu^3_{3s-3p}} \int d\nu \ U_\nu \ \phi_{3s-3p}(\nu) \quad . \tag{10}$$

The coupling of solutions to Eq. (4) to the rate equations is valid provided the plasma sphere is small so that the energy and flux densities, U_ν and Π_ν, adiabatically follow the evolution of the ion densities. When this assumption breaks down, a time derivative must be reinstated in Eq. (4). Solutions to Eq. (4) can be obtained, which accurately describe the progression of the intensity pattern from isotropy at the sphere center to a completely outward directed flux at the surface.[26] The method is illustrated in Figure 2. Rays of varying impact parameter relative to the sphere center are used to construct these patterns. For the calculations of this paper, the sphere is divided into ten shells and ten different flux patterns across these shells are computed.

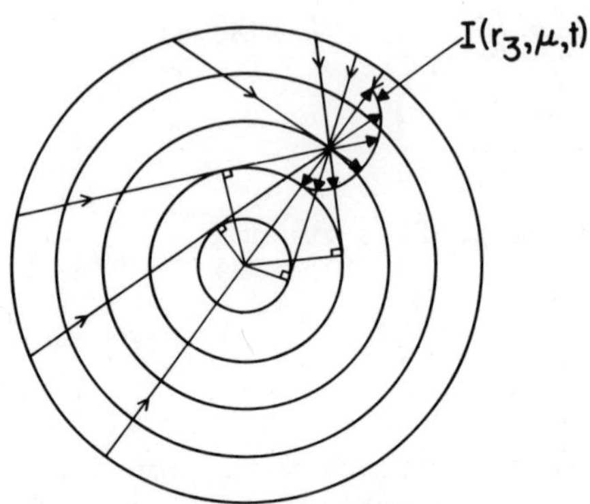

Figure 2. This figure illustrates how the intensity pattern at the third radius position, reading inward, is computed from the four rays of smallest impact parameter in a sphere with a five shell division. μ is the cosine of the angle between the radius and ray vectors ($-1 \leq \mu \leq 1$). The pattern has rotational symmetry about the ray that passes through the center of the sphere.

Each line is divided into ten fixed frequency groups per half line and the assumption of a symmetric line profile is made for the evaluation of the stimulated emission and absorption rate coefficients.

Three moments of Eqs. (3) have special physical significance:

$$N_i = \sum_{\mu} N_{\mu} \quad , \tag{11}$$

$$N_e = \sum_{\mu} Z_{\mu} N_{\mu} \quad , \tag{12}$$

$$E_i = \sum_\mu E_\mu N_\mu \quad . \tag{13}$$

The internal energy E_i, like the ion thermal energy, is generated at the expense of the thermal energy 3/2 $N_e kT_e$ of the electrons; i.e., the total of these stored energies is equal to the balance between laser pump energy S_e absorbed in the sphere, and the radiation and thermal conduction losses from it:

$$\partial_t \left\{ \frac{3}{2} N_e kT_e + \frac{3}{2} N_i kT_i + E_i \right\} = S_e(t) + \frac{1}{r^2} \partial_r \left\{ r^2 \kappa_e \partial_r T_e \right\} - R_i(r,t) - R_b(r,t) \quad , \tag{14}$$

R_i and R_b are the volumetric line plus radiation recombination and bremsstrahlung loss rates respectively.[2] The expression for κ_e was taken from Braginskii:[24]

$$\kappa_e = 12 N_e k \left(\frac{k T_e \tau_e}{m_e} \right) \quad . \tag{15}$$

Equations (1), (3), and (14) were solved using a predictor-corrector algorithm.[27] The radiation field was updated on the integration time scale of the rate equations. The update rate was checked by doubling it in one case. No changes in the results were observed. The choice of the radiation field frequency groups and of the ray impact parameters was also varied. In this case, the results changed slightly but there were no changes of major consequence. The computations were energy, ion number, and photon number conserving.

III. RESULTS

Theoretical calculations have predicted that a steady state population inversion between the 3s and 3p states in carbon-like ions is sustainable for a range of ion densities over which the system could, in principle, be operated as a cw laser.[22] If the ion density is too high, however, the 3s and 3p states collisionally mix faster than the decay of the 3s state and the inversion is lost. These effects are illustrated in Figure 3. The argon plasma is taken to be in collisional-radiative equilibrium at a given electron temperature and the gain coefficient is calculated from steady state solutions to Eqs. (3) for variable N_i. This

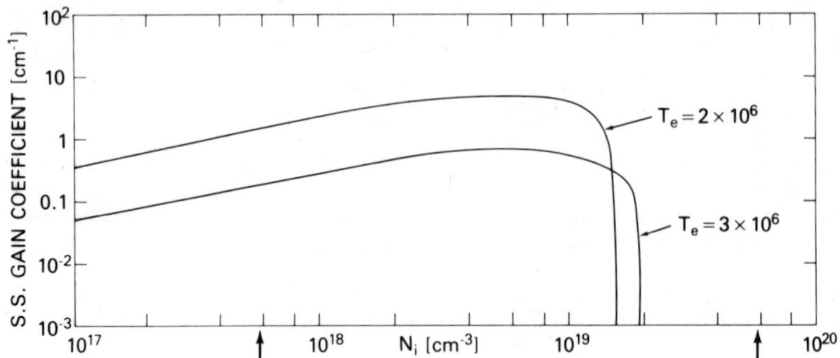

Figure 3. The 3s–3p steady state gain coefficient as a function of ion density for two electron temperatures. The arrows on the ion density axis mark the two fixed densities at which the time dependent calculations were made.

calculation was carried out at electron temperatures of 200 and 300 eV. At the higher temperature, the inversion extends to a higher ion density before cutting off; however, the peak value of the gain coefficient is roughly 8 times smaller than it is at the lower electron temperature. This behavior is demonstrated more clearly in Figure 4. In this case, the steady state gain coefficient and ground state densities are calculated for variable T_e and fixed N_i. The gain reaches its maximum value when the electron temperature is approximately 200 eV and the Ar XIV and XV ground state populations are peaking. Importantly, the magnitudes of the steady state gain coefficients in Figures 3 and 4 agree well with those calculated by Palumbo and Elton.[22]

Because the 3s and 3p levels in carbon-like ions (and in other L-shell ionization stages as well) can be inverted when the plasma is in collisional-radiative equilibrium, these plasmas are potentially attractive sources for obtaining sustained coherent VUV emission. However, even higher gains appear to be achievable when the electrons and ions are not in equilibrium.[21,22] A dramatic illustration of this effect is provided in the Ar XIII system. Consider an argon plasma at a density of 5.52×10^{19} ions/cm^3. When fully ionized, this plasma will have an electron density of 10^{21} cm^{-3}, critical to neodymium laser radiation. According to Fig. 3, at 5.5×10^{19} ions/cm^3, there is no (or negligible) steady state gain in the 3s–3p transition. A sizable transient gain, however, can be generated. In demonstration of this point, a number of calculations were made in which approximately 23 J of energy, in a heating pulse of 200 psec, was

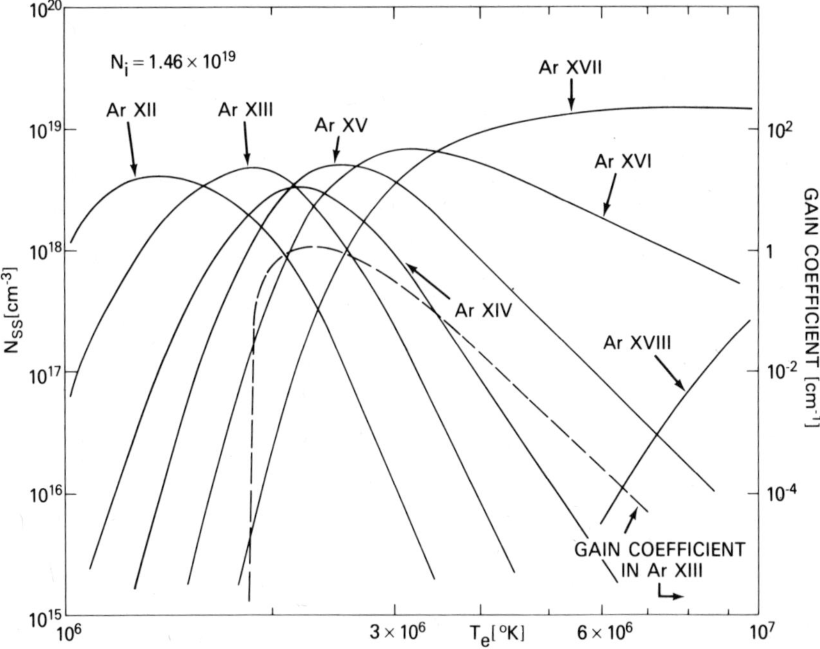

Figure 4. Steady state ground state population densities as a function of electron temperature at a total ion density near the cutoff value where collisional mixing between the 3s and 3p states destroys the population inversion. The dashed curve shows the behavior of the steady state gain coefficient as a function of electron temperature for the same ion density.

uniformly absorbed in a sphere of 314 μ radius containing 5.5×10^{19} argon ions/cm^3.* The results of these calculations are shown in Figures 5-10.

Figure 5 shows the time behavior of the average electron and ion temperatures and the Ar XIII ground state over a 400 psec interval in a calculation done with no radiation transport, i.e., in this calculation (case 1), the plasma was taken to be optically thin. The density of Ar XIII ground states in Figure 5 is also an average over the sphere. During the 100 psec, in which the

*To uniformly heat a plasma of this size and density by inverse bremsstrahlung absorption, one would need to irradiate the plasma uniformly with at least frequency-tripled Nd laser radiation.

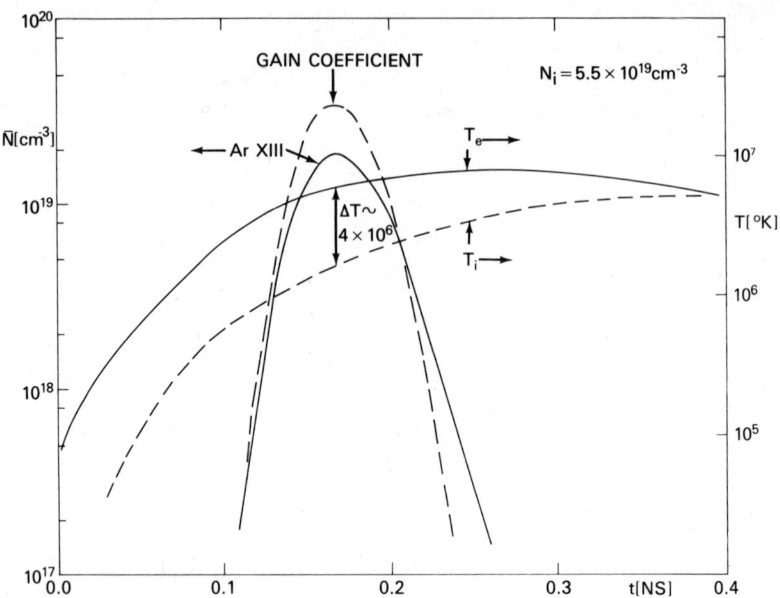

Figure 5. Electron and ion temperatures and the ground state Ar XIII population density as functions of time for the calculation having no radiation feedback. Superimposed over these curves is the gain coefficient curve, number 1, of Figure 8.

Ar XIII system is being excited and ionized, a transient inversion is also generated. The inversion density acquires its peak value at a time when the electron temperature is greater than 500 eV and the 3p and 3d states are being strongly excited. At this time, the ion temperature lags the electron temperature by about 400 eV. The Doppler width of the 3s-3p line is correspondingly smaller and the gain across the line correspondingly larger than it would be were the electrons and ions to be in equilibrium.

When radiation feedback in the 3s-3p, $2p^2$-2p3d, and $2p^2$-2p3s Ar XIII transitions is included in the calculation (case 4) the temperature curves in Fig. 5 remain effectively unchanged. However, the dynamics of the Ar XIII ionization stage are significantly altered. In Figure 6, a comparison is made between the average population densities in the $2p^2$, 2p3s, 2p3p, and 2p3d states of Ar XIII, as functions of time, in cases 1 and 4 (with and without radiation transport). The transport calculation was carried out using Voigt profiles with natural broadening being the source of the Lorentzian falloff in the wings of the lines.

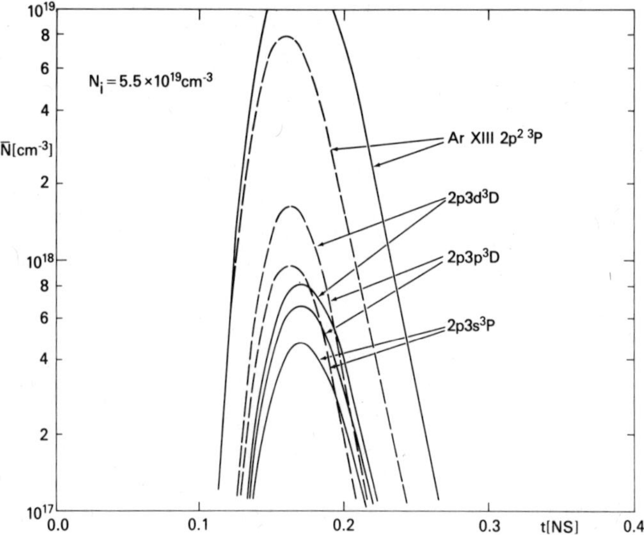

Figure 6. Comparison of the average population densities as functions of time in the Ar XIII ionization stage for cases of no radiation transport (the solid curves) and of a 3-line transport with Voigt profiles (the dashed curves). The solid curve showing the Ar XIII ground state population density is the same one as drawn in Figure 5.

The photo-excitation of the 3s and 3d states increases their population in this case by a factor of 2 and the population of the 3p state by almost a factor of 3. The ground state population is correspondingly decreased. The optical depths of the two resonance lines, the $2p^2$-2p3s and $2p^2$-2p3d, must therefore be self-consistently determined. As shown in Figure 7, with or without radiation feedback, the opacity of the $2p^2$-2p3d line peaks at 200 or 600 respectively. The opacity of the $2p^2$-2p3s line is over 10 times smaller than that of the $2p^2$-2p3d line. Hence, as indicated in Figure 6, when the n = 3 states are photo-excited the 3p state tends to equilibriate more strongly with the 3d than with the 3s state. This statement is true only on the average, however, as illustrated in Figures 8-10.

The four sets of curves in Figures 8-10 provide a comparison of the effects of radiation feedback using the two cases discussed above, plus two other cases where only the 3s-3p line was transported (case 2) or where only the 3s-3p and $2p^2$-2p3d lines were transported (case 3). In both cases 2 and 3, the lines were given

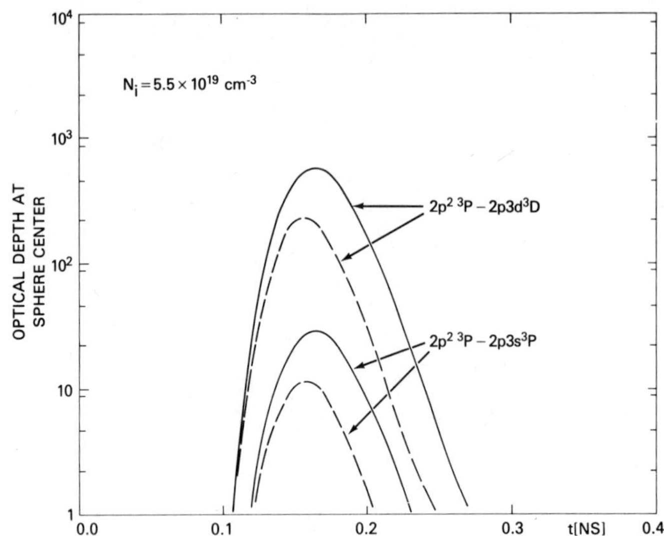

Figure 7. Comparison of optical depths at sphere center as functions of time of the pumping and quenching transitions for the cases of no radiation transport (solid curves) and of a 3-line transport with Voigt profiles (dashed curves).

Doppler rather than Voigt profiles. The computed gain coefficients near line center and at $r = 0$ in these four calculations are shown in Figure 8, the 3s-3p power output curves in Figure 9, and the 3s-3p time integrated line spectral profiles in Figure 10.

The coincidence of the gain curves in cases 1 and 2 occurs because the stimulated emission and absorption rates come close to but never exceed the collisional mixing rate between the 3s and 3p states. When photoexcitation of the 3d state alone is active as in case 3, the peak gain increases by more than a factor of 2 since the 3d state couples more strongly to the 3p than to the 3s state. However, radiation trapping in the $2p^2$-2p3s transition can overcome this pumping effect as curve 4 shows. The peak gain is brought below the case 1 and 2 values by over a factor of 3 at the center of the sphere. The curves in Figure 6 do not suggest that the overall quenching effect of the $2p^2$-2p3s radiation is that strong since they represent the average behavior of the population densities over the sphere, and near its surface; the photo-excitation rate of the 3s state is 2-1/2 times weaker than it is at the center of the sphere while for the 3d state the decline is only by a factor of 2/3. The dip in the gain curve occurs at a time when the $n = 3$ state populations are

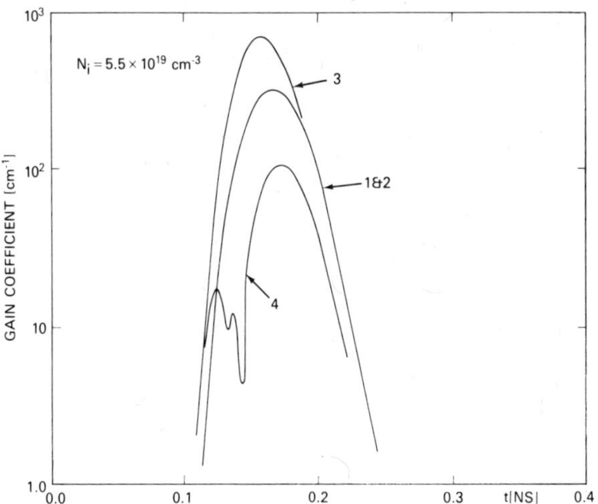

Figure 8. Gain coefficients at sphere center and at line center as functions of time for the following four cases: 1. no radiation transport; 2. transport of the 3s-3p lasing line only; 3. transport of the 3s-3p and $2p^2$-2p3d lines with Doppler profiles; and 4. transport of 3s-3p, $2p^2$-2p3d, and $2p^2$-2p3s lines using Voigt profiles.

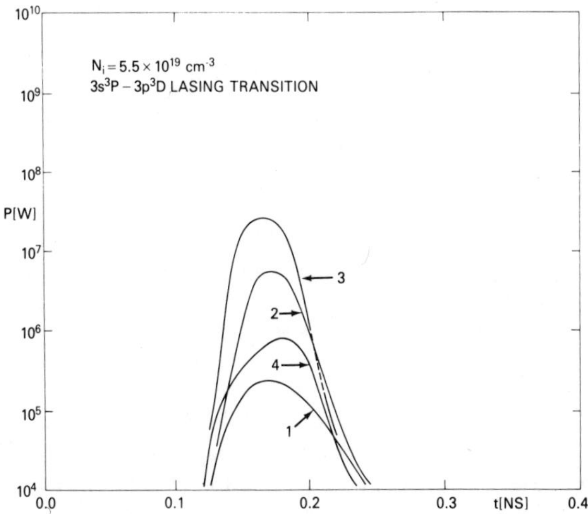

Figure 9. Power outputs in the lasing line as functions of time for the following four cases: 1. no radiation transport; 2. transport of the 3s-3p lasing line; 3. transport of the 3s-3p and $2p^2$-2p3d lines with Doppler profiles; and 4. transport of the 3s-3p, $2p^2$-2p3d, and $2p^2$-2p3s lines with Voigt profiles.

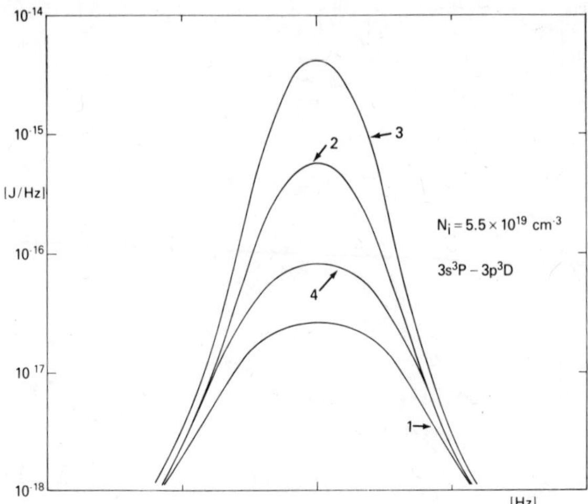

Figure 10. Calculated line profiles of the 3s-3p lasing line for the following four cases: 1. no radiation transport; 2. transport of the 3s-3p lasing line; 3. transport of the 3s-3p and $2p^2$-2p3d lines with Doppler profiles; and 4. transport of the 3s-3p, $2p^2$-2p3d, and $2p^2$-2p3s lines with Voigt profiles.

rapidly increasing and represents a momentary dominance in the photo-quenching rate over the photo-pumping rate. It takes place at a time when the 3s and 3d photo-excitation rates at the center of the sphere are roughly 7 and 6 times, respectively, the collisional-excitation rates to these same states.

In a sphere, stimulated emission from inverted populations also exhibits some properties that are familiar from column geometries. In case 3, for example, as seen in Figure 9, the gain coefficient attains its largest value leading to a 3s-3p power output that is two orders of magnitude larger than the spontaneous output. In case 4, however, when all three lines are transported, a significant portion of this line emission is lost. In Fig. 10, the amount of gain narrowing of the time integrated 3s-3p line profiles associated with the differences in stimulated emission are shown for each of the four cases.

At lower ion densities, the nonequilibrium response of the Ar XIII system will again generate larger inversions between the 3s and 3p states than are achievable in equilibrium. As the ion density is decreased, however, the frequency of the laser radiation that is used to heat the plasma should also be decreased.

For example, if CO_2 laser pulses are used to first breakdown and then uniformly and efficiently heat an argon gas, the atom density must be 100 times less than the corresponding density needed for efficient heating by Nd laser pulses. In Figures 11-14 we present some results from a set of five calculations in which the transient behavior of an argon plasma was investigated at an ion density of 5.5×10^{17} cm^{-3}, for which a steady state inversion will exist between the Ar XIII 3s-3p states. In all of these calculations the radius of the sphere was taken to be 986 μ and the plasma was heated by the net absorption of approximately 3.6 J in a 2 nsec heating pulse. Four of the calculations were similar to the ones described above; namely, in case 1, the plasma was taken to be optically thin and no radiation fields were coupled to the Ar XIII ionization stage; in case 2, only the 3s-3p radiation was coupled to Ar XIII with a Doppler profile; in case 3, both 3s-3p and $2p^2$-2p3d lines were coupled to Ar XIII with Doppler profiles; and in case 5, all three lines 3s-3p, $2p^2$-2p3d, and $2p^2$-2p3s, were coupled to Ar XIII with Voigt profiles that were naturally broadened in the wings of the line. In order to determine the magnitude of the effect that the use of Voigt profiles had on the transport calculations, another case was run (case 4) where all three lines were transported using Doppler profiles.

As before, the electron and ion temperature evolutions that are shown in Fig. 11 were identical in all five calculations. The heating conditions were chosen to optimize the inversion. Thus, the peak in the gain curve was achieved at a time when the electrons were at 700 eV and collisional excitation rates were large. The electron temperature then declined (because of some heat losses from the sphere) so that Ar XIII was maintained as the most highly populated ionization stage. One benefit that is derived when the argon ion density is lowered is evident in Figure 11. Since the electron-ion heating rate decreases as the ion density is decreased, the difference in electron and ion temperatures at the time of peak gain is increased. The equilibriation of the electrons and ions at later times is then partly responsible for the decline in the gain coefficient toward its steady state value.

Another advantage of lowering the ion density is shown in Figures 12 and 13. The trapping of the $2p^2$-2p3d line radiation will again pump the 2p3p state and increase the 3s-3p population inversion, and again the trapping of $2p^2$-2p3s line radiation will counteract this effect. However, their combined effect now yields a net in p-state pumping rather than a net quenching. The Voigt profiles also allow more radiation to escape in the wings of the

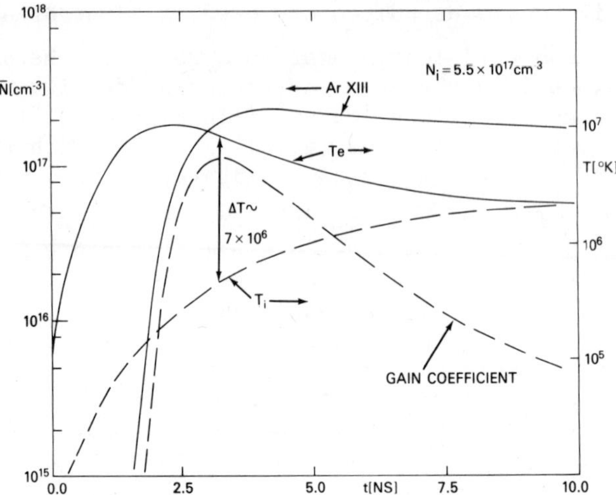

Figure 11. Average electron and ion temperatures and the ground state Ar XIII population density as functions of time for the case of no radiation feedback. Superimposed over these curves is the gain coefficient curve, number 1, of Figure 12.

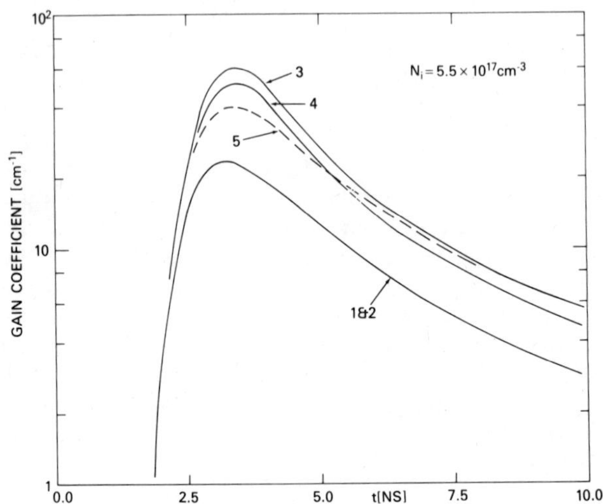

Figure 12. Gain coefficients at sphere center and line center as functions of time for the five cases: 1. no radiation transport; 2. transport of the 3s-3p lasing line; 3. transport of the 3s-3p and $2p^2$-2p3d lines with Doppler profiles; 4. transport of the 3s-3p, $2p^2$-2p3d, and $2p^2$-2p3s lines with Doppler profiles; and 5. transport of the 3s-3p, $2p^2$-2p3d, and $2p^2$-2p3s lines with Voigt profiles.

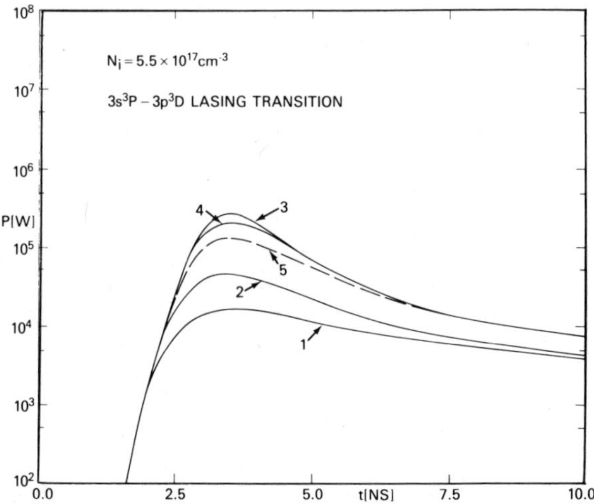

Figure 13. Power outputs in the lasing line as functions of time for the five cases: 1. no radiation transport; 2. transport of the 3s-3p lasing line; 3. transport of the 3s-3p and $2p^2$-2p3d lines with Doppler profiles; 4. transport of the 3s-3p, $2p^2$-2p3d, and $2p^2$-2p3s lines with Doppler profiles; and 5. transport of the 3s-3p, $2p^2$-2p3d, and $2p^2$-2p3s lines with Voigt profiles.

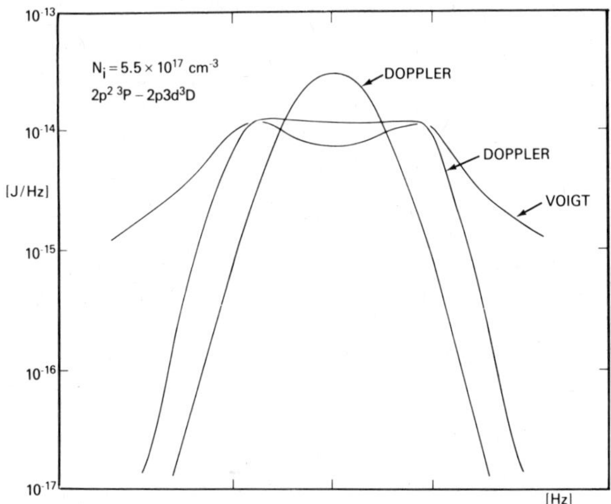

Figure 14. Calculated $2p^2$-2p3d line profiles for the three cases: 1. no radiation transport; 2. transport of the 3s-3p, $2p^2$-2p3d, and $2p^2$-2p3s lines with Doppler profiles; and 3. transport of the 3s-3p, $2p^2$-2p3d, and $2p^2$-2p3s lines with Voigt profiles.

lines but the gain increase due to $2p^2$-2p3d photo-pumping remains. This behavior can be partly accounted for in the magnitudes of the line opacities. The maximum optical depth of the $2p^2$-2p3d line at the center of the sphere is approximately 20, while the optical depth of the $2p^2$-2p3s line never exceeds 1.

In spite of this reduction in line opacities from those shown in Figure 7, photo-excitation of both the 3s and 3d states remains nonetheless the dominant process by which these states are excited. At the center of the sphere, for example, when all three lines are transported with Voigt profiles the photo-excitation rates of the 3s and 3d states reach peak values of 9×10^8 and 7.5×10^{10} sec^{-1} respectively at a time when the collisional excitation rates of these states are only 4×10^7 and 1.5×10^9 sec^{-1}. We have an example, therefore, of a physical situation in which the self-absorption of an optically thin resonance line can still play an important role in the excitation of an ionization stage. In this case, when both $2p^2$-2p3s and $2p^2$-2p3d resonance lines are self-absorbed, the number of 3p excited states is increased by a factor of 5/3, while the number of 3s states goes up by a factor of 7/2. Moreover, the net increase in the 3s-3p inversion occurs at a time when the 3d photo-excitation rate is larger than the rate of collisional mixing between the 3s and 3p states.

The weakness of collisional-relative to photo-excitation processes is also manifested in the shape of the time integrated line profile of the emitted $2p^2$-2p3d radiation, which is shown in Fig. 14. When the line is transported with a Doppler absorption profile, the output profile flattens at line center and broadens as absorbed and reemitted photons diffuse into the wings of the line. In the calculations for which $N_i = 5.5 \times 10^{19}$ cm^{-3}, this profile was flattened but not broadened; i.e., more photons were quenched before they could diffuse into the line wings. The dip at line center in the profile that was calculated using the Voigt function was obtained because our calculation had better resolution of the transition region between the optically thick and thin regions of the plasma. The same dip is, therefore, expected to be present when the lines are transported with Doppler absorption profiles and the same resolution of the transition region.

IV. SUMMARY

The idea that optical lasers can be used to pump VUV or x-ray lasers is very attractive since the optical laser energy can be easily focused or concentrated, the pulse duration can be controlled, and different laser wavelengths can be selected to tailor pump energy delivery to the different needs of different ion laser plasma schemes. For the argon ion laser studied in this paper, our calculations have shown that population inversions between the 3s and 3p states of Ar XIII can be generated that are one to two orders of magnitude larger than the inversion densities achievable when the plasma is in collisional-radiative equilibrium. Furthermore, if the gain medium has a cylindrical shape our calculations have shown that it is desirable to choose the radius of the cylinder to be less than 300 μ if the average argon ion density is near the critical plasma density for Nd laser radiation, or to be of the order of 1000 μ if the average ion density is near the critical value for CO_2 laser radiation. In the former case, the quenching effects due to self-absorption of the $2p^2$-2p3s resonance line are avoided; in the latter case some pumping benefits from the self-absorbed $2p^2$-2p3d resonance line are gained. Our calculations also suggest what the lengths of the plasma columns and the absorbed pulse energies should be to observe amplified spontaneous emission of the 3s-3p line.

The importance of photo-excitation to the dynamics of laser-heated plasmas was demonstrated by the fact that, in both sets of our calculations, the photo-excitation rates of the 3s and 3d states were larger than the collisional excitation rates over a significant period of the Ar XIII system's evolution. This phenomenon occurred even when the $2p^2$-2p3s line was optically thin. The importance of these processes suggests a way to increase the 3s-3p inversion. The 3s photo-excitation rate can be reduced if argon is mixed with another gas that will absorb but not reemit the $2p^2$-2p3s line radiation. For example, two of the $2p^2$-2p3s Ar XIII lines have wavelengths of 31.92 and 31.86 Å, which lie close to the Lyman series limit in CV. In an argon-methane gas mixture of the right proportions, $2p^2$-2p3s line radiation may preferentially ionize CV rather than photo-pump the 3s state of Ar XIII.

Lines with gain in laser-produced plasmas are also potentially valuable sources of diagnostic information. In contrast to an optically thick line, which allows one to see beyond the plasma corona only by resolving the wings of the emitted line profile, a stimulated line allows one to view the plasma *a fortiori* at line core. Angular distributions and pinhole pictures of a line that is amplified rather than attenuated by its passage through a laser heated plasma should be of particular value in diagnosing temperature and density gradients. As the length of the transit path increases, other interesting phenomena will also occur since photon transit times will no longer be short compared to other collisional or photo-excitation and ionization times. In our calculations at

$N_i = 5.5 \times 10^{19}$ argon ions/cm^3, in fact, these transit times across a sphere diameter were comparable to or longer than the collisional mixing times of the Ar XIII excited states. This represents one of the weak points of these calculations. The plasma diameters were chosen, however, so that the maximum 3s-3p gain across them were, in all cases, larger than 4 and of order 10. In this way, effects like the gain narrowing of the 3s-3p line profile and the order of magnitude increases in power output could be calculated for a physical situation in which the laser medium did not have an axis of maximum gain.

References

(1) P. J. Mallozzi, H. M. Epstein, R. G. Jung, D. C. Applebaum, B. P. Fairand, and W. J. Gallagher, "Fundamental and Applied Laser Physics," edited by M. S. Feld, A. Javan, and N. A. Kurnit (Wiley, New York, 1971), p. 165.

(2) K. G. Whitney and J. Davis, J. Appl. Phys. 45, 5294 (1974).

(3) F. E. Irons, J. Phys. B 8, 3044 (1975).

(4) L. F. Chase, W. C. Jordan, J. D. Perez, and J. G. Pronko, Appl. Phys. Lett. 30, 137 (1977).

(5) P. Kaw, G. Schmidt, and T. Wilcox, Phys. Fluids 16, 1522 (1973).

(6) M. D. Feit and J. A. Fleck, Jr., Appl. Phys. Lett. 28, 121 (1976); M. D. Feit and D. E. Maiden, Appl. Phys. Lett. 28, 331 (1976).

(7) N. G. Loter, D. R. Cohn, W. Halverson, and B. Lax, J. Appl. Phys. 46, 3302 (1975).

(8) K. Lee, D. Forslund, J. Kindel, and E. Lindeman, Phys. Fluids 20, 51 (1977).

(9) B. H. Ripin, J. M. McMahon, E. A. McLean, W. M. Manheimer, and J. A. Stamper, Phys. Rev. Lett. 33, 634 (1974); B. H. Ripin, Appl. Phys. Lett. 30, 134 (1977), W. M. Manheimer, D. G. Colombant, and B. H. Ripin NRL Report 3426 (1976) to be published.

(10) D. W. Forslund, J. M. Kindel, K. Lee, E. L. Lindman and R. L. Morse, Phys. Fluids 11, 679 (1975); E. G. Estabrook; E. J. Valeo, and W. L. Kruer, Phys. Fluids 18, 1151 (1975).

(11) G. J. Pert and S. A. Ramsden, Opt. Commun. 11, 270 (1974).

(12) E´ Ya Kononov and K. N. Koshelev, Sov. J. Quant. Electron 4, 1340 (1975).

(13) B. A. Norton and N. J. Peacock, J. Phys. B 8, 989 (1975).

(14) A. G. Molchanov, Sov. Phys. Uspekhi 15, 124 (1972).

(15) F. E. Irons and N. J. Peacock, J. Phys. B 7, 1109 (1974).

(16) R. J. Dewhurst, D. Jacoby, G. J. Pert, and S. A. Ramsden, Phys. Rev. Lett. 37, 1265 (1976).

(17) P. K. Cheo and H. G. Cooper, J. Appl. Phys. 36, 1862 (1965); Appl. Phys. Lett. 7, 202 (1965).

(18) R. Pappalardo, J. Appl. Phys. 45, 3547 (1974).

(19) W. B. Bridges and A. N. Chester, IEEE J. Quant. Elect. QE-1, 66 (1965); W. B. Bridges and A. N. Chester, IEEE J. Quant. Elec. QE-1, 66 (1965); W. B. Bridges, A. N. Chester, A. S. Halstead, and J. V. Parker, Proc. IEEE 59, 724 (1971).

(20) C. K. Rhodes, IEEE J. Quant. Elec. QE-10, 153 (1974).

(21) J. Davis and K. G. Whitney, Appl. Phys. Lett. 29, 419 (1976).

(22) L. J. Palumbo and R. C. Elton, JOSA (to be published).

(23) J. Davis and K. G. Whitney, J. Appl. Phys. 47, 1426 (1976); J. P. Apruzese, J. Davis, and K. G. Whitney, J. Appl. Phys. 48, 667 (1977).

(24) S. I. Braginskii, "Reviews of Plasma Physics," edited by M. A. Leontovich (Consultants Bureau, New York, 1965), p. 205.

(25) S. Chandrasekhar, "Radiative Transfer" (Clarendon Press, Oxford, 1950), p. 23.

(26) D. G. Hummer, C. V. Kunasz, and P. B. Kunasz, Comp. Phys. Comm. 6, 38 (1973); P. B. Kunasz and D. G. Hummer, MNRAS 166, 19 (1974).

(27) T. R. Young and J. P. Boris, NRL Report No. 2611, 1973 (unpublished).

BEAM-PROFILE EFFECTS IN SELF-INDUCED TRANSPARENCY: ON-RESONANCE SELF-FOCUSING OF COHERENT OPTICAL PULSES IN ABSORBING MEDIA**

F. P. Mattar** and M. C. Newstein

Laboratory for Laser Energetics

The University of Rochester, Rochester, New York 14627

Abstract: Analytic and numerical solutions of the Maxwell-Bloch equations, including transverse and time-dependent phase variations, predict on-resonance self-focusing and elucidate its formation due to the combined effects of diffraction and inertial response of the medium. This self-focusing can be characterized by a single parameter in terms of the beam and medium parameters. Recently, two independent experiments in sodium[+] and neon[++] demonstrated this new self-focusing effect of spatially non-uniform self-induced transparency pulses propagating in thick resonant absorbers. Comparison of the experimental results with the theoretical analysis will be presented.

*All the computations performed from June '73 to January '77 were jointly supported by F. P. Mattar, the joint services Electronic Program, the office of Naval Research, the International Division of Mobil Oil Company and the University of Montreal.

ˣGratitude and appreciation are extended to the Experimentalists for allowing their observed data to be used.

**The theoretical work reported here is part of a dissertation submitted by F. P. Mattar in December 1975 as a partial fulfillment for the Ph.D. degree to Professor M. C. Newstein at the Department of Electrical Engineering and Electrophysics, Polytechnic Institute of New York, Farmingdale, N.Y. 11735.

+The first experimental demonstration in Na was jointly performed by B. Bolger, L. Baede (at Philips Research Lab in Eindhoven - The Netherlands) and H. B. Gibbs (Bell Labs - Murray Hill, N.J.).

++The second observation was done in Ne by G. Forster and P. E. Toscheck at the Institute of Applied Physics, University of Heidelberg, Heidelberg, West Germany.

I. INTRODUCTION

When strong optical pulses propagate through resonant media, significant exchange of energy between the radiation field and the material system can occur in a time short compared to the thermal relaxation time of the medium. When this occurs, the pulses are said to be coherent and this type of propagation originated with the experimental and theoretical work of McCall and Hahn[1] for an uniform plane-wave. Their investigation led to the discovery of self-induced transparency (SIT): the absorbed energy from the leading edge of a pulse (whose time-integrated area is 2π) is completely re-radiated back into the trailing edge of the pulse through stimulated emission. Their conclusion has been thoroughly verified by Gibbs and Slusher[2]. Extensive studies have evolved extending the range of application to allow frequency and phase modulation, the inclusion of degeneracies in the medium and the effect of host dispersion. G. Lamb[3], Kryukow and Letokov[4] and Courtens[5], published reviews of these coherent phenomena. Since experimental arrangements often do not satisfy the uniform plane-wave conditions, attempts[6] have been made for the modification of pulse behavior on account of transverse variations. Newstein and Wright[7] undertook the first full nonlinear study. They attempted to understand and quantitatively describe the nonlinear evolution of coherent pulses in inhomogeneous (sharp-line), non-degenerate resonant media. They treated mainly amplifiers and found significant phase modulation along with transverse spreading; this may partially account for certain mechanisms that limit the useful output of long amplifiers. Because of the complexity and essential nonlinearity of the governing coupled field-matter equations, they rely heavily on numerical methods. Reflection of the field at the boundary wall ρ_{max} (that limits the extent of the transverse region over which the numerical solution is to be determined) being amplified[8], turned out to play a substantial role in the calculation since it tended to obscure the emergence of new effects in this three-dimensional analysis.* This suggested a similar calculation for absorbers since reflection would not become amplified and hence boundary effects are reduced to a negligible minimum. They showed that a new form of self-focusing would occur[9]. This focusing appeared to be a *consequence of strong coherent interactions between field and matter*. We believed this problem was representative of a class of propagation problems which

---

*Private discussion with E. Courtens during the third conference of Quantum optics at Rochester University, Rochester, New York (June 1972).

were not satisfactorily understood; it fell in the category of *transient self-focusing associated with the inertia of the coherent nonlinear response*. It was therefore appropriate to examine this promising initial result to see whether it could withstand close scrutiny, and was indeed a real physical phenomenon. Finally, *having thoroughly eliminated the possibility of this new type of self-focusing being a numerical artifact, the understanding of the detailed features and consequences of this new effect seemed to be a desirable goal*. This motivated the development of a more accurate numerical algorithm that would attempt to reproduce the new phenomenon, verify its authenticity and evaluate its importance[10,11]. Wright and Newstein's basic observations were reproduced with some modifications in results due to our more accurate numerical procedure[12]. *The study of this phenomenon was then pursued with the goals of determining its dependence on the different pulse and medium descriptive parameters and of understanding its physical basis. A systematic perturbation analysis was carried out to reach physical insight and provide a guide curve to the thorough three-dimensional numerical analysis.* Transverse derivatives in the field equation provide the coupling leading to the development of radially dependent frequency variation and phase curvature. The study of the transient transverse reshaping (that occurs during the propagation through the resonant medium) led to the identification of the transverse energy flow as the leading mechanism to significant coherent self-focusing. The study of the origin and evolution of this energy current determined the appropriate conditions responsible for the onset and development of the focusing phenomenon. This coherent self-action (self-phase modulation and self-focusing) arises through the combined effects of diffraction and the nonlinear inertial response of the media[12]. In the SIT reshaping regime, little focusing occurs while each annular ring of the input Gaussian profile evolves according to the uniform plane-wave theory. Since the more intense rings propagate more rapidly, the tail of the pulse has more intensity in the outer rings (and the appearance of a hole near the axis). Diffraction into the forward central region (still in an amplifying condition) may result in an inward flow of energy. Summing up, the diffraction of light from the trailing edge of the pulse towards the axis will fill up the hole and will be shown to be equivalent to the transversal motion of solitons.*

---

*Private discussion with H. M. Gibbs, B. Bolger and J. Marburger during the IX International Quantum Electronics Conference in Amsterdam (June 1976)(also illustrated in Fig. 27): *The total transient field can be considered as made up of individual pulses*

The evolution (development) of this new self-focusing effect for a given pulse shape and beam profile, its location, the sharpness of its threshold and the spacing between multiple foci (when they occur) depend on the Fresnel number, absorption coefficient, relaxation times, laser-absorber frequency mismatch and input on-axis time integrated electric field area. The effect *also* occurs in the presence of inhomogeneous broadening[14].

Two independent experiments, with non-uniform plane-waves, demonstrating coherent self-focusing were reported with as much as a factor of two increase of axial fluency (energy per unit area) on-resonance and a factor of 4.5 off-resonance. The first experiment was made by Gibbs et al in inhomogeneously broadened sodium[15]; self-focusing and temporal reshaping were also studied as a function of laser detuning. The (careful) observation of focusing on resonance for coherent pulses[16] (but not for CW) clearly illustrates that coherent transient self-focusing is different from previous self-focusing involving resonant interactions investigated either in the rate-equation[17] or in the adiabatic-following[18] approximations. The second experiment carried out by Toschek et al[19], involves the investigation of the on-resonance self-focusing in quasi-degenerate inhomogeneously

---

of different strength for different radii. If one looks at the peak of the output curves in time for fixed radii and propagation distance as one goes downstream, one has the situation where the less intense rim of the beam is delayed with respect to the center, since pulses of lower intensity travel slower. That is, one finds that the off-center peak intensity comes later in time than does the on-axis peak intensity. Light diffracting towards the axis from the trailing edge of the pulse in the rim of the beam (at larger radius) interacts with those atoms which were excited by the preceding pulse (which was close to the axis) and can experience net amplification for the case illustrated in Fig. 27. This boosting phenomenon continues as energy flows towards the axis, making a positive contribution to the center of the beam. The leading portion of the pulse in the rim of the beam essentially sees an absorber and thus will experience net absorption.

broadened neon and its dependence on input pulse on-axis area and on pulse duration. In summary, coherent transient focusing has been convincingly demonstrated. It greatly alters the transverse and temporal evolution of optical pulses propagating in thick resonant absorbers when sufficient diffraction is present[20]. Coherent transient self-focusing may explain previously not understood transverse effects in SIT experiments[35-38, 46].

II. EQUATIONS OF MOTION

For the class of problems of interest it is permissable to describe a transverse component of the electric field, E, and of the medium polarization density, P, as a product of a slowly varying envelope and a high frequency carrier, namely

$$E = \text{Re}\ \xi' \exp\left(i[\omega t - (\omega\eta/c)\ z]\right) \quad , \tag{1}$$

$$P = \text{Re}\ i p' \exp\left(i[\omega t - (\omega\eta/c)\ z]\right) \quad . \tag{2}$$

Assuming that the complex amplitudes ξ' and ρ' change by a small fractional amount, temporally in the optical period $2\pi/\omega$ and spatially in the optical wavelength $2\pi c/\omega$, the field equations become first order in z and t. Introducing the dimensionless variables

$$\xi = (2\mu/h)\tau_p\ \xi' \quad , \tag{3}$$

$$p = (2/\mu)\ p' \quad , \tag{4}$$

the dimensionless field-matter equations which describe our systems are:

$$-i\ F\ \nabla_\rho^2\ \xi + \frac{\partial\xi}{\partial\eta} = p \quad , \tag{5}$$

$$\partial p/\partial\tau = \xi W - [i\Delta\Omega + 1/\tau_2]\ p \quad , \tag{6}$$

and

$$\partial W/\partial \tau = -1/2\ (\xi^* \mathcal{p} + \xi \mathcal{p}^*) - (W - W_o)/\tau_1 \quad . \tag{7}$$

The complex field amplitude ξ, the complex polarization density $\mathcal{p}$, and the energy stored per atom W, are functions of the transverse coordinate ρ, the longitudinal coordinate η, and the retarded time τ. The time scale is normalized to a characteristic time τ_p of the input pulse, and the transverse dimension scales to a characteristic spatial width r_p of the input pulse. The longitudinal distance is normalized to the effective absorption length $(\alpha_{eff})^{-1}$, where[21]

$$\alpha_{eff} = \left[\frac{\omega\ \mu^2 N}{nhc}\right]\tau_p \quad . \tag{8}$$

In this expression ω is the angular carrier frequency of the optical pulse, μ is the dipole moment of the resonant transition, N is the number density of resonant molecules, and n is the index of refraction of the background material. Thus in terms of the physical coordinates, ρ, z, and t, we have:

$$\tau = [t - zn/c]/\tau_p \quad ,$$

$$\eta = z\ \alpha_{eff} \quad ,$$

$$\rho = r/r_p \quad . \tag{9}$$

The dimensionless quantities

$$\Delta\Omega = (\omega - \omega_o)\tau_p \quad ,$$

$$\tau_1 = T_1/\tau_p \quad , \quad \tau_2 = T_2/\tau_p \tag{10}$$

measure the offset of the optical carrier frequency ω from the central frequency of the molecular resonance ω_o, the thermal relaxation time T_1, and the coherence relaxation time T_2, respectively.

The dimensionless parameter, F, is given by

$$F = \frac{(\alpha_{eff})^{-1}\lambda}{4\pi\, r_p^2} \quad . \tag{11}$$

The reciprocal of F is the Fresnel number associated with an aperture radius r_p and a propagation distance $(\alpha_{eff})^{-1}$. The magnitude of the parameter F determines whether it is possible to divide up the transverse dependence of the field into "pencils" (one pencil for each beam radius ρ) which may be treated in the plane-wave approximation; i.e., the parameter F plays an important role in the transverse propagation behavior[22]. If F is small, each pencil may be treated independently; the situation is equivalent to the geometrical optics approximation. The beam distortion is solely due to self-induced nonlinear interaction with the active medium. On the other hand, if F is larger, diffraction will strongly couple the different pencils preventing this separation across the beam; we are essential dealing with the wave-optics limit where the nonlinear interaction can be neglected. Clearly, the behavior of the beam varies smoothly between these two limits. Inspection of the normalized scalar wave-equation suggests that $F = 1$ forms a threshold boundary between the two types of beam distortion mechanisms. Indeed when $F = 1$, the diffraction term and the resonant polarization due to the nonlinear interaction are of the same order. When the Fresnel number differs from unity, the diffraction coupling term or the nonlinear interaction alternately dominate depending on whether $F > 1$ or $F < 1$.

III. ENERGY CONSIDERATIONS

From the field-matter relations (5)-(7) one obtains the energy current equations:

$$\nabla_T J_T + \partial_\eta J_z = -2\left[\partial_\tau W + \frac{(W - W_o)}{\tau_1}\right] \quad ,$$

or

$$\nabla\cdot\underline{J} = -2\left[\partial_\tau W + \frac{(W - W_o)}{\tau_1}\right] \quad , \tag{12}$$

where, using the polar representation of the complex envelope, we have

$$\xi = A \exp[i \phi] \quad , \tag{13}$$

$$J_z = A^2 \quad , \tag{14}$$

and

$$J_T = 2F \left[A^2 \frac{\partial \phi}{\partial \rho} \right] \quad . \tag{15}$$

The components J_z and J_T represent the longitudinal and transverse energy current flow. Thus, the existence of transverse energy is clearly associated with the radial variation of the phase of the complex field amplitude ξ. When J_T is negative [i.e., $\partial\phi/\partial\rho < 0$], self-induced focusing dominates diffraction spreading.

One may rewrite the continuity equation (12) in the *laboratory frame* to recover its familiar form

$$\nabla \cdot \underline{J} = - \frac{\partial}{\partial \tau} \left[2W + \frac{n}{c\tau_p \alpha_{eff}} A^2 \right] - 2 \frac{(W - W_o)}{\tau_1} \quad . \tag{16}$$

IV. PERTURBATION APPROACH TO THE INITIAL BEHAVIOR

We expect an incident field with sufficient transverse intensity variation to undergo significant alterations of transverse structure as it propagates. This will occur even within the near-field region of the effective input aperture. The region surrounding each light pencil is expected first to propagate like the corresponding one-dimensional input field of the same amplitude. However, after a certain distance longitudinal reshaping will cause different pencils to propagate with *different* speeds leading to transverse shaping. When this occurs, the radial derivatives in the scalar-wave equation (the source of coupling between the different pencils) lead to significant phase variations and transverse energy flow. In the following example, a comparison of the distinct temporal evolution of two separate pencils will be made for short propagation distances. One of these is on axis where the intensity is at maximum, and the second is just off-axis where the intensity is smaller. The nonlinear polarization that results will make the group velocity of the pulse peak at the center pencil exceed the corresponding off-axis group velocity. This is sketched in Fig. 1. The situation at the input plane is pictured in diagram (a). It is seen that the peaks for the two pencils coincide in time. The situation at $\eta = \eta_1$ (in the direction of propagation) is indicated in diagram (b); here it is seen that the peaks no longer coincide, rather the peak corresponding to the on-axis pencil precedes the peak of the outer pencil. For a certain instant of time τ_0, the value of the field is the same for both pencils. The third diagram (c) illustrates the situation which occurs at a distance η_2, still further along the direction of propagation, where the temporal separation between peaks is even greater. For the particular instant of time τ_0, the off-axis field is now larger than the on-axis field.

Associated with this relative motion between adjacent pencils, we have a variation in the sign for the on-axis transverse coupling (transverse Laplacian) term. At the input plane its contribution is negative. As the pulse propagates along η, for a later instant of time τ_0 the transverse coupling term eventually vanishes. Still further away, its contribution for lagging times τ_0 becomes positive. Thus, the sign of the transverse Laplacian is a function of time differing at the leading and lagging portions of the pulse. Along with these changes in amplitude, the phase also changes. These amplitude (A) and phase (ϕ) delays lead to buckling (larger phase accumulation on-axis). The combined amplitude and phase variations will affect the pulse evolution in the nonlinear resonant absorber in different ways as a function of time. Transverse variation of the phase will induce a transverse energy current, defined by $J_T = 2F\,[A^2\ \partial\phi/\partial\rho]$, flowing inwards at some times and outwards at others. This interrelation can be understood in terms of a perturbation treatment of a coupled system of field matter equations (5)-(7). The idea behind this treatment is that for a small enough propagation distance, the behavior along each pencil can be described using a uniform plane-wave approximation. Here we will restrict ourselves to the on-resonance situation with infinite relaxation times.

Expanding the relevant variables in powers of the small quantity F, and keeping only the first order terms, we obtain

$$\xi = \xi_o + F\ \xi_1 = \xi_o + F(\xi_{1r} + i\ \xi_{1i}) \quad ,$$

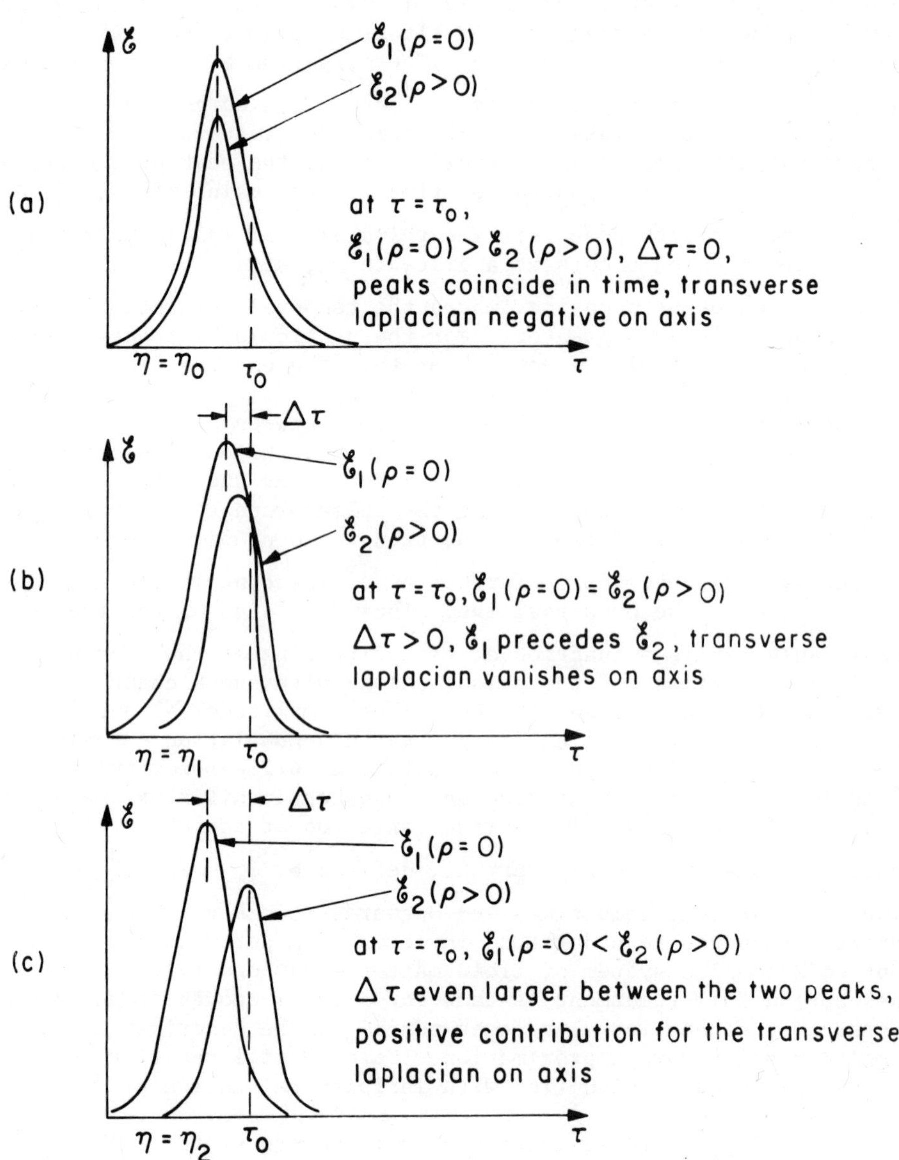

Figure 1. The relative motion among adjacent pencils propagating coherently in the nonlinear resonant absorber.

$$p = p_o + F\,p_1 = p_o + F(p_{1r} + p_{1i}) \quad ,$$

$$W = W_o + FW_1 = W_o + FW_{1r} \quad , \tag{17}$$

where ξ_0 and p_0 are the (real) solution of the uniform plane-wave propagation problem (obtained by integrating the one dimensional equations). At the input plane $\xi_1 = 0$, $p_1 = 0$, and $W_1 = 0$; as the pulse propagates, the imaginary part of the three-dimensional field grows. The rate of growth of ξ_{1i} and its sign depend on the radial variation of ξ_0. From the known uniform plane-wave properties, these transverse variations can be calculated assuming each pencil propagates independently of the other. It turns out that the quantities ξ_{1r}, p_{1r} and W_{1r} remain zero for all η. From this, one can calculate ξ_{1i} and hence obtain the phase

$$\phi = \arctan\,(F\xi_{1i}/\xi_0) \quad . \tag{18}$$

The validity of this procedure is limited to the range of propagation distances where one-dimensional pulses do not differ significantly from their three-dimensional counterparts. A significant prediction of the perturbation theory is the development of a substantial focusing curvature of the wavefront in the tail of the pulse. This occurs well within the reshaping region before any substantial focusing begins to occur. The fact that the phase is smaller on the wings of the beam than on the axis is indicative of an inward radial flow of energy current. The results of the perturbation theory, using the plane-wave solution as the unperturbed behavior, are in good agreement with the corresponding three-dimensional results obtained from the rigorous numerical calculations.* This is illustrated in Fig. 2 which refers to a situation well within the reshaping region (substantial focusing has not yet occurred). Nevertheless, there is already substantial phase variation on the tail of the pulse. This is indicated by

*For details see Ph.D. thesis, "Transverse Effects Associated with the Propagation of Coherent Optical Pulses in Resonant Absorbing Media," by F. P. Mattar, Polytechnic Institute of New York (Distributed by University Microfilms, Ann Arbor, Michigan).

the curves labeled $\phi(3D;\rho = 0)$ and $\phi(3D;\rho = \Delta\rho)$. These give the phase variation for the three dimensional case (3D) as determined by the full numerical calculation for two values of ρ. In the input plane, the pulse area is 2π at $\rho = 0$ and is 1.8 at $\rho = \Delta\rho$. The fact that the phase is smaller at $\rho = \Delta\rho$ than at $\rho = 0$ shows that energy is flowing radially toward $\rho = 0$ in the tail of the pulse. The results of the perturbation theory, using the one-dimensional solution as the unperturbed behavior, are shown by the curves labeled $\xi(ID;\ \rho = 0)$ and $\xi(ID;\ \rho = \Delta\rho)$. They compare well with the corresponding three-dimensional results from the numerical calculation. Fig. 2 also gives the comparison of the perturbation theory (called 1-D) with corresponding numerically derived (called 3-D) field amplitudes. The perturbation theory breaks down if these curves differ significantly from one another. The agreement is seen to be good over the domain of validity of the perturbation theory.

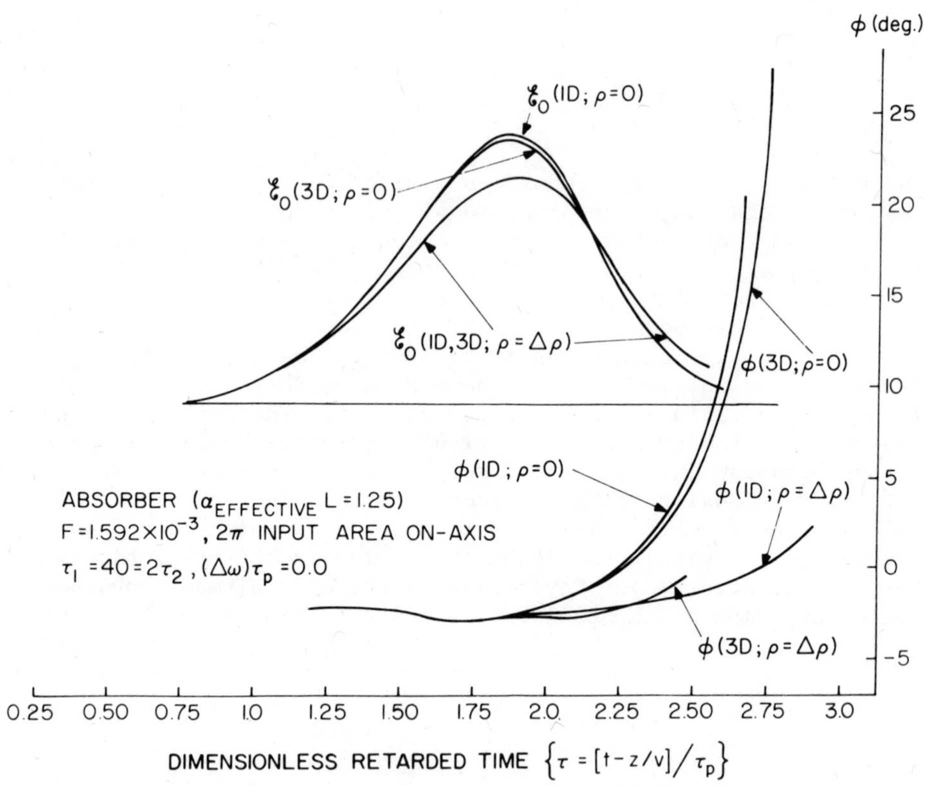

Figure 2. The comparison of the perturbation theory results concerning the development of longitudinal and transverse phase variations with the 3-D numerical results.

Furthermore, if one assumes that for small propagational distances the induced polarization is driven by the field at the input plane, an analytic expression for the one-dimensional field is obtained from the energy equation

$$\partial_\eta \xi_o^2 = 2\xi_o p_o \tag{19}$$

as follows:

$$\xi_o \cong [\xi_a^2 - 2\eta \sin \theta_a]^{1/2} \quad , \tag{20}$$

with ξ_a and θ_a the field and its time-integrated area at the aperture (input plane) respectively.

For large values of τ, ξ_a tends to zero causing ξ_o to be bound and finite. From the previous pencils approach, one obtains a (completely) analytic expression for the phase that emerges due to transverse coupling, as well as for J_T, the transverse energy current.* The resulting expression for J_T is a power series in η with coefficients of alternate sign. For small enough η (where higher order terms of η can be neglected), the energy current is positive indicating an outward transverse energy flow in agreement with the spreading due to a near-field diffraction effect. As η increases the second term must be considered also. Since its contribution is a negative quantity, J_T eventually changes sign and begins flowing inwards towards the axis forming a converging lens. The latter may counteract and overcome the diffraction, giving rise to the coherent self-focusing. If η increases further, the third term contribution becomes important; its effect tends to reduce the inward transverse energy flow due to the second term. The third term may be interpreted as the manifestation of additional diffraction due to the narrowing of the optical beam, which is the result of propagating through the nonlinear converging lens-like resonant absorber.

---

*For details see Ph.D. thesis, "Transverse Effects Associated with the Propagation of Coherent Optical Pulses in Resonant Absorbing Media," by F. P. Mattar, Polytechnic Institute of New York (distributed by University Microfilms, Ann Arbor, Michigan).

Due to the approximation used, the expressions obtained can at best suggest trends of what really occurs, which can be only studied by rigorous three-dimensional numerical computations. However, the present results were found to be consistent with the full three-dimensional calculation results.

Finally, one sees that the phase variations developed due to transverse coupling are not only a function of the spatial (longitudinal and radial) coordinates but also evolve in time. As a result, the time derivative $\partial\phi/\partial\tau$ will appear as a non-uniform frequency offset in the Bloch's equations (when expressed in a rotating coordinates system)[23]. The *absorptive* (as well as the *dispersive*) properties of the resonant attenuator, which depend on the frequency offset, *will* in turn *be altered for the various radii along the direction of propagation*. Consequently, the rate of energy exchange between field and matter will vary spatially as well as temporally. As a result of the radially different temporary energy storage in the two-level system, the pulse velocity of the various radii will not be uniformly retarded; this translates eventually into a distortion of the entire beam. Furthermore, the curvature acquired by the wave-front will induce a transverse energy current flowing either inwards or outwards; consequently, either a coherent self-focusing or self-defocusing will eventually ensue.

The above discussion refers to the situation where the input field is on resonance. An offset of the input field frequency relative to the medium resonance can account for additional phase modulation during propagation[24-27]. This arises because the dispersive component of the polarization develops right at the input plane, leading to a further contribution in the evolution of the imaginary part of the field envelope.*

V. SIMPLE PHYSICAL PICTURE**

The diffractional expansion of the optical pulses is balanced or overcome by the refractive effect of the resonant medium. The transient absorption is calculated by use of the full Bloch equations, rather than the particular use of the rate-equation approximation or the adiabatic-following condition, since the pulses are short compared to the medium relaxation times. The resulting transverse dependence of the absorption may lead to self-focusing.

---

*Op. cit., p. 151.

**Op. cit., p. 142.

The Self-Induced Transparency (SIT) phenomenon deals with the unattenuated propagation of an ultra-short pulse in a medium containing resonant centers which would normally absorb the electromagnetic energy. When the absorbed energy from the leading edge of the 2π hyperbolic-secant pulse is re-radiated into the trailing edge of the pulse, an undisturbed steady-state pulse is caused to propagate. In this process, no net energy is lost from the pulse. It is clear that field-atom interactions are large in this regime. The field energy is transitionally stored in the excitation of the medium, causing (among other effects) a *substantial* reduction of the pulse envelope velocity.

Following the intensity profile of the beam we have distinct resonant interactions for different beam radii. Consequently, the *absorptive and dispersive parts of the induced nonlinear polarization will vary across the beam* (hence, also the associated index of refraction) making the transfer of energy (between the field and the atomic system) radially dependent. Accordingly, the temporal delay of the pulse will increase differently along the direction of propagation depending on the beam radius; in other words, relative motion between neighboring radii will result.

Summing up, we expect to see different SIT effects for different beam radii. The various pencils will experience different SIT effects and therefore have different delays; this causes distortion of the wave-front and the appearance of a dip in the profile near the axis. However, linear diffraction will fill up the dip as shown in Fig. 3. Consequently, part of the beam shows inward energy flow (hence, self-focusing); whereas, another part shows outward transverse energy flow (i.e., self-defocusing).

To illustrate these ideas, we consider a fairly simple equivalent situation where we begin with a family of 2π hyperbolic-secant pulses, where the pulse duration τ_p changes across the beam. It is well known that when a 2π hyperbolic-secant pulse passes through a resonant attenuating medium, it will be substantially delayed in time[1]. This time delay is a direct consequence of atomic coherence. We can see that there are some particularly marked effects on the delay times as one varied the input time duration. From Fig. 4, the relative motion is evident.

Furthermore, the transverse coupling causes a temporal and radially dependent phase, leading to the flow of a transverse energy current. Since the more intense inner rings propagate more rapidly, the tail of the pulse has more intensity in the outer rings than in the center. *Diffraction into the forward*

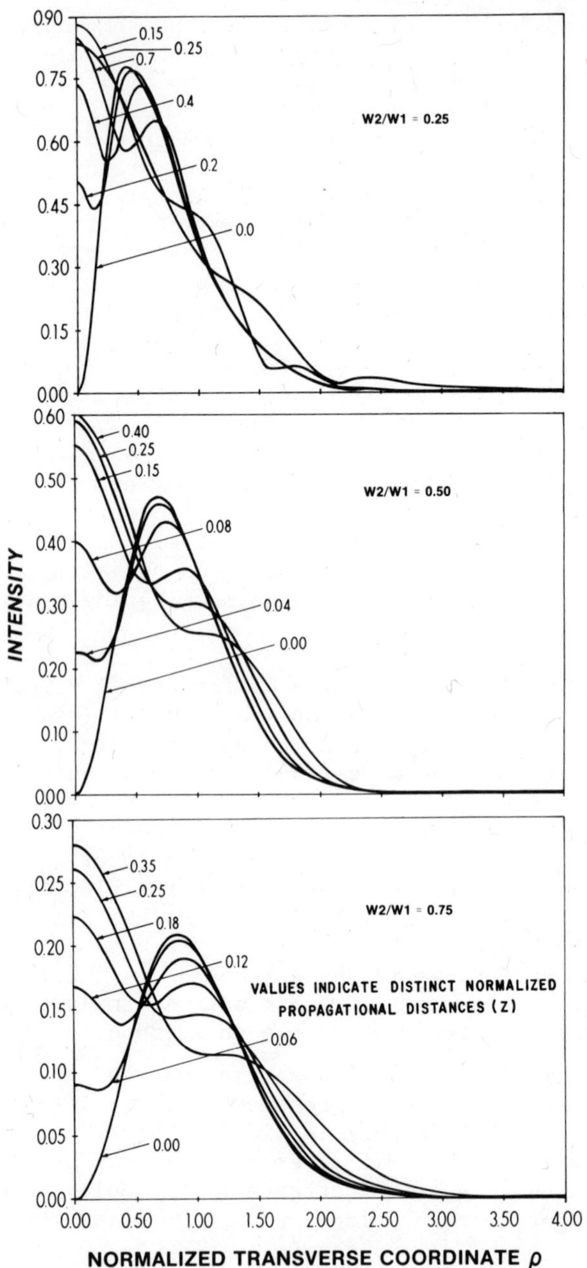

Figure 3. Illustrative effects of linear diffraction on the propagation of an intensity profile with a hole near the axis. This input profile is achieved by a subtraction of two Gaussians with different beam width. The propagation follows the analytical work of Kogelnik and Li.

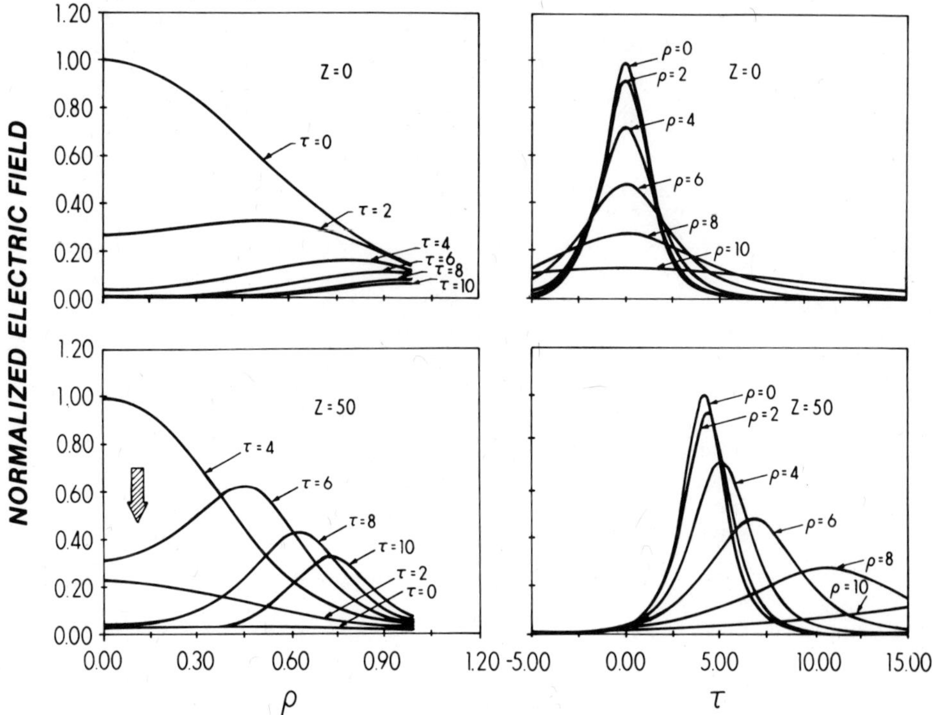

Figure 4. Plots of a family of 2π hyperbolic-secant pulses with radially-dependent pulse-width, at both the input plane and after a distance of propagation η. This graph illustrates the relative motion between adjacent pencils. At the new location, η, the profile will develop a hole near the axis for certain instants of the retarded time.

central region, still in an amplifying mode, may result in a continuous boost of an inward flow of energy. Because of this always-present interaction with population-inverted atoms, the forward flow to the axis is being amplified. The amount arriving at the axis is significantly higher than it would be if the remaining medium was linear. *Had the remaining propagation been in free space, the self-focusing distance associated with the phase curvature $\partial\phi/\partial\rho$ (hence with the transverse energy current $A^2\ \partial\phi/\partial\rho$) would have been substantially longer; whereas, the presence of the active absorber triggers the self-lensing mechanism* *sooner*. In addition, temporal and spatial variations of the phase also develop, leading to a dispersive contribution as well.

VI. PERTINENT PARAMETERS

Since F (the inverse Fresnel number per characteristic length), the "area" of the input pulse on axis, the relaxation times and the off-line center frequency shift are the pertinent parameters describing the temporal and transverse evolution of these coherent pulses in the absorbing media, we studied the dependence of the focusing characteristics on these parameters. The location of the focusing, the sharpness of its threshold, as well as the appearance of the multiple foci on-axis along the direction of propagation were altered by the specific choice of the following characteristics:

a. The reciprocal of the Fresnel number per characteristic absorbing length, F, through its different constituents;

   (1) the temporal length of the pulse τ_p,

   (2) the spatial beamwidth of the Gaussian profile r_p,

   (3) the "effective" absorption length/τ_p = $(\alpha')^{-1}$ (effect of medium gain),

   (4) The carrier wavelength λ.

b. The material relaxation times τ_1 and τ_2.

c. The frequency offset $\Delta\Omega$.

d. The input on-axis (time-integrated field amplitude) area.

VII. SCALING

In exploring the parametric dependence of the focusing phenomenon, we have studied the variations of the significant propagation characteristics with respect to F, $\Delta\Omega$, τ_1 and τ_2, the dimensionless parameters which appear in Eqs. (5)-(7). Some of these investigations have been done under the conditions that the input field (at $\eta = 0$) was taken to have a fixed Gaussian form in both the transverse coordinate ρ, and the time coordinate τ, with the additional input condition $\int_{-\infty}^{\infty} d\tau\, \xi(\rho = 0, \eta = 0, \tau) = 2\pi$, i.e., a "$2\pi$" pulse on axis. Thus, the solution for a particular value of F can be related to a set of properly scaled physical solutions corresponding to varying the parameters that enter into F in such a way that F, and the input conditions, remain constant.

On the other hand, the set of dimensionless solutions for fixed input conditions (and varying F) can be related to *physically* different input widths by assuming that the F variations are associated with variations in r_p, the transverse width of the input pulse. Similarly, physically different input pulse durations τ_p (with field amplitude modified so that they continue to satisfy the 2π-pulse condition) can be described in terms of the dimensionless solutions for fixed input (and varying F), when one takes into account the variation of α_{eff} with τ_p (both in F and in the scale of the longitudinal distance η).

VIII. NUMERICAL RESULTS

The results of the numerical integration of Eqs. (5)-(7) will now be discussed. The effect of coherent self-focusing is illustrated in Fig. 5. The integrated pulse energy per unit area (Fluence) is plotted, for various values of the transverse coordinate, as a function of the propagation distance. At the input plane, on the left of the figure, the transverse distribution of Fluence is Gaussian. After a reshaping period, during which the relative retardation of adjacent pencils leads to the development of curvature of the phase fronts, strong focusing starts to occur which leads to substantial magnification of the energy per unit area on-axis. There appears to be a rapid attenuation of the total pulse energy after the focal plane is passed. Part of this attenuation is a calculational effect in that (for the later distances) not all the pulse energy falls within the time interval over which the numerical integration was performed.

Fig. 6 gives the pulse characteristics (as a function of the propagating distance) for the on-axis time-integrated field energy per unit area, the total field energy (integrated over the beam cross-section) and a quantity ρ_{eff} which characterizes the beam cross-section. The quantity ρ_{eff} is the ratio of the total field energy to the on-axis field energy per unit area,

$$\rho_{eff} = \left[\frac{\int_0^{\rho_{max}} d\rho \; \rho \int_{-\infty}^{\infty} |\xi(\rho,\eta,\tau)|^2 \, d\tau}{\int_{-\infty}^{\infty} |\xi(0,\eta,\tau)|^2 \, d\tau} \right]^{1/2} . \qquad (21)$$

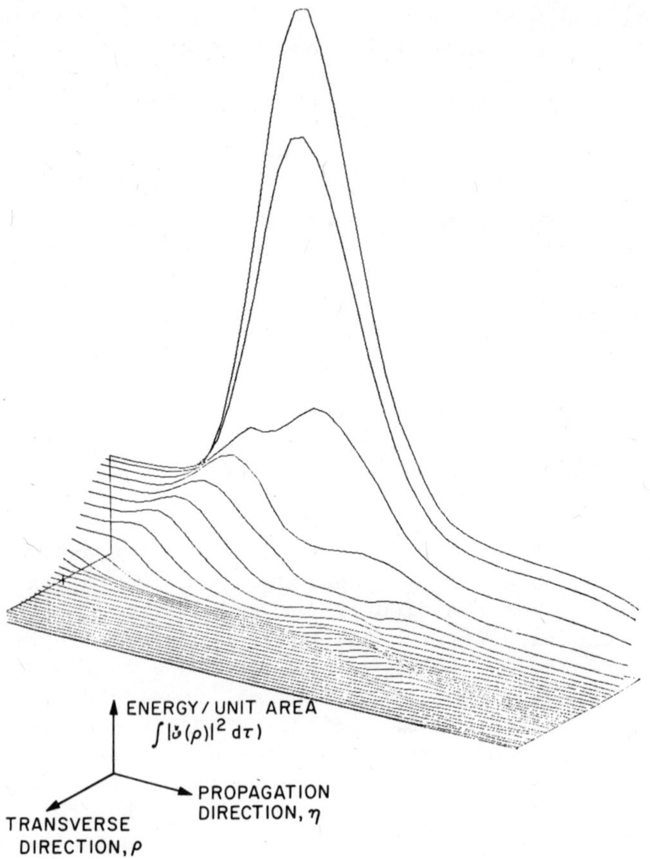

Figure 5. The energy per unit area $\{\int_0^{\tau} |\xi(\rho,\eta,\tau')|^2 \, d\tau'\}$. The fluency is displayed as a function of the distance in the direction of propagation for various values of the coordinates transverse to the direction of propagation.

It is noteworthy that the total field energy is a smooth decaying function of the propagating distance in the resonant absorber. This is in agreement with the physics of the problem even in the presence of strong magnification on-axis and provides additional confidence in the computation.

Fig. 7 contrasts the transverse distribution of the time integrated energy per unit area at the focal plane with that at the input plane in order to illustrate the coherent narrowing of the beam. The development of self-focusing is clearly evident.

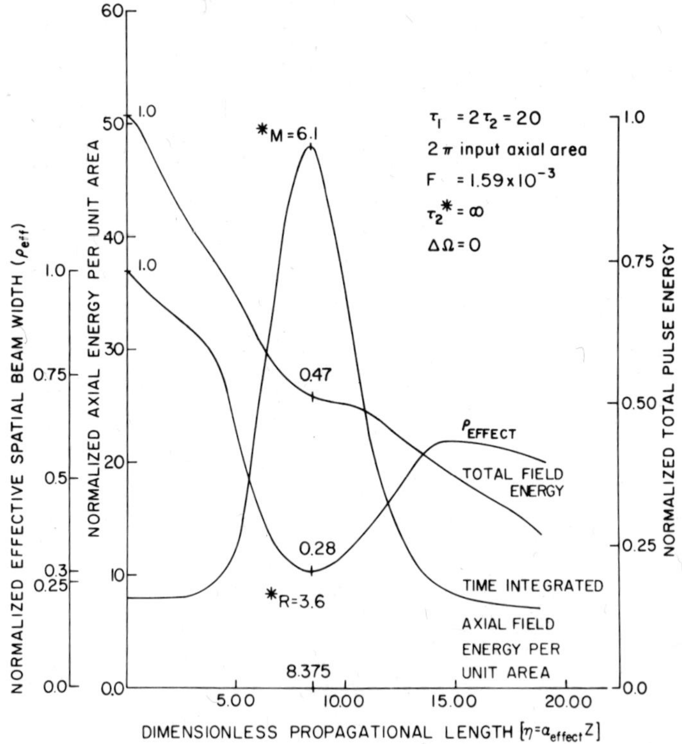

Figure 6. The principal characteristics plotted against the dimensionless propagation distance for a particular value of F: the on-axis energy density, the total field energy and an effective radius defined as the square root of the total field energy divided by the on-axis energy density.

In Figs. 8(a), 8(b), and 8(c) the field amplitude is plotted versus the retarded time for three stages of the propagation process: (a) the build-up region; (b) the reshaping region; and (c) the focal region. The transverse energy current is plotted versus the retarded time for the same three distances in Figs. 8(d), 8(e), and 8(f). In each case the plots are given for several values of the transverse coordinate ρ. Positive values of the transverse energy flow correspond to outward flow, and negative values to inward flow.

Figs. 8 clearly illustrate the following features of the self-focusing process. In the earliest stages of the propagation (Figs. 8(a) and 8(d)), the near axis energy current is outward for most of the pulse time, but becomes inward (self-focusing) toward the rear ($\tau \approx 2.4$). For this value of τ the field

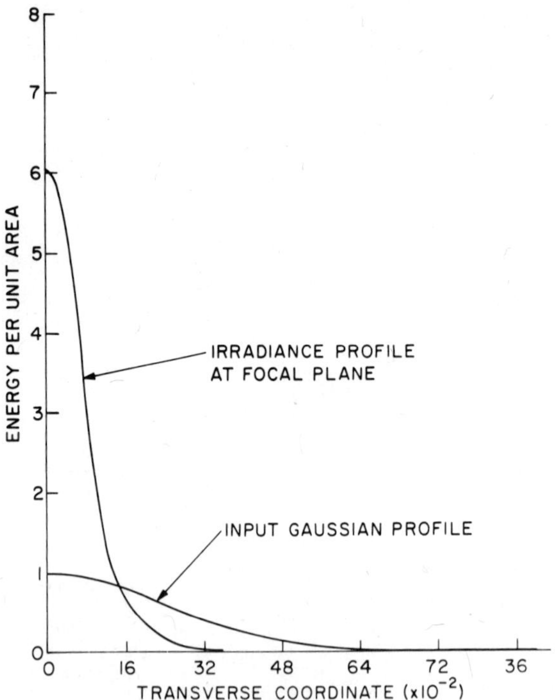

Figure 7. The profile of the energy per unit area for both the input and focal planes.

amplitude (Figure 8(a)) is already past its peak and has a small value. As we proceed into the reshaping region (Figures 8(b) and 8(e)), the near axis peak amplitude moves back in time (corresponding to the fact that the group velocity is less than c/η) while the temporal location of the change from focusing to defocusing energy flow remains the same. This leads to a large increase in the value of the transverse energy flow. In Figs. 8(c) and 8(f) (in the focal plane) the peak amplitude occurs at $\tau = 2.4$. The energy current flow in the earlier stages of the pulse is now outgoing, corresponding to power which has already been focused and is now diverging.

The results of the earlier stage (Figures 8(a) and 8(d)) are in quantitative agreement with the analytic predictions of the perturbation theory presented earlier.

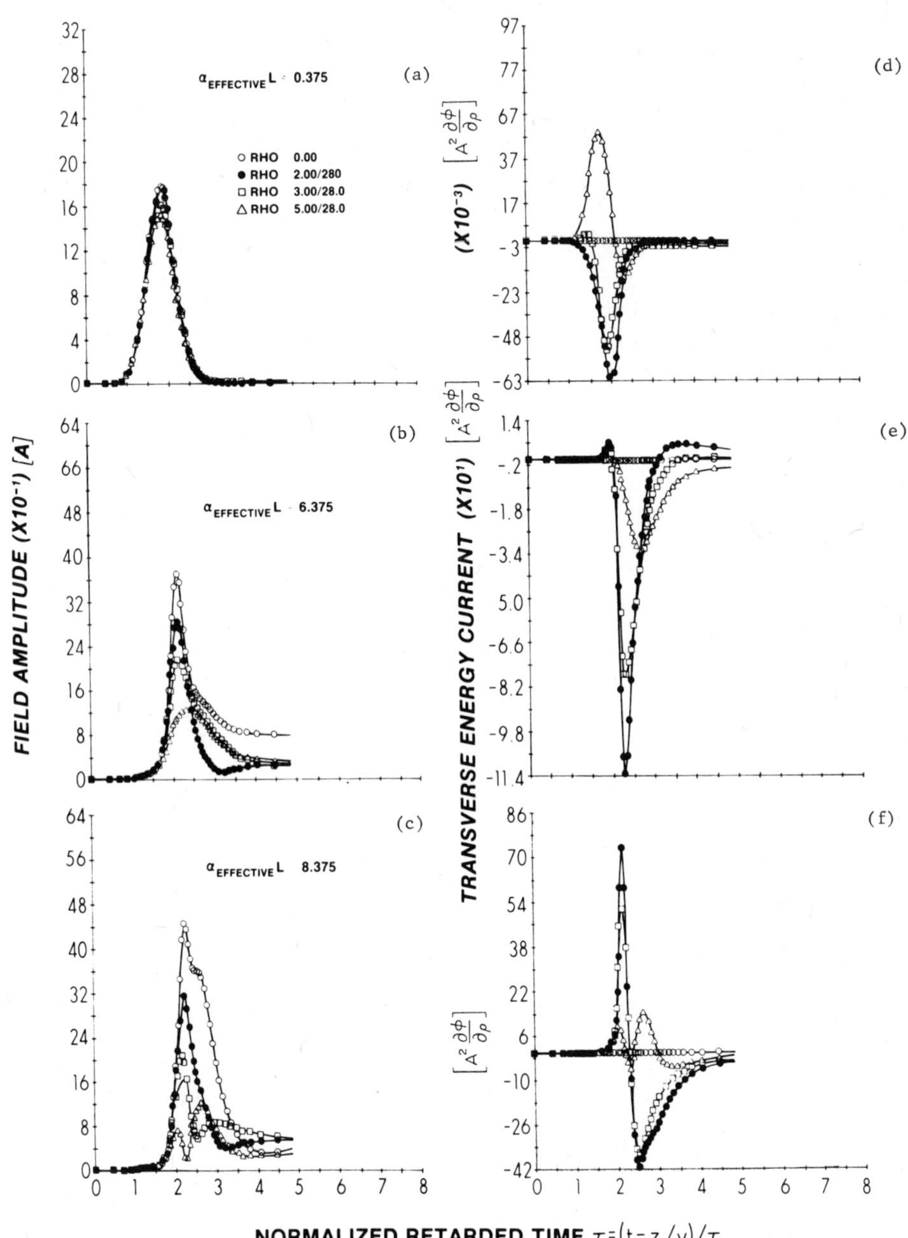

Figure 8. The field amplitude (a,b,c) and the transverse energy current (d,e,f) for several radii versus the retarded time for three stages of the propagation: the reshaping region, the build-up region and the focal region (as a function of the transverse coordinate).

In graphs (a), (b), and (c) in Figs. 9 and Figs. 10, the profile of the field amplitude is plotted for several instants of time in the leading part of the pulse or in the lagging parts of the pulse, respectively, for the same three stages of the propagation process. In the plots of Figs. 8(a,b,c), one can notice the distinct deviation of the profile from the input Gaussian shape. The beam splits into more than one lobe, indicating the different concentration of energy in more than one ring around the axis (outwardly) due to diffraction. The quantitative agreement (of the hole formation, then its filling up) with the simplified physical picture predictions presented earlier is quite evident. Graphs (d), (e), and (f) in Figs. 9 and Figs. 10 display the transverse energy current for the earlier and later instants of time, respectively, at the main stages of the propagation process. In the earlier stage of propagation, the energy current was flowing outwardly. A substantial transverse energy current flows inwardly leading to a build-up of field energy on-axis which is the new coherent self-focusing phenomenon. Immediately after the focal plane, both inward and outward transverse energy flow occur within the same pulse at different times; the outward flow occurs in the leading portion resulting from the diffraction of the focused beam, while the inward flow occurs later and prepares the second (but weaker) focusing.

In Fig. 11 we display four characteristics of the phenomenon as functions of the parameter F. These are: (1) the dimensionless focal length $(\alpha_{eff} \cdot L(\text{focus})) = \Lambda_{focus}$; (2) the ratio, M, of the axial energy/unit area at the focal plane to that at the input plane; (3) the time delay at the focal plane of the peak of the pulse on axis; and (4) a number which characterizes the pulse cross-sectional area at the focal plane (the ratio of the total field energy to the axial energy/unit area, $\pi\rho^2_{eff}$). The general monotonic decrease of Λ as F increases is consistent with the expectation that increasing the importance of the transverse derivatives in Eq. (5) will decrease the distance to the focal plane. The corresponding decrease in the time delay is because of the shorter distance over which the time delay accumulates (the pulse velocity is relatively insensitive to variation in F).

The magnification factor M, decreases monotonically as F increases for $F \tilde{>} 0.4 \times 10^{-3}$. This decrease is associated with the decreasing distance to the focal plane and the decreasing retardation. As F decreases from $\sim 0.4 \times 10^{-3}$, the distance to the focal plane increases rapidly. The phenomenon of retardation leads to the numerical effect referred to previously, namely not all the pulse energy falls within the preset time interval for numerical integration. Thus, part of the decrease in M as

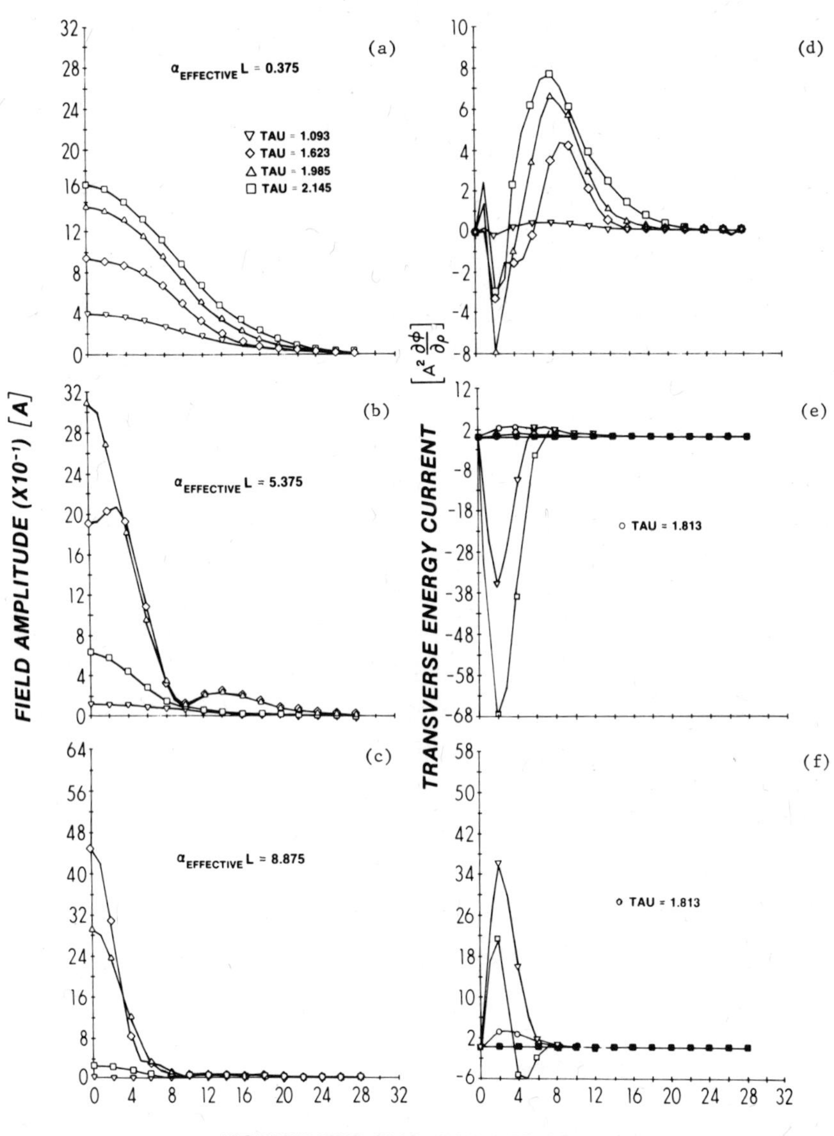

Figure 9. The profile of the field amplitude (a,b,c) and the profile of the transverse energy current (d,e,f) for several earlier (small) instants of time for three stages of the propagation: the reshaping region, the build-up region and the focal region (as a function of earlier retarded time).

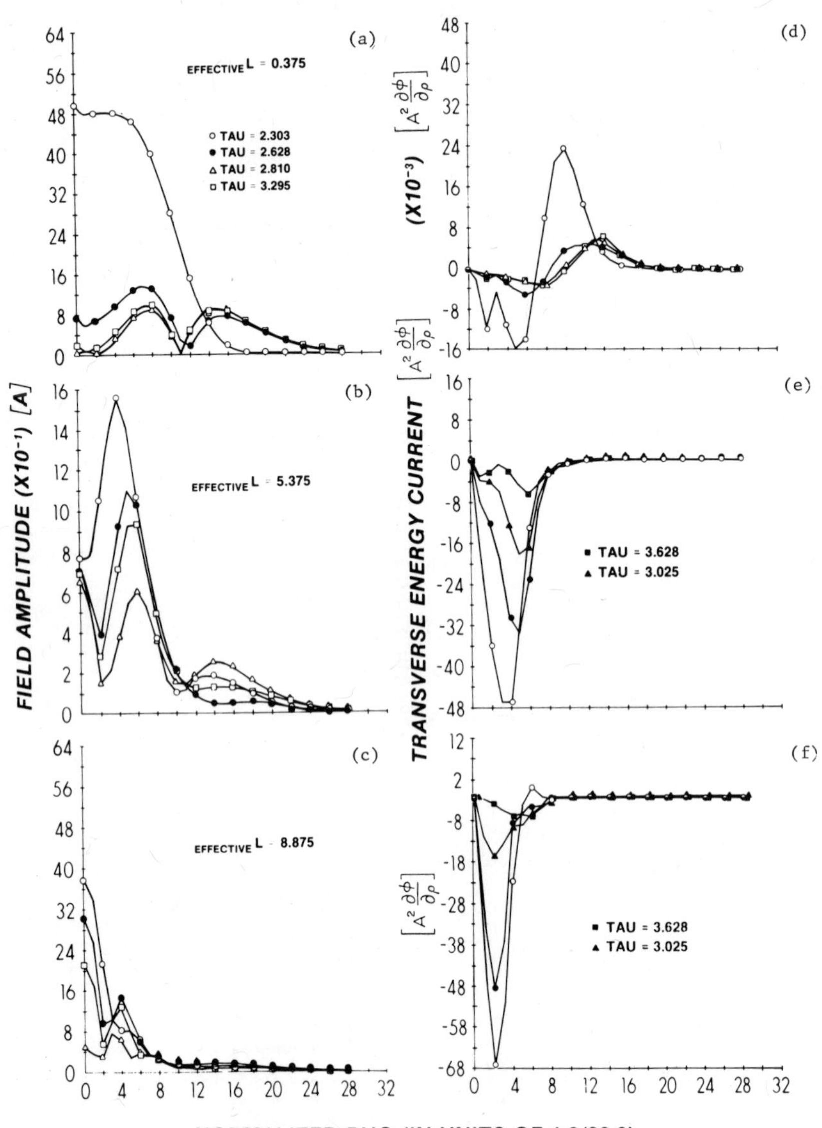

Figure 10. The profile of the field amplitude (a,b,c) and the profile of the transverse energy current (d,e,f) for several later (subsequent) instants of time for the three stages of the propagation: the reshaping region, the build-up region and the focal region (as a function of subsequent retarded times).

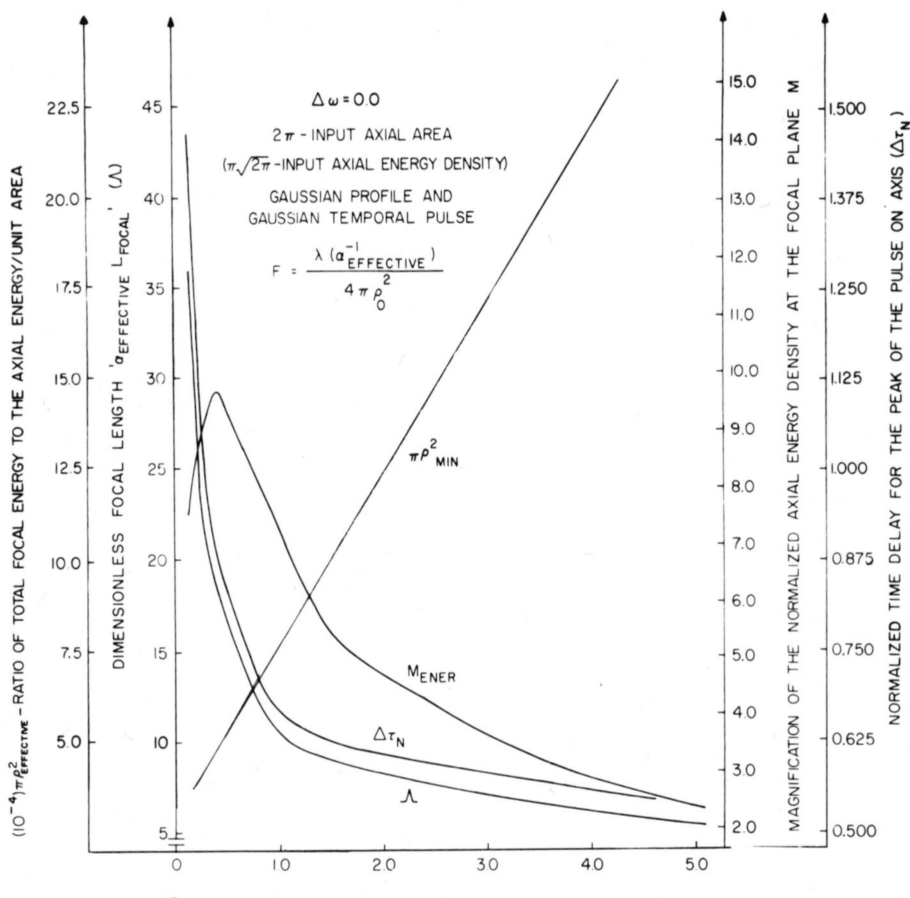

Figure 11. The principal characteristics of the focal plane as a function of the parameter F.

$F \rightarrow 0$ is calculational. The effective focal plane area $\pi\rho^2_{min}$ appears to be a linear function of the F parameter. This is not a consequence of the similarity laws for this system; however, one may make the following observation. If one scales the F parameter into the transverse coordinate of Eq. (5), one notes that doubling F is equivalent to keeping F fixed and doubling the transverse area of the input field. The fact that the effective area at the focal plane also seems to approximately double is an unexplained feature of our numerical results.

Fig. 12 gives the axial energy density/unit area, $\int_0^\infty d\tau \, |\xi(\rho = 0,\eta,\tau)|^2$, as a function of the normalized propagation distance η for various values of the dimensionless parameter F. We see that the reshaping interval, before the relatively sharp onset of the focusing phenomenon, increases as the F parameter decreases. Since these curves are shown for $F > 0.4 \times 10^{-3}$, their behavior up to the focal plane does not show the loss associated with the numerical effect referred to in connection with Fig. 11 for $F < 0.4 \times 10^{-3}$.

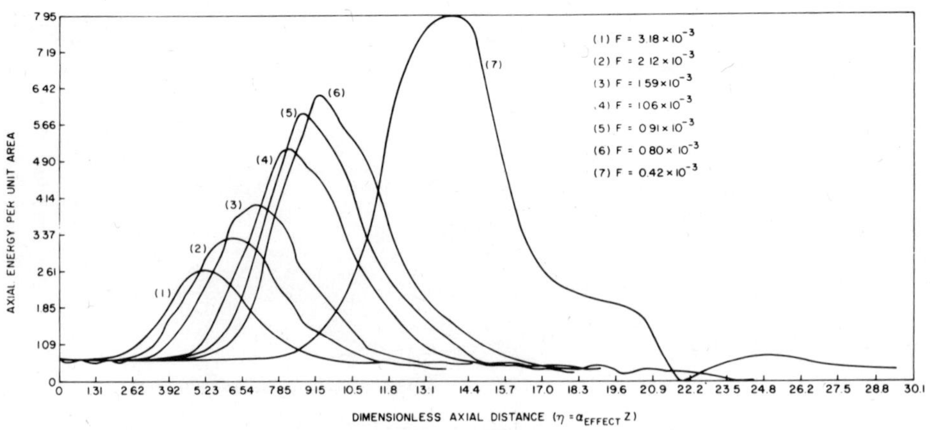

Figure 12. Universal plots of the axial energy per unit area versus the dimensionless propagation distances for a family of the parameter F.

Fig. 13 illustrates the dependence of the distance to the focal plane Λ and of the focal plane axial energy density $\int_{-\infty}^{\infty} d\tau \, |\xi(\rho = 0,\eta = \Lambda,\tau)|^2$ on the offset $\Delta\Omega$. In the $\Delta\Omega$ interval over which the focusing is significant, $-0.3 \lesssim \Delta\Omega \lesssim 1.5$, the distance to the focal plane is an increasing function of the magnitude of the deviation from resonance $|\Delta\Omega|$. The strongest focusing occurs when the carrier frequency is greater than the resonance frequency, such that $\Delta\Omega \approx 0.2$.

As one increases the offset, the axial energy/unit area at the focal plane decreases until at $\Delta\Omega \approx 4$ it is below the value at the input plane. As one decreases the offset, it rapidly reaches the input plane value at $\Delta\Omega = -0.38$. For sufficiently

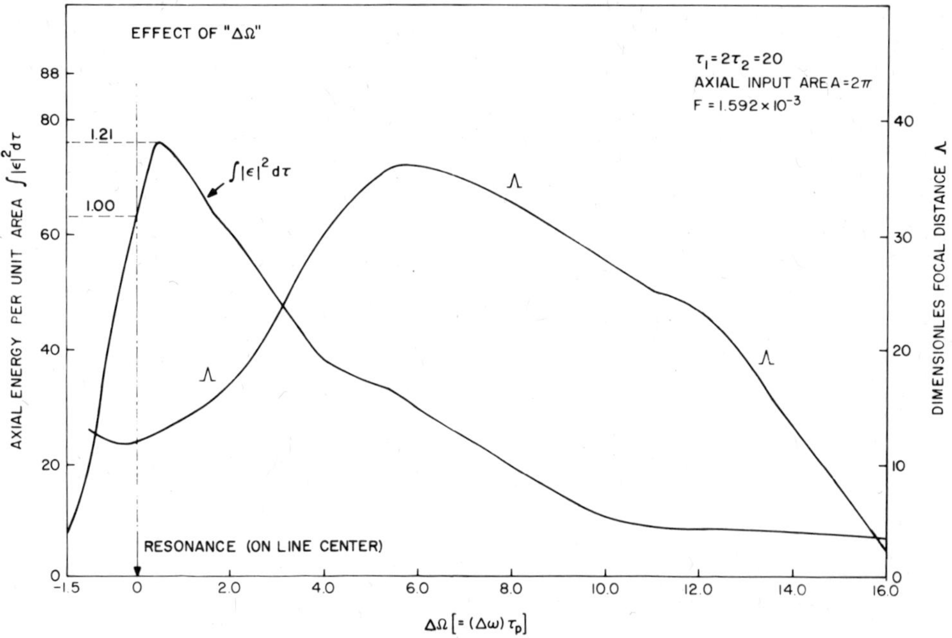

Figure 13. The frequency detuning effect on the coherent self-focusing.

large offsets in either direction one might expect a connection to the adiabatic following model. The sense of the change with $\Delta\Omega$ is in agreement with the predictions of that model.

The on-axis energy per unit area is plotted as a function of longitudinal distance for varying values of the frequency offset $\Delta\Omega$ in Fig. 14. Focusing occurs over the whole range of offsets from -1 to +4. From $\Delta\Omega = -1$ to $+0.5$ the location of the plane of peak axial energy density is fairly constant corresponding to the flat minimum near resonance in Fig. 13. As $\Delta\Omega$ is increased beyond 1.0, the focal plane rapidly retreats and the magnification factor decreases.

Two different aspects of the medium polarization must be related to the observed behavior. The absorptive properties of the medium are a resonant function of the frequency offset. A maximum rate of energy exchange between field and matter occurs at $\Delta\Omega = 0$. The dispersive properties of the medium become important as $|\Delta\Omega|$ increases from zero[23]. The development of a dispersive part of the polarization leads to self-refraction directly. A secondary effect is the evolution of phase modulation[27] which, in turn, leads indirectly to self-refraction.

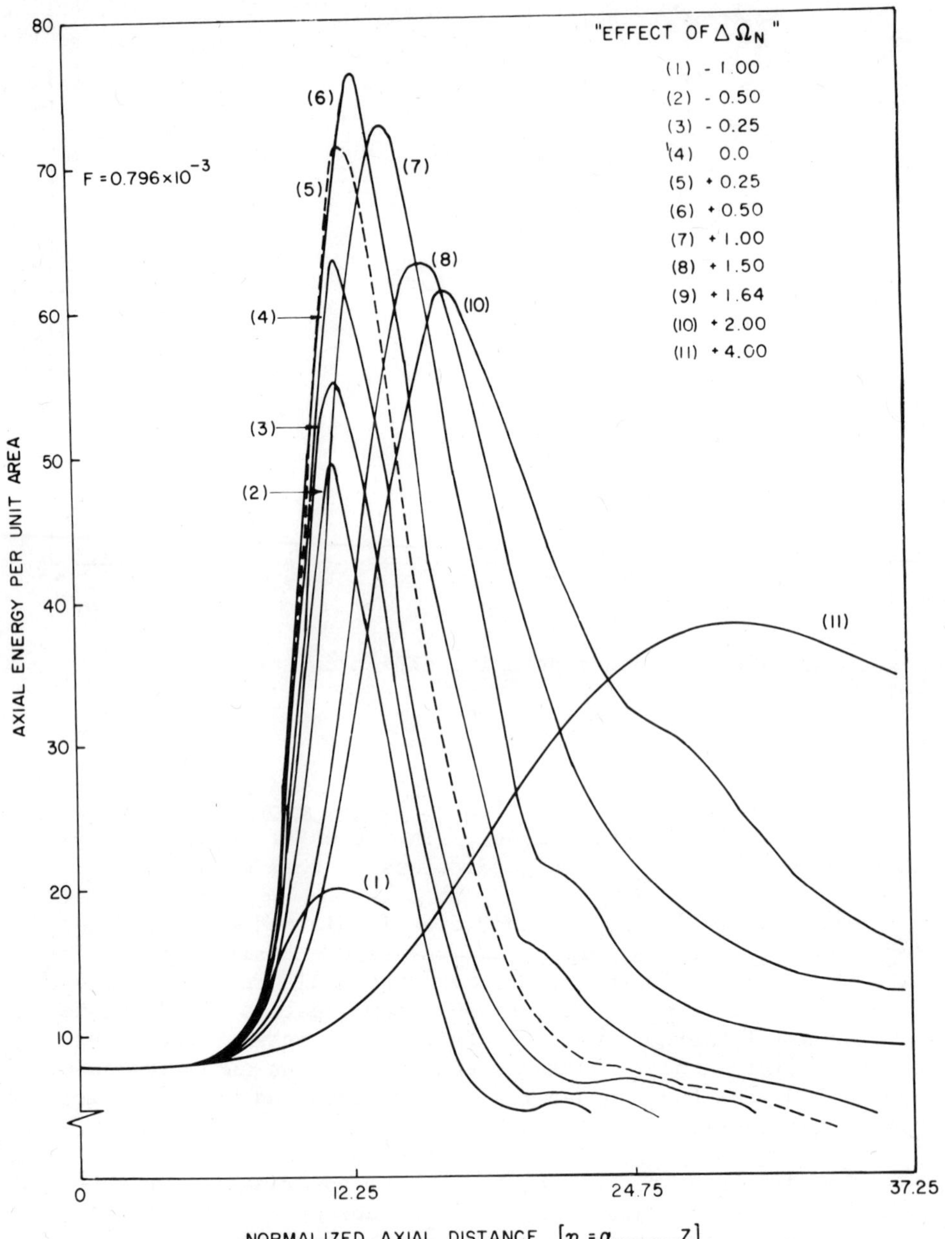

Figure 14. The on-axis energy per unit area versus the dimensionless propagational distance for various frequency-offset at the input plane.

Fig. 15 gives the dependence of the strength of the focusing phenomenon on the dimensionless relaxation time τ_1. For the cases considered $\tau_1 = 2\tau_2$. As expected, the strongest effect, as measured by the dependence of the axial energy per unit area on distance, occurs for $\tau_1 = \infty$. The peak value decreases (and occurs at later values of distance) as the relaxation time is decreased.

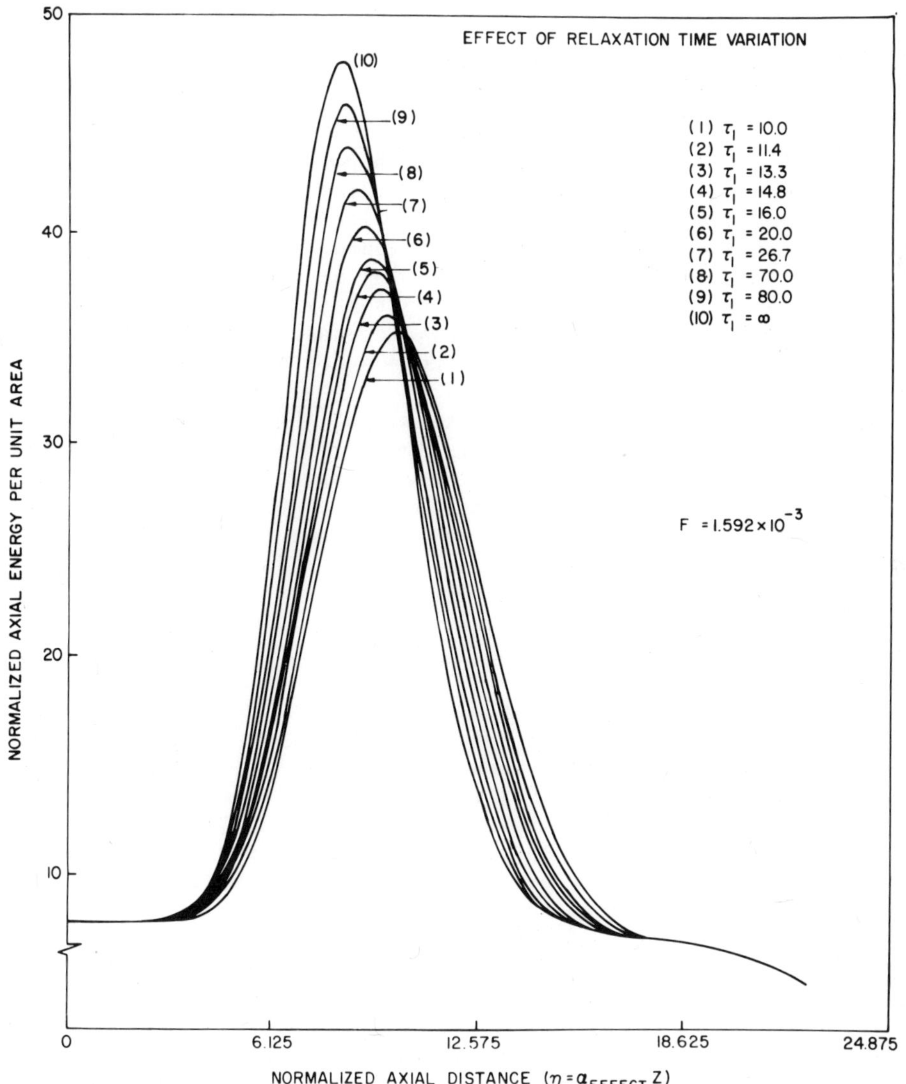

Figure 15. The effect on the focusing of varying the ratio of the relaxation times τ_1 and τ_2 to the pulse length.

Figs. 16 and 17 show the effect of varying the input axial pulse area, $\int_{-\infty}^{\infty} d\tau \, |\xi(\rho = 0, \eta = 0,\tau)|$. Our previous examples were all for an input area of 2π. Fig. 16 gives three quantities as a function of input axial area: the delay of the peak of the pulse from the input to the focal plane, $\Delta\tau$; the distance to the focal plane, Λ; and the axial energy per unit area $\int_{-\infty}^{\infty} |\xi(\rho = 0,\Lambda,\tau)|^2 \, d\tau$ at the focal plane $\eta = \Lambda$. Fig. 17 shows the axial energy per unit area as a function of η for various values of the input area. The observation of a strong dependence of delay times on the input value of the area (with pulses of small area exhibiting much larger delay times than those for larger area) is in agreement with Hopf and Scully's study on the transient pulse behavior in S.I.T.[28,29]. These figures show a rapid strengthening of the energy magnification as the input area is increased from 2π, associated with an increasing distance to the focal plane and a decreasing retardation of the peak of the pulse (the pulse speed is increasing at a faster rate than the focal plane is receding). As the input axial area is decreased toward π, the magnification rapidly decreases as does the distance to the focal plane. The focusing ceases for areas less than 1.1π.

A factor of importance in determining the effect of input area on pulse behavior is the relative fraction of energy carried by the field and matter systems. For 2π pulses, the material system returns to the tail of the pulse the energy it has received from the front. For π pulses, after the major portion of the pulse has passed, the material system is left in an excited state. For pulse areas in the neighborhood of π some energy is returned to the field at a rate insensitive to the input pulse shape.

Figs. 18 and 19 display the magnification of the on-axis normalized field energy as a function of the propagational distance for various small and large input areas, respectively. The interesting and noticeable point is the appearance of a small primary focusing for large area at the same (exact) propagational distance; a substantially larger secondary focusing appears as the pulse propagates further in the absorber. As would be expected, the weaker the field the more efficient is the magnification. This remark agrees well with parametric calculations for a uniform plane-wave pulse propagation by Iscevgi and Lamb[30].

Fig. 20 illustrates the effect of a frequency offset on the propagating characteristics of a $1.1\,\pi$ pulse on-axis. The latter was selected for this investigation since it is the largest area pulse that does not exhibit the self-focusing phenomenon for $\Delta\Omega = 0$. The most striking feature of Fig. 20 is the separation

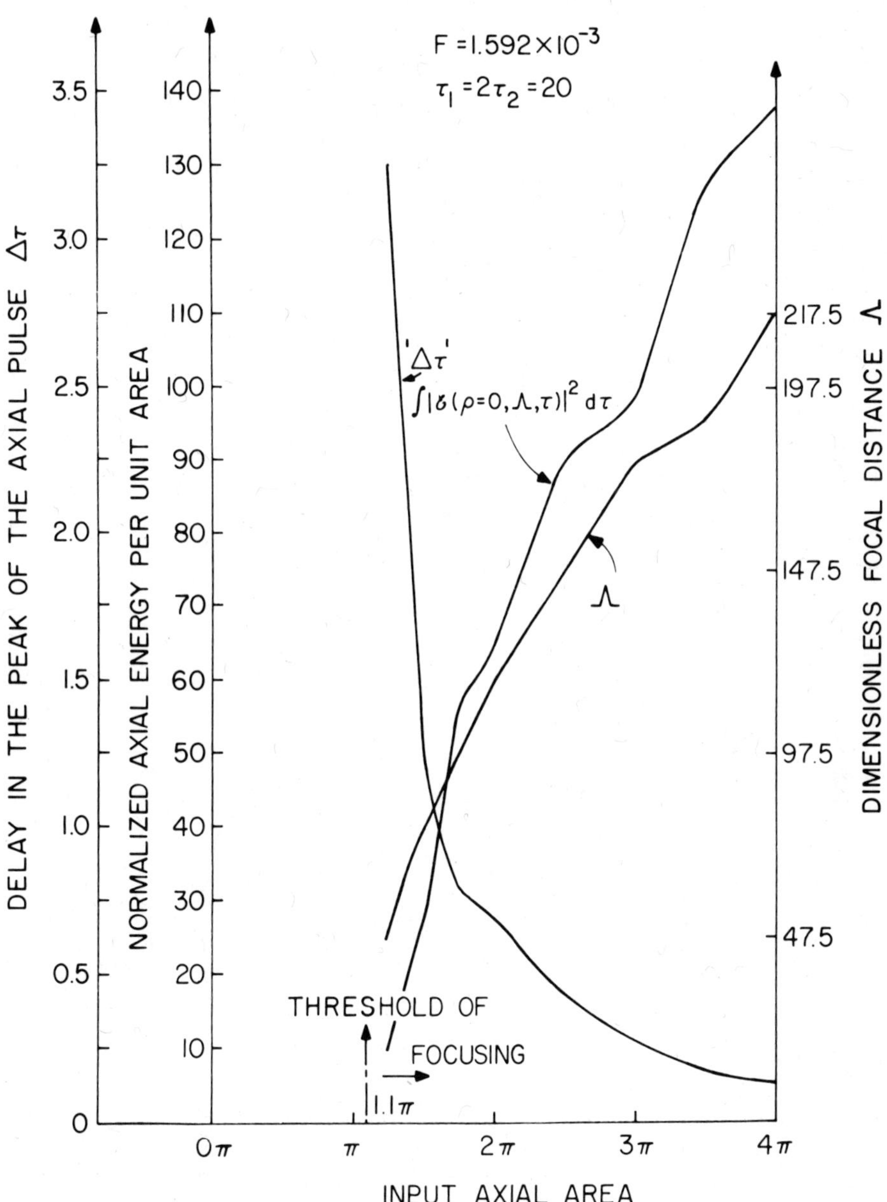

Figure 16. The effect of the input area on the focusing phenomenon.

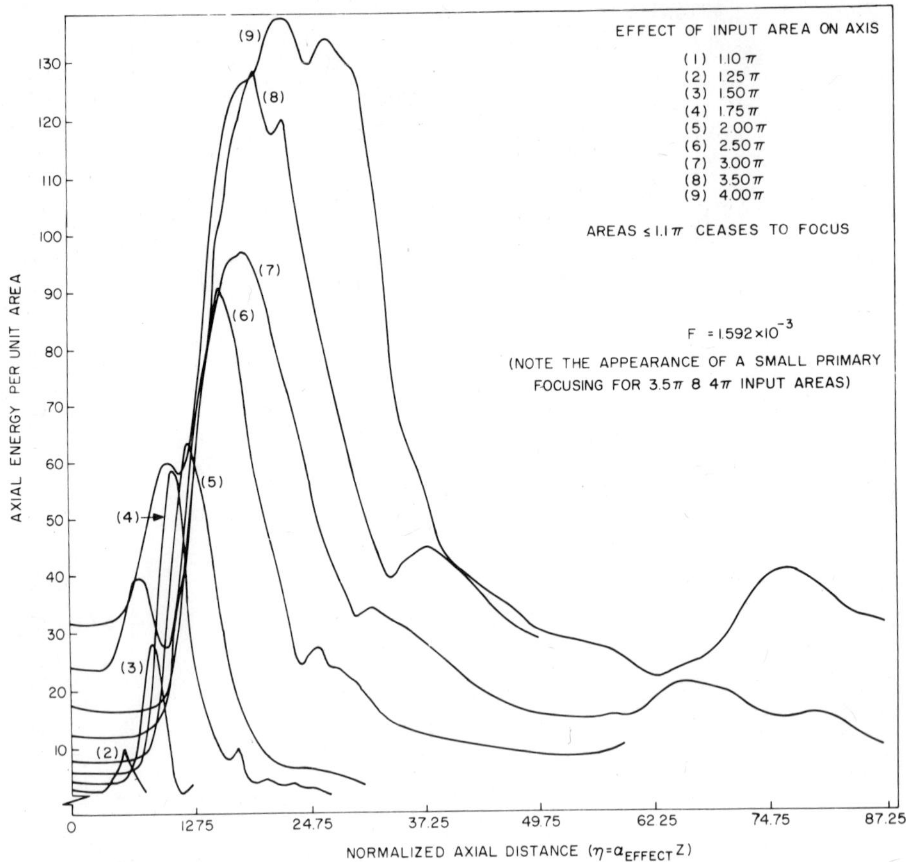

Figure 17. The on-axis field energy per unit area versus the dimensionless propagational distance for various values of the input pulse area (area under pulse time envelope).

of the curves into two classes: those corresponding to negative and small positive frequency offset and those for larger positive offset. The former corresponds to a propagating behavior whereby the energy decays rapidly as a function of longitudinal distance. The latter shows no tendencies to decrease, but rather a mild initial tendency towards focusing seems to develop. The general decaying behavior for the first group is consistent with our previous observation that for a zero offset pulse (curve #4), the 1.1 π area ceases to exhibit focusing (as in Fig. 17).

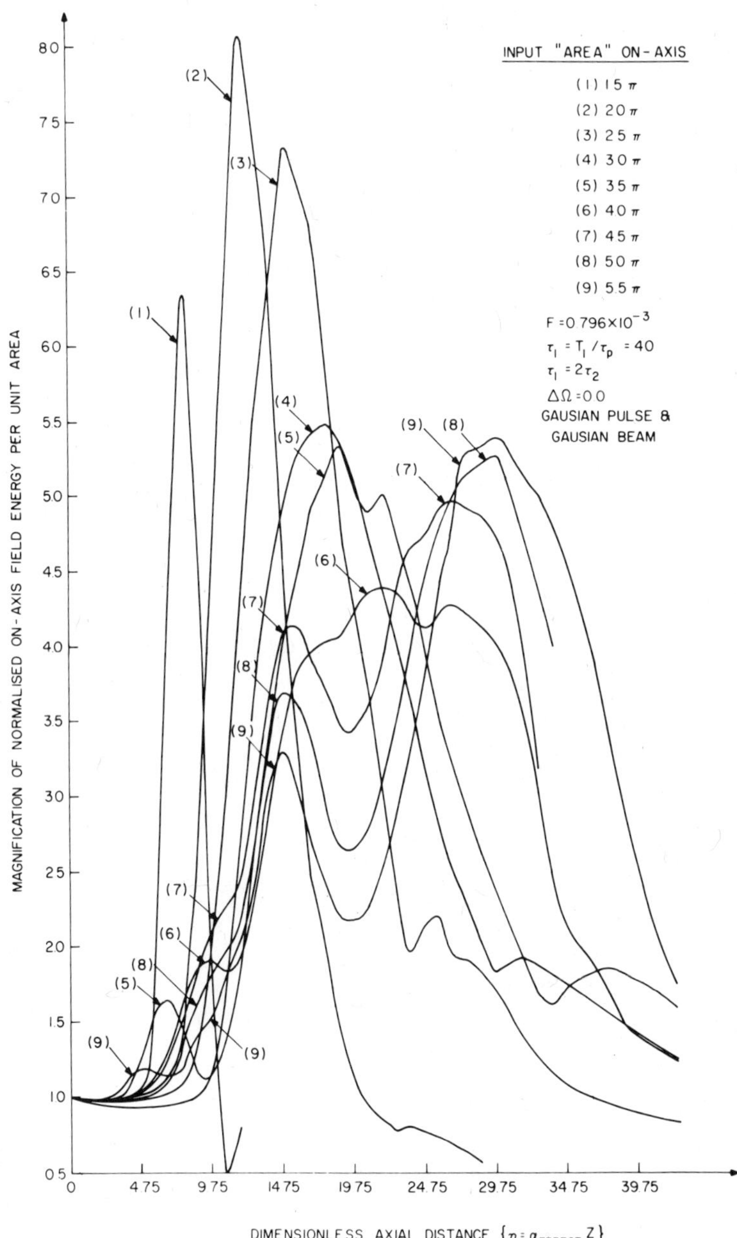

Figure 18. The on-axis magnification for the field energy per unit area versus the dimensionless propagational distance for various small values (1.5π–5.5π) of the input area.

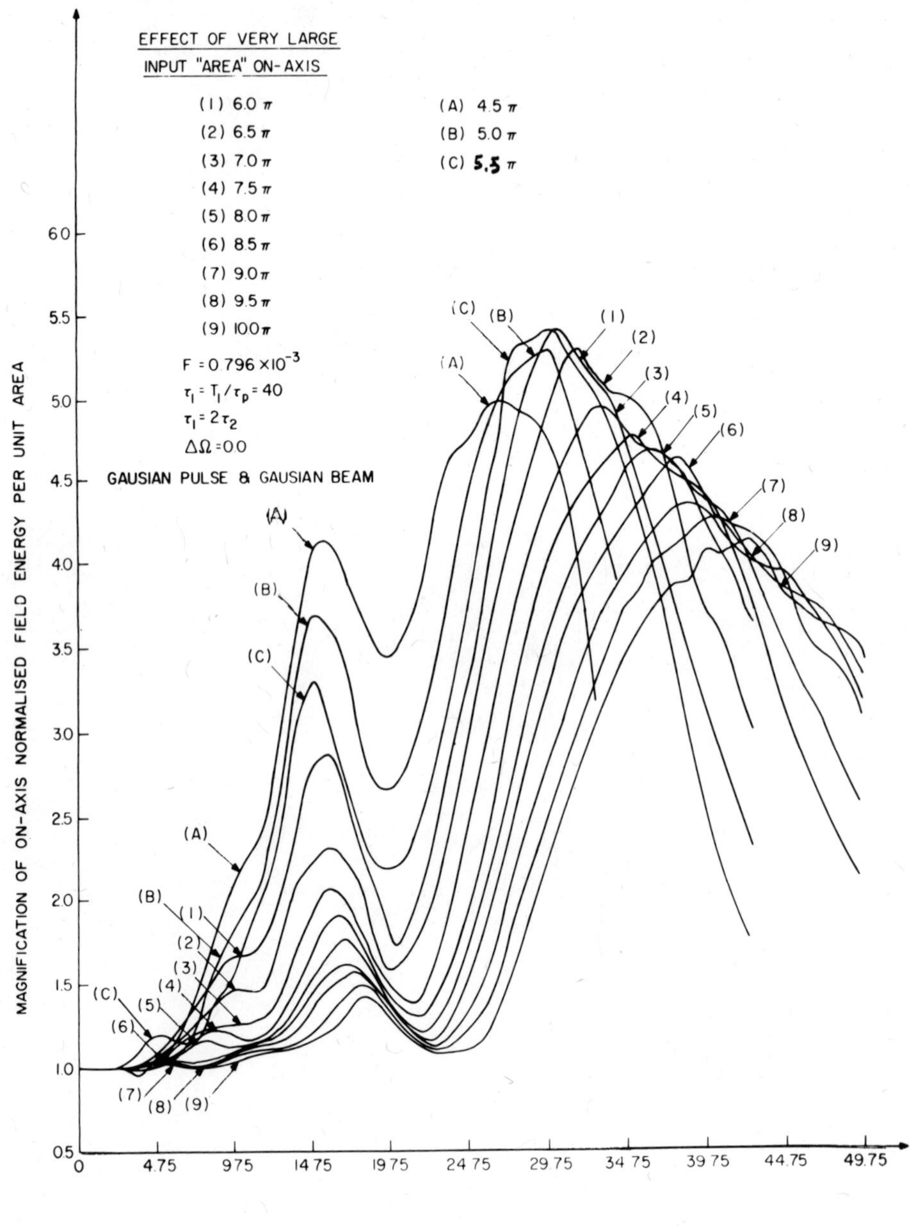

Figure 19. The on-axis magnification for the field energy per unit area versus the dimensionless propagational distance for various large (6.0π-10.0π) values of the input area.

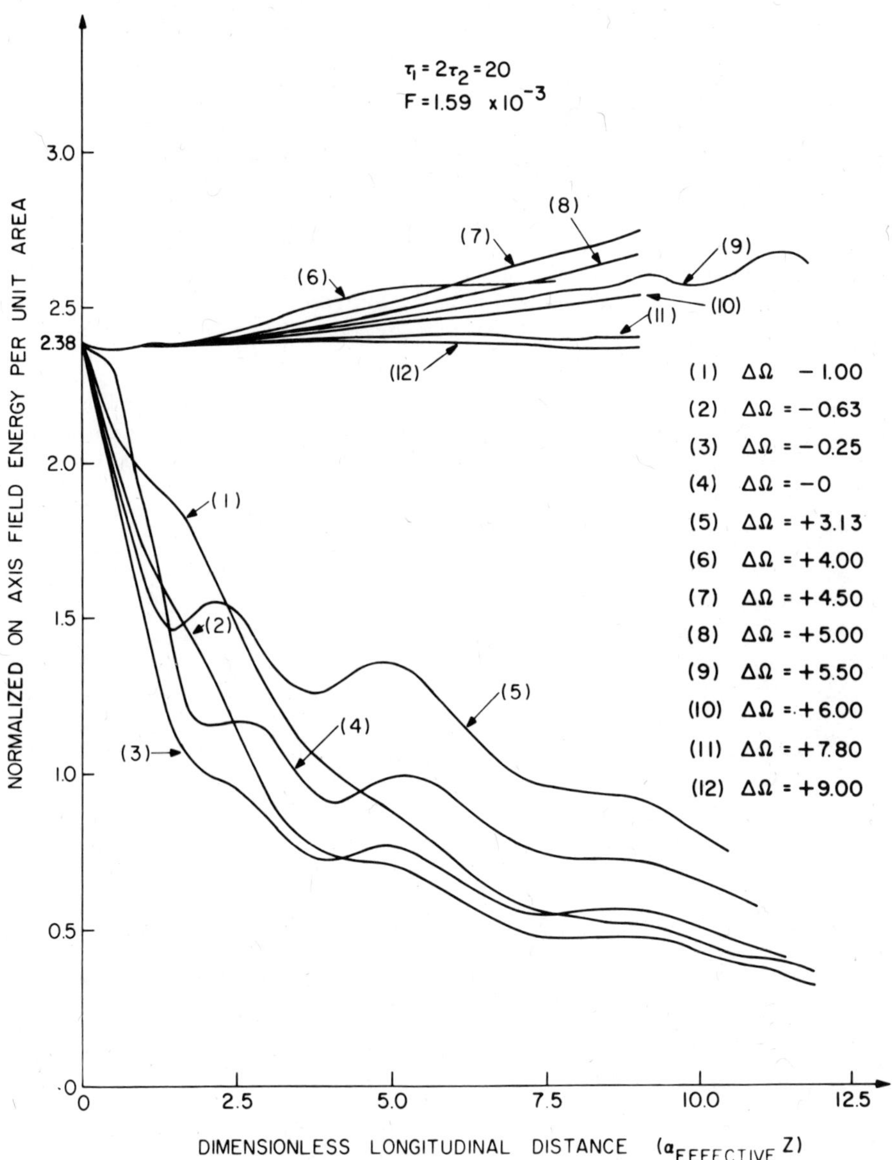

Figure 20. The frequency offset effect on the first area 1.1 π that ceases to focus, to illustrate the importance of nonlinear dispersion.

IX. EXPERIMENTAL VERIFICATION

Recently, two experiments demonstrated this new self-focusing (SF) that does not vanish on resonance and requires coherent propagation of spatially nonuniform pulses. Under experimental conditions (similar to the theoretical optimum parameters) an increase in the Fluency (axial energy per unit area), by a factor of 2 on-resonance and a factor of 4.5 off-resonance is observed. Self-Induced Transparency outputs illustrate that SF can dominate pulse reshaping in thick absorbers.

The first experimental observation[15] of this SIT near-resonance SF and self-defocusing was carried out at Philips Laboratories in Na. Transform-limited 2-ns pulses are produced by a CW dye laser and N_2-laser-pumped dye amplifier. The pulses are spatially filtered, expanded to 5 mm diam, and focused by a 800-mm lens to a 125-μm-diam Gaussian-profile equiphase plane wave in the 11-mm-long Na cell. The circularly polarized light interacts with only the nondegenerate D_1 transition in a high magnetic field (all M_1 transitions have the same dipole moments, so their 0.44-GHz splittings have the effect of inhomogeneous broadening and their dispersions cancel at line center). Frequency tuning of the transition is affected by the magnetic field. Plane-wave SIT (large-diameter beam; $\alpha L \approx 5$) was used to calibrate the input area.

Focusing, defined here as an increase in energy per unit area in the center of the beam, was monitored by 12.5X imaging of the cell exit upon a Philips TV camera and 280-μm-diam Si avalanche photodiode. The TV monitor permitted visual observation in two dimensions of the dramatic reduction in size and enhancement of center intensity from self-focusing. Quantitative beam profiles were obtained by photographing an oscilloscope display of a TV scan through the beam center (2-3 μm resolution at the cell exit); see Fig. 21. This 2X on-resonance magnification was insensitive to the position at which the beam passed through the cell, and to whether the beam was parallel or slightly divergent or convergent through the cell. Thus, it is likely that diffraction in the cell rather than initial phase variations initiated this SF. Drastic changes in spatial and temporal profiles for pulses undergoing Self-Induced Transparency with transverse energy flow are shown in Figs. 22, for the maximum off-resonance (a and a_1) and on-resonance (b and b_1) self-focusing. The smaller curves are the inputs in all cases. The same pattern of pulse compression and beam narrowing, as predicted by the computations in graphs (b) and (c) of Figs. 8, 9, and 10, is clearly evident.

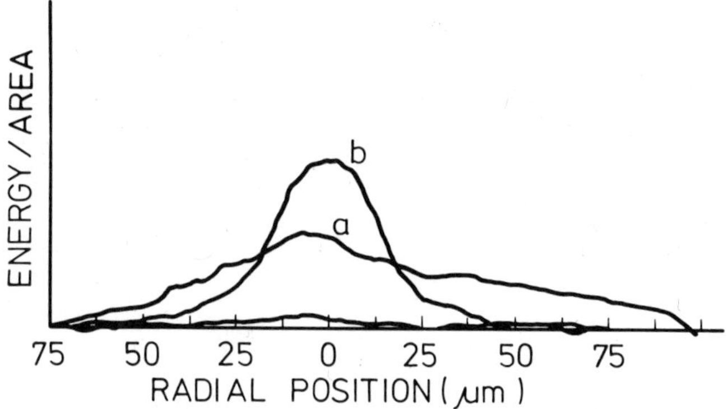

Figure 21. Cross-section of beam at cell exit in the first experiment: curve a (without Na) and curve b (with Na on-resonance), with input area of 3π to 4π and magnetic field of 3.5 kG.

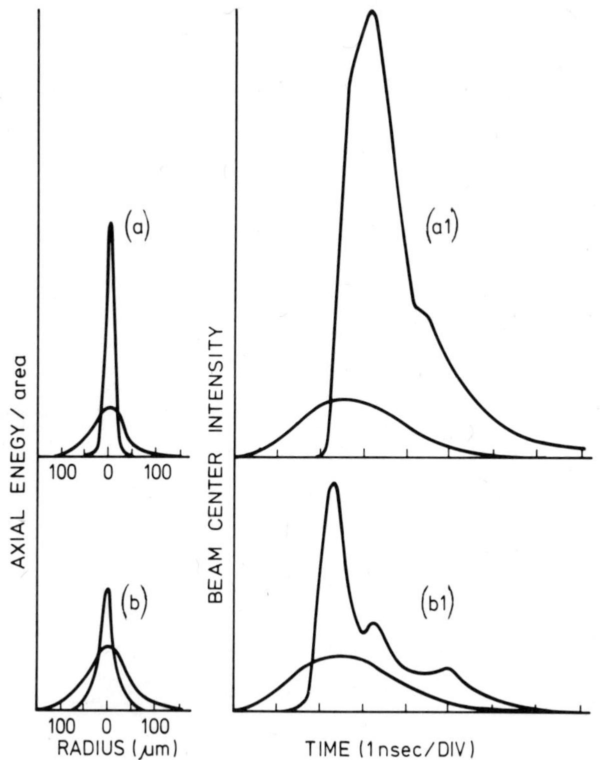

Figure 22. Changes in spatial and temporal profiles for pulses undergoing Self-Induced-Transparency with Transverse Energy flow that lead to coherent Self-Focusing. For the maximum off-resonance: curves b and b_1.

With uniform plane-wave inputs, one can normally see 4π break-up with peaking of 30 to 50%, Fig. 23(b). With non-uniform plane-wave inputs the peaking by on-resonance self-focusing was over 3X as shown in Fig. 23(a). These data illustrate the dominating effect on-resonance self-focusing can have upon spatial and temporal evolution; even larger effects occur off-resonance.

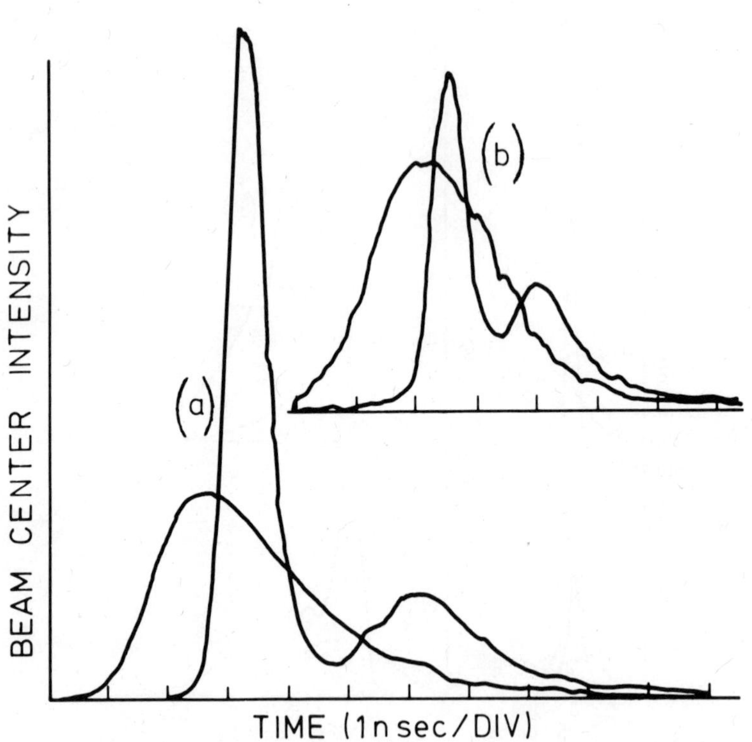

Figure 23. Temporal behavior of pulses. Break-up of a 4π input pulse is shown under (a) self-focusing, and (b) uniform plane-wave conditions in Na. In (a) the integrated output is 34% larger than the input.

The experimental dependence of coherent transient SF upon absorption and magnetic detuning in Na is shown in Figs. 24. By a simple change from an off-resonance frequency (where there is no interaction with the atoms) to resonant interaction with the atoms, one can see a drastic change in the output diameter and the peak intensity of the transmitted pulse. CW light has no transient SF and focuses for $\Delta\omega < 0$ and defocuses for $\Delta\omega > 0$ (Fig. 24, curve g)[31]. For 2-ns pulses and $\alpha L < 3$, slight SF occurs peaked on resonance (Fig. 24, curve a). At higher absorption the maximum SF occurs for $\Delta\omega < 0$, but SF is still seen on resonance (Fig. 24, curves a-d). Only for very high absorption ($\alpha L \approx 20$) is there no magnification on resonance (Fig. 24, curves e and f); presumably absorption destroys the pulse after it passes its focus. *The observation of focusing on resonance for coherent pulses and not for CW light clearly illustrates that coherent transient self-focusing is different from previous index-saturation self-focusing*[17,18].

Beam diameter (Figs. 24) is not a useful criteria of SF because the weak outer portions of the beam may be stripped away without inward energy flow. Losses from temporal reshaping and diffraction help reduce the energy magnification to far less than the square of the ratio of input to output diameters. SF drastically alters pulse shapes, too. With uniform plane-wave inputs, peaking in 4π is typically 30 to 50%. With on-resonance SF, breakup still occurs, but the peaking is more than 3X; still larger effects occur off resonance[32].

The experimental F value of 10^{-2} corresponds to a Fresnel number of about 3, i.e., strong enough diffraction to provide radial communications, but weak enough that SF is able to compete with diffraction. Numerical simulations predict maximum magnification of almost 10 for $F = 4 \times 10^{-4}$ and $\alpha L \approx 18$ ($F \approx 44$)[10,14]. Attempts to observe on-resonance magnifications greater than 2 with large-diameter beams resulted in hot spots (filaments) in the output even for $\alpha L < 15$. These result from focusing of smaller diameter regions with $F \simeq 10^{-2}$, initiated by input phase and intensity variations (small ripples)[33] differing from the theoretically assumed perfect Gaussian input. Consequently, *self-focusing may be unavoidably important in the propagation of coherent optical pulses through thick absorbers* ($\alpha L \gtrsim 15$) *even when large-diameter beams (large F) are used on resonance.*

The second experimental observation was carried out in Heidelberg. The researchers investigated the propagation of linearly polarized 1.15-μm light pulses in an absorbing neon

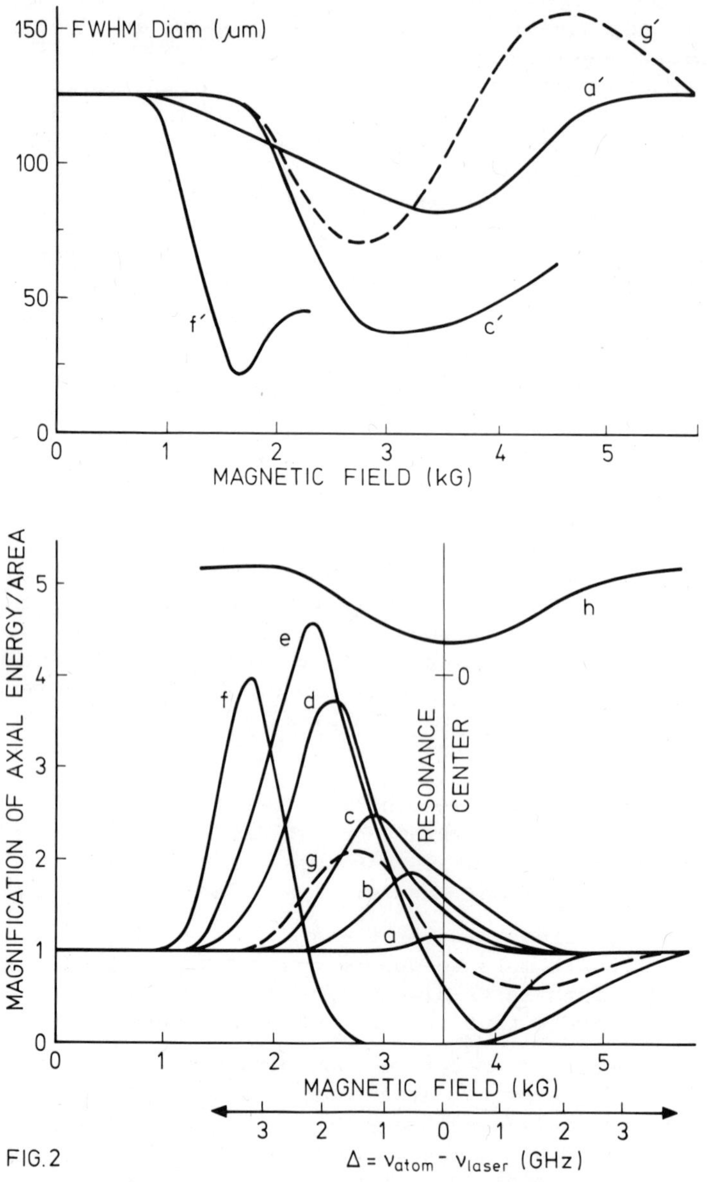

Figure 24. Experimental energy-density magnification and diameter reduction in Na as a function of detuning. Curves a to f are for 2-ns, 5 pulses of 125 μm diam in an 11-mm cell. The absorption increases from curve a to curve f; curve g is self-focusing of CW light; curve h shows the atomic absorption. Above, Curve a is the diameter for the conditions of curve a, etc.

discharge ($\alpha L < 9$) and observed both SIT[34] and SF[35]. A single-frequency, 1-mW, CW, He-Ne laser is kept *on resonance* with the $2s_2$-$2p_4$ transitions by a short intracavity low-pressure Ne absorber discharge. Pulses of 20-W, 1-ns are generated by a pulsed optical amplifier consisting of a fast transverse Ne discharge of the Blumlein type. The amplifier output pulses pass a demagnifier, spatial filter, an offset-free and deviation-free precision attenuator, and the 110-cm-long absorber (ρ = 0.8 torr, i_c = 0.5 A). Finally, the central 0.5 mm of the 2-mm-diam beam is apertured, and its power is detected by an avalanche photodiode and oscilloscope (150-ps risetime). The transmission T is determined by graphic integration of about forty superimposed pulses. Particular care is taken to produce a perfect Gaussian beam, with less than 3% variation of diameter along the absorber to avoid small-scale self-focusing. Comparison with numerical simulations[19] of uniform-plane-wave Maxwell-Schrödinger equations with T_1 = 33 ns and T_2' = 10 ns calibrates the squared pulse area, θ^2, in agreement with power measurements.

Data and simulations are compared in Fig. 25[19]. For τ_p = 3 ns, experimental T's slightly exceed those from uniform-plane-wave theory; for 1-ns pulses the discrepancy is even larger. *This increase in SF with increased T_2'/τ_p agrees with three-dimensional calculations* and emphasizes the coherent nature of the effect.* Magnifications up to 40%, and more recently up to 60%[49], were seen (see Fig. 26). Note that the experimental transmissions increase and the uniform plane-wave simulations decrease with increasing αL.

Simulations with all of the Na or Ne experimental parameters (and complications, such as inhomogeneous broadening and M_I splittings and relaxation times) have not yet been made. A family of calculations was made to verify that the coherent self-focusing predicted in sharp-line will remain in broad-line atomic systems. The purpose was to see ahead of time if the proposed experiment[47] in Na, would work. This calibration curve allows one to extrapolate the various parameters studied to the case of broadening. Future joint theoretical and experimental efforts are planned to refine the study of the pertinent features of the dynamic self-action phenomena due to the coherent interaction. Nevertheless,

---

* <u>Op. cit.</u>, p. 151.

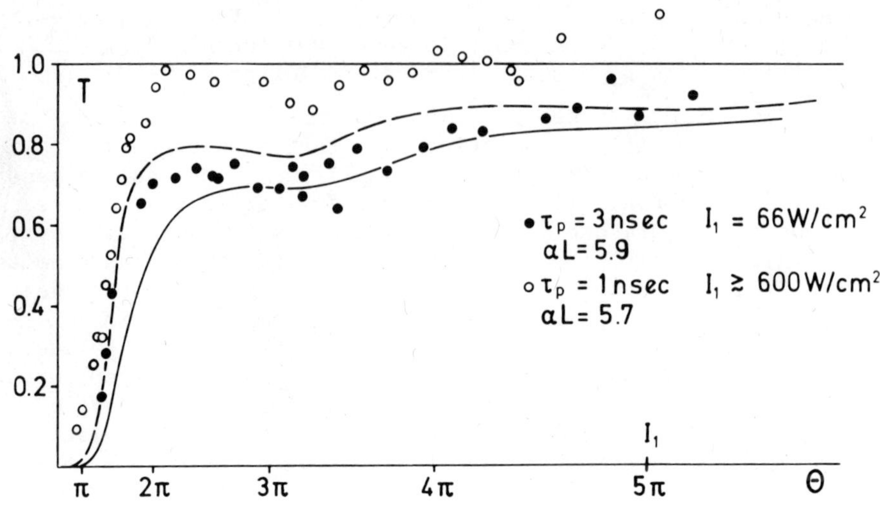

Figure 25. Pulse energy transmission (output per input) in Ne versus squared pulse area for 3-ns (full dots) and 1-ns (open dots) pulses. Curves are the corresponding plane-wave computer simulations.

there is a *surprisingly good semi-quantitative* agreement between the predictions of the sharp-line, single transitions numerical calculations and the experimental data, in particular, for the observed magnification of $\alpha L \approx 10$ and $F \approx 10^{-2}$. Data and theory agree that coherent self-focusing can be as large on resonance as off, can occur for a wide range of input areas above π, and is most effective when the relaxation times are long compared to the pulse length. *There is no doubt that the experimentally observed self-focusing is the new mechanism recently predicted numerically*[14].

In summary, coherent transient self-focusing greatly alters the transverse and temporal evolution of optical pulses propagating in thick ($\alpha L > 5$) resonant absorbers if $[F = \lambda/4\pi r_\rho^2 \alpha]$ is between 10^{-4} and 10^{-2}. Even large-diameter beams may break up into SF filaments of smaller diameters. Coherent transient SF may explain previously not understood observed transverse *effects*.

Perhaps the reason this effect has not been seen clear before is that it becomes significant only *after* the reshaping region, i.e., $\alpha L \leq 5$, on which most SIT experiments have concentrated. Much higher absorption ($\alpha L \approx 25$) was used by McCall

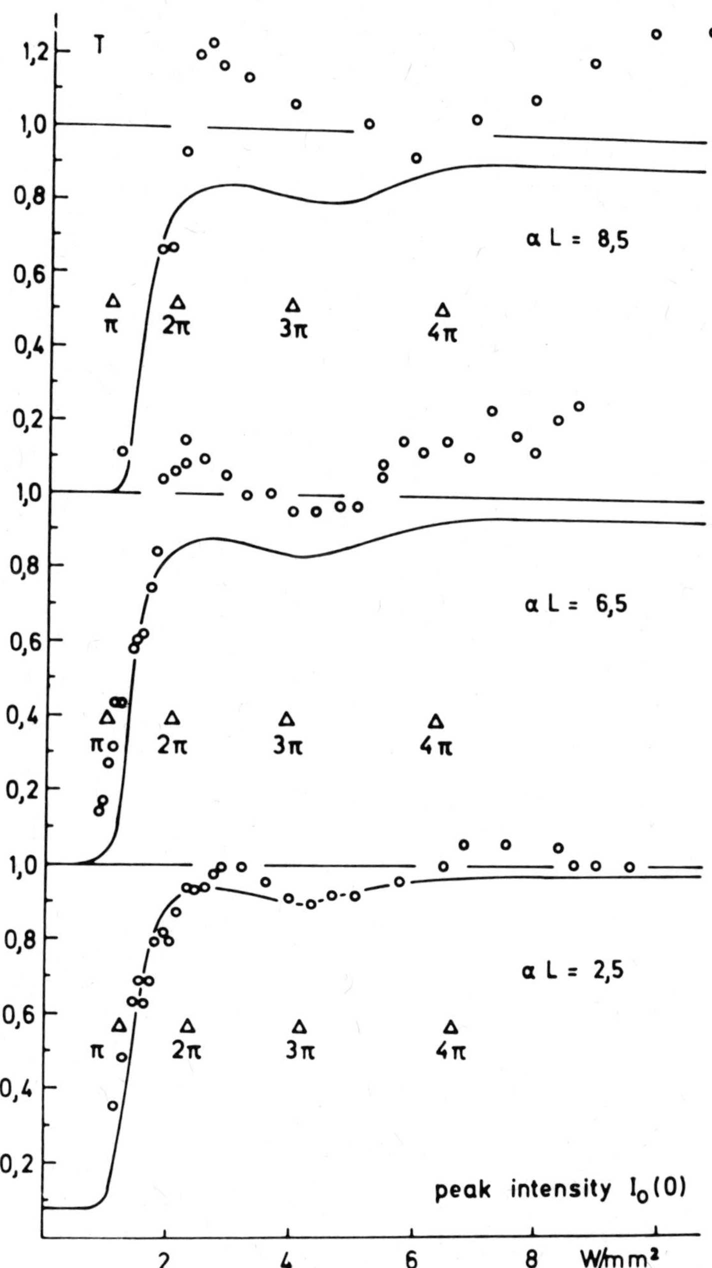

Figure 26. Energy-density magnification versus peak intensity for 0.8-ns pulses, showing increase of self-focusing with Ne absorption, αL. Curves are plane wave computer simulations with $T_2' = 10$ ns.

and Hahn[36] in the first SIT experiment; they reported hot spots in the output (also seen in the Na experiment for large input diameters) which they attributed to transverse instabilities. Zembrod and Gruhl[37] used $\alpha L \approx 11$ and $F \approx 10^{-2}$, so self-focusing should be beginning; their Fig. 1c has an output about 14% higher than the input and most of their data have higher transmissions than expected by uniform plane-wave simulations. Rhode and Szöke[38] also reported transverse effects seen in SF_6 with $\alpha L \approx 20$, which may have resulted from self-focusing.

Furthermore, if one adjusts (in the numerical simulation for a sharp-line two-level atomic system) the absorption is such a manner that it is equal to the effective absorption that one gets in the broad-line case (which is the situation of most of the experiments), one would find that on-resonance coherent SF does indeed occur at an αL of 7.5 with a magnification of 2.4.

The importance of coherent transient SF in amplifying media remains to be worked out.

X. DISCUSSIONS AND SUMMARY

Studies of plane-wave coherent propagation show a sensitivity of evolved characteristics to the temporal behavior of the input pulse. When *transverse* variations are allowed in the input plane, the complex one-dimensional effects in adjacent pencils are coupled by diffraction (various parts of the transverse profile can communicate with each other) and refraction[14]. This leads to substantial deviations from the one-dimensional behavior and to the development of a variety of self-action phenomena. A particularly interesting effect, revealed by our numerical analysis of the propagation of a coherent pulse in a resonant absorber, is the self-focusing of an input pulse with transverse Gaussian variation and 2π area on-axis. This *dynamic* coherent behavior cannot be described in terms of a closed-form nonlinear susceptibility, but requires the solution of the full Bloch equations for the material system.

Several separate physical[40-45] effects seem to play important roles in the focusing process. These effects, such as *coherent reshaping, self-phase modulation,* and the spontaneous development of *adiabatic following,* come into play during different space and time intervals as the pulse evolves from the input plane.

The initial behavior, near the input plane, can be understood in terms of a theory in which the *transverse coupling terms of the wave equation perturb the one-dimensional propagation characteristics. This leads to the development of radially dependent*

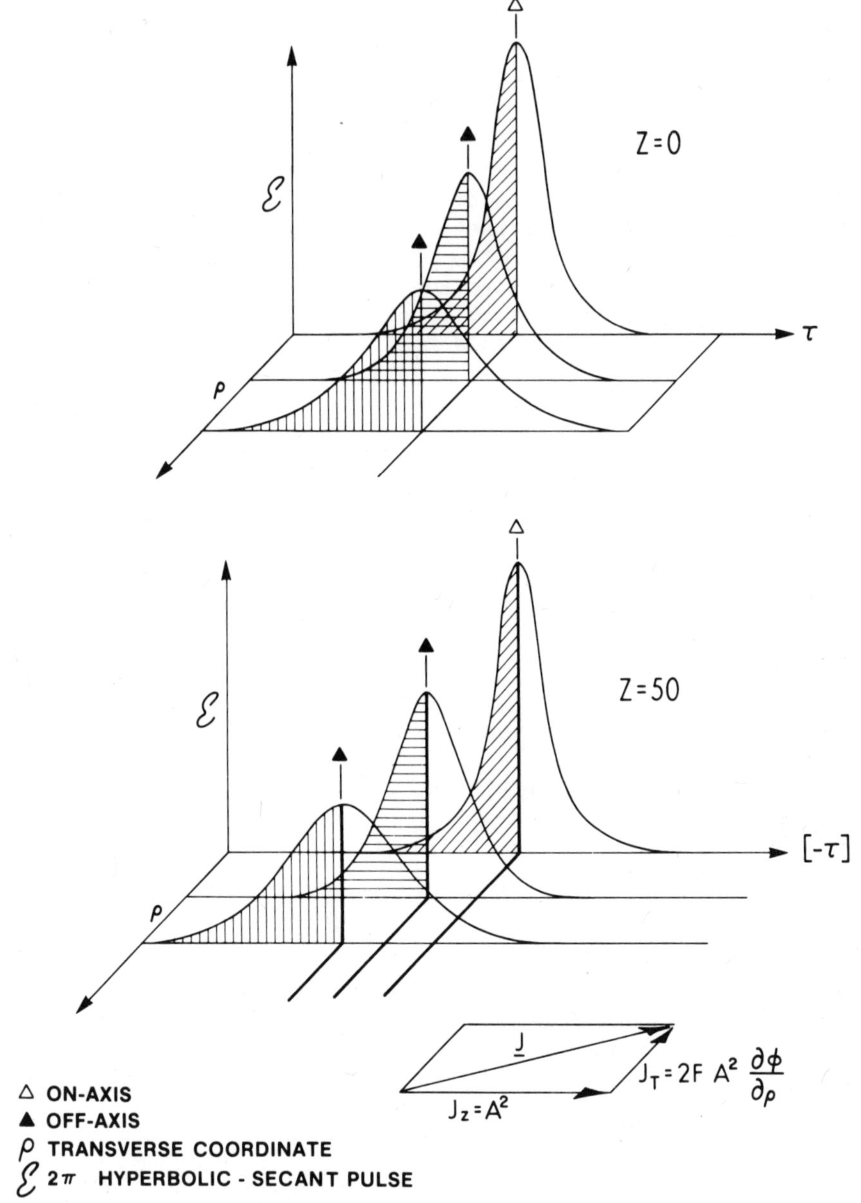

Figure 27. An isometric time evolution for three distinct radii representing three 2π hyperbolic-secant pulses (with radially dependent pulse-width) after a propagation distance η. This plot illustrates the relative motion as well as the boosting operation that the light, diffracted from the tail of the pulse on the rim of the beam, experiences while flowing toward the axis.

self-phase modulation in the temporal tail of the pulse. The radial phase variations lead, in turn, *to inward transverse energy flow in the later time portions of the pulse.* The sites of significant radial phase variation propagate with greater velocity than the loci of peak intensity.

The former sites are characterized by strong self-phase modulation and hence weaker interaction between field and material systems. This part of the pulse tends to propagate at the velocity of light in the background medium. At the loci of peak intensity there is negligible phase modulation, and the field-matter exchange of energy is of greater importance in determining the propagation velocity. This leads to a lagging of the peak of the pulse toward the sites of strong transverse energy flow. When a significant portion of the pulse intensity lags into the same site as strong radial phase variation, the inward transverse energy current rapidly increases in importance and dominates the subsequent propagation characteristics, causing self-focusing.

We have studied these processes numerically. Change in shape, temporal length, spatial width and amplitude of the input pulse were considered. Pulse development when the inhomogeneous broading τ_2, the characteristic inverse length α_{eff} and the frequency-offset are varied was observed. We have compiled universal output curves illustrating the expected output from the absorber as one modifies the characteristics of the input field and the active media.

As an example, from Fig. 11, we predict the observation of coherent self-focusing (an on-axis magnification of axial energy per unit area of 10) when the F parameter has the value 0.4×10^{-3}. The distance to the focal plane will then be $30(\alpha_{eff}^{-1})$.

Until quite recently, the work whose physical parameters most closely match the situation in our computations was the experiment of Gibbs and Slusher[46]. Unfortunately, the length of their cell and the range of operating pressures were such that the distance to the focal plane exceeded their cell length.

More recently, at Philips Research Laboratory, H. M. Gibbs, B. Bölger, and L. Baede[15,32] have observed on-resonance coherent self-focusing in sodium. Independently, G. Forster and P. E. Toschek observed[19,35] coherent self-focusing in a neon discharge. A description of these experimental results and their relationship to our theoretical model has been presented by H. M. Gibbs et. al.[20]

This work is not as a model of any specific situation. Rather, what we have attempted to study is a situation where coherent interaction leads to strong deviation from the conditions in plane-wave propagation. We first dealt with a simplified model (scalar-wave equation coupled to a two-level sharp-line atomic system without degeneracy) where transverse effects will enhance inhomogeneites, and lead to nonlinear dispersion and nonlinear absorption. The significant role of the dynamic transverse energy flow is expected to play the same physical role in more realistic situations, where it might be somehow modified by other effects but still will not be washed out. *Coherent phenomena are not confined just to the situation where the pulse duration is much shorter than the relaxation time: They will also appear whenever the field is large enough so that significant exchange of energy between the light pulse and matter takes place in a time short compared to a relaxation time.* When this situation is combined with a sufficiently rapid spatial variation in the input plane, significant self-action phenomena are expected.

ACKNOWLEDGEMENTS

F. P. Mattar would like to express his gratitude and appreciation to Professors: L. Bergstein, M. Bloom, H. A. Haus, J. T. LaTourrette, M. C. Newstein, A. A. Oliner, P. E. Serafim, J. Teichmann and to Dr. H. M. Gibbs for their encouragement and their help in obtaining the computer time necessary for the completion of this work (from June 1973 until January 1977).

Furthermore, he has benefited from enlightening discussions with his thesis advisers M. C. N. and J. T. L., as well as with H. A. H., P. E. S. and H. M. G., that led to additional clarity in understanding the subtleties of the Physics. He is indebted to H. M. G. for his faith in the work that led to the first experimental verification of the coherent on-resonance self-focusing.

Last, but not least, F. P. M. would like to thank Dr. J. Soures and D. Brown for their encouragement to pursue this work during the tenure of his post-doctoral fellowship at the Laboratory of Laser Energetics.

The editing efforts of Dr. C. Bowden (which clarify the manuscript) as well as his patience are joyfully appreciated.

References

(1) S. L. McCall and E. L. Hahn, "Self-Induced Transparency by Pulsed Coherent Light:, Phys. Rev. Letters, 18, 308 (1967) and "Self-Induced Transparency", Phys. Rev., 183, 457 (1969).

(2) H. M. Gibbs and R. E. Slusher, "Optical Pulse Compression by Focusing a Resonant Absorber", Appl. Phys. Letters, 18, 505 (1971); "Self-Induced Transparency in Atomic Rubidium", Phys. Rev., A5, 1634 (1972); "Sharp-Line Self-Induced Transparency", Phys. Rev., A6, 2326 (1972).

(3) G. Lamb, Jr., "Analytic Descriptions of Ultra-Short Optical Pulse Propagation in a Resonant Media", Rev. Mod. Phys., 43, 94 (April 1971).

(4) P. Kryuokow and Y. Letokov, "Propagation of a Light Pulse in a Resonantly Amplifying (Absorbing) Medium", Rev. Sov. Phys. Uspekhy, Vol. 12, #5 (March-April, 1970).

(5) E. Courtens, "Coherent Resonant Propagation Phenomenon in Absorbers," Proceeding of 1969 Chania Conference (unpublished); "Nonlinear Coherent Resonant Phenomena", Laser Handbook, Ed., F. Arecchi, North-Holland Pub. Co., (1972).

(6) S. L. McCall and E. L. Hahn, "Pulse Area-Energy Description of a Travelling-Wave Laser Amplifier", Phys. Rev. A, Vol. 2, 870-71 (1970).

(7) M. C. Newstein and N. Wright, "Transverse Behavior of Pulses Propagating Through Resonant Media", Bull. Am. Phys. Soc., Series 11, 16, 12 (December 1971).

(8) M. C. Newstein, N. Wright and F. P. Mattar, "The Spatial and Temporal Evolution of Optical Pulses in Resonant Media", Progress Report #37 to JSTAC, Polytechnic Institute of Brooklyn (1972).

(9) N. Wright and M. C. Newstein, "Self-Focusing of Coherent Pulses", Optics Communications, 1, 8-13 (1973).

(10) M. C. Newstein and F. P. Mattar, Selected Annual Review to JSTAC of the Microwave Research Institute of the Polytechnic Institute of New York (December 1974); Laser TAC Meeting of "Advances in Radiation Technology", M.I.T., (May 1975); and Optical Society of America, Fall Meeting in Boston (October 1975) J. Opt. Soc. Am. 65, 10, 1181 (1975) and IX International Conf. of Quantum Electronics, Amsterdam (June 1976), Opt. Comm. 18, 70-72 (July 1976).

(11) M. C. Newstein and F. P. Mattar, Proceedings of the Seventh Conference on Numerical Simulation of Plasma, Courant Institute of Mathematical Studies, N.Y. University, (June 1975).

(12) F. P. Mattar, "Application of the McCormick Scheme and the Rezoning Technique to the Transient Propagation of Optical Beam in Active Media", Laboratory of Laser Energetics, U. of Rochester, Technical Note #63 (1976) (to be published).

(13) F. P. Mattar, "Self-Focusing of Coherent Pulses in Resonant Absorbers", Proceedings of the 10th Congress of the International Commission for Optics, Prague, Czechoslovakia, (August 1975).

(14) F. P. Mattar and M. C. Newstein, "Transverse Effects Associated with the Propagation of Coherent Optical Pulses in Resonant Media", (to be published in IEEE J. of Quantum Electronics).

(15) H. M. Gibbs, B. Bölger and L. Baede, "On-Resonance Self-Focusing of Optical Pulses Propagating Coherently in Sodium", Opt. Commu. 18, 199-200 (1976).

(16) J. E. Bjorkholm and A. Ashkin, "CW Self-Focusing and Self-Trapping of Light in Sodium Vapor", Phys. Rev. letters 32, 28-32 (1974).

(17) A. Javan and P. L. Kelley, "Possibility of Self-Focusing Due to Intensity Dependent Anomalous Dispersion", IEEE J. of Quantum Electronics 2, 9, 470-473 (1966), and J. A. Fleck, Jr., and R. L. Carman, "Laser Pulse Shaping Due to Self-Phase Modulation in Amplifying Media", Appl. Phys. Lett. 22, 10, 546-548 (1973).

(18) D. Grischkowsky, "Self-Focusing of Light by Potassium Vapor", Phys. Rev. Letters 24, 866-869 (1970); D. Grischkowsky and J. Armstrong, "Adiabatic Following and the Self-Defocusing of Light in Rubidium", 3rd Conference of Quantum Optics, University of Rochester (June 1972), pp. 829-831.

(19) W. Krieger, G. Gaida and P. E. Toschek, "Experimental and Numerical Study of Self-Induced Transparency in a Neon Absorber", (to be published in Zeitschrift für Physik B).

(20) H. M. Gibbs, B. Bölger, F. P. Mattar, M. C. Newstein, G. Forster and P. E. Toschek, "Coherent On-Resonance Self-Focusing of Optical Pulses in Absorbers," Phys. Rev. Letters, 37, 1743-1746 (December 1976).

(21) The concept of "effective absorption coefficient has been introduced by Gibbs and Slusher in their discussion of coherent propagation in the regime where the inhomogeneous width $(T_2^*)^{-1}$ was greater than the homogeneous width $(T_2)^{-1}$ but less than the inverse pulse duration $(\tau_p)^{-1}$. Under these conditions, the pulse behavior corresponds to that in a medium with a homogeneously broadened line.

(22) S. L. McCall, "Instabilities in Continuous-Wave Light Propagation in Absorbing Media," Phys. Review A 9, 4, 1515-1525, (1974).

(23) R. P. Feynman, F. L. Vernon, Jr., and R. W. Hellwarth, "Geometrical Representation of the Schrödinger's Equations Solving Master Problems", J. Appl. Phys. 28, 43-52 (1957).

(24) S. Chi, "Theory of the Generation and Propagation of Ultra-short Pulses in Laser Systems", Ph.D. Thesis, Polytechnic Institute of Brooklyn (1972).

(25) E. Courtens, "Coherent Resonant Propagation Phenomenon in Absorbers", Proceedings of 1969 Chania Conference, IBM Research Zurich Tech. Report RZ 333 #12642, October 30, 1961; "Frequency of Steady 2π Pulses", VI IEEE Intl. Conf. of Quantum Electronics, Kyoto, Japan, pp. 298-299 (1970).

(26) R. Marth, "Theory of Chirped Optical Pulse Propagation in Resonant Absorbers", Ph.D. Thesis, Polytechnic Institute of Brooklyn (1971).

(27) J. C. Diels and E. L. Hahn, "Carrier-Frequency Distance Dependence of Pulse Propagating in a Two-Level System", Physical Review A 8, 2, 1084-1110 (1973), and "Phase-Modulation Propagation Effects in Ruby", Phys. Rev. A 10, 6, 2501-2509 (1974).

(28) F. A. Hopf and M. O. Scully, "Transient Pulse Behavior and Self-Induced Transparency", Phys. Rev. B1, 1, 50-53 (1970).

(29) F. A. Hopf, "Amplifier Theory", Higher Energy Lasers and Their Applications, Ed. by S. Jacobs et al, pp. 77-176, Addison Wesley (1975).

(30) A. Icsevgi and W. E. Lamb, Jr., "Propagation of Light Pulses in a Nonlinear Laser", Phys. Rev. 185, 517-547 (1969).

(31) J. E. Bjorkholm and A. Ashkin, Phys. Rev. Lett. 32, 129 (1974). If coherent on-resonance self-focusing exists it is small relative to the maximum off-resonance CW SF.

(32) B. Bolger, L. Baede and H. M. Gibbs, "Production of Tunable 300 Watt, Nanosecond, Transform Limited Optical Pulses and Their Application to Coherent Pulse Propagation," Opt. Commun. 18, 67-68 (1976); "Production of 300W, Nanosecond Transformed Limited Optical Pulses," Opt. Comm. 19, 346-349 (1976).

(33) B. R. Suydam, "Self-Focusing of Very Powerful Laser Beams", Laser Induced Damage in Optical Material, NBS special publication 387, p. 42-48 (1973); and "Self-Focusing of Very Powerful Laser Beam II", IEEE J. Quantum Electronics 10, 837-843 (1974); and "Effects of Refractive-Index Nonlinearity on the Optical Quality of High-Power Laser Beams", IEEE J. Quantum Electronics 11, 225-230 (1975).

(34) W. Krieger and P. E. Toschek, "Self-Induced Transparency on the 1.5 m Line of Neon," Phys. Rev. A 11, 276-279 (1975).

(35) G. Forster and P. E. Toschek, paper presented at Quantum Optics Spring Meeting of the German Physical Society, Hannover, Germany, 1976 (unpublished).

(36) S. L. McCall, "Self-Induced Transparency by Pulsed Coherent Light", Ph.D., Physics, U. California, Berkely (1967) (Section V-A, pp. 34-39, XI-F, pp. 118-120, XII, pp. 121-122).

(37) A. Zembrod and T. Gruhl, "Self-Induced Transparency of Degenerate Transitions with Thermally Equilibrated Levels", Phys. Rev. Lett. 27, 287-290 (1971).

(38) C. K. Rhodes and A. Szoke, "Transmission of Coherent Optical in SF_6," Phys. Rev. 184, 25-37 (1969) and C. K. Rhodes, A. Szoke and A. Javan, "The Influence of Degeneracy on the Self-Induced Transparency Effect", Phys. Rev. Lett. 21, 1151-1155 (1968); and "Influence of Degeneracy on Coherent Pulse Propagation in an Inhomogeneously Broadened Attenuator", F. A. Hopf, C. K. Rhodes and A. Szoke, Phys. Rev. B 1, 7, 2833-2842.

(39) F. A. Hopf and M. O. Scully, "Theory of Inhomogeneous Amplifiers", Phys. Rev. 179, 399-416 (1969).

(40) T. Gustafson, J. Taran, H. A. Haus, J. Lifsitz, P. Kelley, "Self-Modulation, Self-Steepening, and Spectral Development of Light in Small Scale Trapped Filaments", Phys. Rev., 177, 306-313 (1969).

(41) F. Demartini, C. Townes, T. Gustafson and P. Kelley, "Self-Steepening of Light Pulses", Phys. Rev., 164, 312-323 (1967).

(42) D. Grischkowsky, E. K. Courtens and J. Armstrong, "Observation of Self-Steepening of Optical Pulses with Possible Shock Formation", Phys. Rev. Lett. 31, 422-425 (1973).

(43) A. J. Campillo, J. E. Pearson, S. L. Shapiro and N. J. Terrell, Jr., "Fresnel Diffraction Effects in the Design of High-Power Laser Systems", Appl. Phys. Lett. 23, 85-87 (1973).

(44) R. A. Fisher and W. Bishel, "Numerical Studies of the Interplay Between Self-Phase Modulation and Dispersion for Intense Plane-Wave Laser Pulses", J. Appl. Phys. 46, 4821-4834 (1973).

(45) D. Grischkowsky, "Adiabatic Following", Laser Applications to Optics and Spectroscopy, Ed. S. Jacobs et al. Addison Wesley (1975).

(46) H. M. Gibbs, R. G. Slusher, "Self-Induced Transparency in Atomic Rubidium", Phys. Rev. A5, 1634-1659 (1972).

(47) H. M. Gibbs (private Communications to F. P. Mattar and M. C. Newstein, December, 1975).

(48) H. M. Gibbs, "On-resonance Self-Focusing Phenomenon", Joint seminar sponsored by LLE and the Physics and Astronomy Department at the University of Rochester (October, 1975).

(49) F. P. Mattar, G. Forster, and P. E. Toschek, "On-resonance Self-Focusing In Neon," Spring Meeting of the German Physical Society, Mayence, West Germany (28 February-4 March, 1977).

MAXWELL-BLOCH EQUATIONS AND MEAN-FIELD THEORY FOR SUPERFLUORESCENCE*

R. Bonifacio, M. Gronchi, L. A. Lugiato**, and

A. M. Ricca† §

**Istituto di Scienze Fisiche dell'Università

Milano, Italy

†CISE, Milano, Italy

Abstract: In this paper we show that under proper initial conditions numerical solutions of Maxwell-Bloch equations confirm the existence both of pure and oscillatory superfluorescence in agreement with Mean-Field theory and with very recent experimental results.

I. INTRODUCTION

Superfluorescence (SF) has been described in terms of Mean Field (MF) theory[1] and of numerical solutions of Maxwell-Bloch (MB) equations[2,3]. The MF approach gives an average description of propagation effects and field inhomogeneity inside the active volume leading to an analytical description of classical and quantum mechanical properties. In particular, MF theory predicts the existence of pure (single-pulse) SF if $L/L_c \ll 1$ and ringing if $L/L_c \gtrsim 1$, where L is the length of the "pencil-shaped" active volume and L_c is the cooperation length[4]. Furthermore, the time scale is predicted to change from the superradiant characteristic time τ_R[1,2,3] to the cooperation time $\tau_c = L_c/c = \sqrt{L\tau_R/c}$ depending on whether the ratio L/L_c is smaller or larger than 1. This feature has been confirmed also by the numerical results of Ref. 3. On the other hand the authors of Refs. 2, 3 on the basis of a numerical analysis of MB equations predict always ringing as due to propagation effects. However, very recent experimental results obtained by Gibbs and Vrehen[5] have proved the existence of pure SF and the appearance of ringing depending on the value of the ratio L/L_c. Hence, one can think that MB equations are in contrast with the above experiment and with MF theory, as regard to the existence or non-existence of pure SF.

In this paper we show that this contrast arises only if one solves MB equations under the initial conditions used by the authors of Ref. 2, whereas the authors of Ref. 3 did not consider the case $L/L_c \ll 1$. On the contrary, we find that there are other initial conditions for which there is no contradiction between MB equations and MF theory. More precisely, we solve (numerically) lossless MB equations and we simulate the noise polarization source by tilting the Bloch vector from the north pole by a small angle θ_o, as done also in Refs. 2,3. This angle should be considered as an effective parameter to fit experimental results (one finds that all physical quantities depend only logarithmically on θ_o). We find perfect agreement between MF theory and MB equations if $B = |\log(\theta_o/2\pi)|/4 \lesssim 1$. In the opposite case, i.e., $B \gg 1$, MF theory is still quantitatively correct for $L/L_c \gtrsim 1$, whereas for $L/L_c \ll 1$ it predicts time scales within a numerical factor $|\log \theta_o|$. In any case, the fundamental dependence of the time scale from the quantity L/L_c predicted by MF theory turns out to be correct at most within a numerical factor. Furthermore, if $B \lesssim 1$, MB equations confirm the existence of pure SF when $L/L_c \ll 1$, and of ringing when $L/L_c \gtrsim 1$ as predicted by MF theory. If $B \gg 1$, which is the only case considered in Ref. 2, MF theory breaks down because there is a residual ringing due to inhomogeneity which does not disappear when $L/L_c < 1$. This inhomogeneity ringing must not be confused with the one due to stimulated processes which appears only when $L/L_c > 1$, as described by MF theory.

Furthermore, since the factor B is a constant in each experiment, one can understand the different features of the experiment of Ref. 2 in which ringing has been practically always observed and of the experiment of Ref. 5 where pure SF has been observed, taking $B \gg 1$ in the first case and $B \lesssim 1$ in the second case. Finally, we give a best fit formula for the delay time which fits all numerical data for a large set of values of B and L/L_c. Such a formula contains as particular cases previous numerical and analytical results, showing explicitly their limits of validity.

II. MAXWELL-BLOCH EQUATIONS AND MEAN-FIELD THEORY

Short pulse propagation in a homogeneously broadened two-level system is described by the following equations

$$\frac{\partial \mathcal{P}(z,t)}{\partial t} = E(z,t)\,\Delta(z,t) \quad , \tag{2.1}$$

$$\frac{\partial \Delta(z,t)}{\partial t} = -E(z,t)\,\mathcal{P}(z,t) \quad , \tag{2.2}$$

$$\frac{\partial E(z,t)}{\partial t} + c\,\frac{\partial E(z,t)}{\partial z} = \frac{1}{\tau_c^2}\,\mathcal{P}(z,t) \quad , \tag{2.3}$$

where: 1) $E(z,t) = \mathcal{E}(z,t)p_{1,2}/\hbar$ is the Rabi frequency, $\mathcal{E}(z,t)$ is the field and $p_{1,2}$ is the matrix element of the dipole moment between the two states; 2) $\mathcal{P}(z,t)$ is the polarization per atom; 3) $\Delta(z,t) = (N_2 - N_1)/N$ is the population inversion per atom; 4) c is the velocity of light in vacuum; 5) τ_c is the characteristic time of atom-field interaction

$$\tau_c = \sqrt{\frac{8\pi\tau_o}{\rho c\lambda^2}} \quad , \tag{2.4}$$

where τ_o is the Wigner-Weisskopf decay time, ρ is the atomic density and λ is the wavelength of the radiation field (τ_c does coincide with the so called "cooperation time"[4]).

MB Equations (2.1) - (2.3) describe the propagation of a pulse which is exactly resonant and shorter than all atomic relaxation times. Through the well-known transformation

$$\mathcal{P}(z,t) = \sin\theta(z,t) \quad , \tag{2.5}$$

$$\Delta(z,t) = \cos\theta(z,t) \quad , \tag{2.6}$$

one obtains from MB Eqs. the so called Sine-Gordon equation[6]

$$\frac{\partial^2\theta(z,t)}{\partial t^2} + c\,\frac{\partial^2\theta(z,t)}{\partial t\partial z} = \frac{1}{\tau_c^2}\sin\theta(z,t) \quad , \tag{2.7}$$

with

$$E(z,t) = \frac{\partial\theta(z,t)}{\partial t} \quad . \tag{2.8}$$

Here, we neglect effects due to opposite waves propagation and diffraction losses. Due to the presence of the space derivative, only numerical solutions can be given which satisfy some boundary conditions. On the contrary, a quantum mechanical scheme of calculation has been recently proposed[1] in terms of a quasi-mode expansion of the field which gives an approximate analytical description of both directions of propagation, off-axial propagation and diffraction losses. Roughly speaking, these quasi-modes give a space-average description of the field and <u>not</u> a "one-mode" description of the field. In fact, in the semiclassical approximation the quasi-mode approach leads to the pendulum equation

$$\ddot\theta + \kappa\dot\theta = \frac{1}{\tau_c^2}\sin\theta \quad , \tag{2.9}$$

with $\dot\theta = \overline{E}(t)$, and $\overline{E}(t)$ has now to be interpreted as a space average of the field inside the cavity. In this sense we speak of "mean field" theory. In Eq. (2.9) κ is a constant which depends on the geometry of the active volume and takes into account both propagation and diffraction losses. Its order of magnitude for a Fresnel number $\mathfrak{F} \simeq 1$ is

$$\kappa \simeq c/L \quad . \tag{2.10}$$

We stress that since κ takes into account very complicated mechanisms, we cannot pretend to evaluate it exactly. We can only estimate its order of magnitude. In other words, MF theory is valid if one can describe experiments using an effective loss parameter whose order of magnitude is given by (2.10).

One could ask for the relation between the MF approximation (2.9) and the Sine-Gordon equation (2.7) derived from MB equations. Contrary to what is claimed in Ref. 2, the MF approximation cannot be derived from MB Equations just by neglecting propagation, i.e., putting $\partial E/\partial z = 0$. In fact, in this way one would obtain a pendulum equation <u>without</u> the linear loss term. On the contrary, <u>pure (single-pulse) SF exists just when κ is the fastest relaxation mechanism</u>[7].

Actually, MF can be related to MB equations assuming in (2.7) an exponential variation of the field on the space scale L,

$$\frac{\partial E}{\partial z} \sim \frac{1}{L} E \quad , \quad \text{so that} \quad c \frac{\partial^2 \theta}{\partial t \partial z} \simeq \frac{c}{L} \frac{\partial \theta}{\partial t} \quad . \tag{2.11}$$

In this way one obtains the pendulum equation (2.9) which in the limit of very large c/L leads to the hyperbolic secant solution.

We stress that the approximation (2.11) is not necessarily the limit of validity of the MF approximation. This is only a rough way to understand the relation between the one direction MB problem and the MF approach to SF.

III. COMPARISON BETWEEN MF PREDICTIONS AND NUMERICAL DISCUSSION OF MB EQUATIONS

A. MF Predictions

The pendulum equation leads very simply to the following predictions[1]:

1) $\kappa\tau_c \gg 1$ (i.e., $L \ll L_c$). One has pure (single-pulse) SF. In this case Eq. (2.9) reduces to

$$\dot{\theta} = \frac{1}{\kappa\tau_c^2} \sin\theta \quad .$$

The pulse has a hyperbolic-secant shape with a characteristic time $2\tau_R$ where τ_R is defined by[2]

$$\tau_R = \frac{c}{L}\tau_c^2 = \frac{8\pi\tau_o}{\rho L \lambda^2} = \frac{L_c}{L}\tau_c \quad , \tag{3.1}$$

which is inversely proportional to the atomic density. The full time width of the pulse at half maximum T_W is of the order of $2\tau_R$ (exactly 1.76 τ_R).

2) $\kappa\tau_c \lesssim 1$ (i.e., $L \gtrsim L_c$). One has oscillatory SF, that is ringing starts appearing as L becomes of the order of or greater than L_c.

3) $\kappa\tau_c \ll 1$ (i.e., $L \gg L_c$). One has strong oscillatory SF with many lobes of radiation. However, the superradiant character is lost. In fact, in this case, the time duration is ruled by $\tau_c \propto \rho^{-1/2}$ and the peak intensity goes as τ_c^{-2} which is proportional to the density and not to the squared density of atoms as in pure SF.

These predictions on the change of time scale, the pulse amplitude and the oscillatory or non-oscillatory character of SF can be derived very simply from the pendulum equation and have been confirmed by the numerical analysis of MB equations worked out in Ref. 3. If such predictions turn out to be experimentally correct, MF theory is a powerful tool to understand at least qualitatively the crucial physics of SF. The change of time scale for different values of L/L_c can be described by the following intermediate formula:

$$T_W \simeq 2\tau_R \left(2 + \frac{L}{L_c}\right) = 2(2\tau_R + \tau_c) \quad . \tag{3.2}$$

This formula gives $T_W \propto \tau_R$ for $L \ll L_c$, whereas when $L \gg L_c$ it is easily seen from Eq. (3.1) that one has $T_W \propto \tau_c$. Now, T_W denotes the time width of the first pulse. The factor 2 in parenthesis arises from the fact that there is best quantitative agreement between MF theory and MB equations choosing for K the value of $2c/L$[8]. This gives a $\tau_R = K\tau_c^2$, twice as large as the one defined in Eq. (3.1). Hence, we take (3.2) as a qualitative formula which should be used to compare MF predictions with numerical or experimental results. In the same way we use the following formula for the delay time T_D (which corresponds to the peak of the first radiated pulse):

$$T_D = \tau_R \left(2 + \frac{L}{L_c}\right) |\log \theta_o| \quad , \qquad \theta_o \ll 1 \quad , \tag{3.3}$$

where θ_o is the initial value of the Bloch angle. Equation (3.3) can be proved analytically only in the case of pure SF (i.e., $L \ll L_c$).

Let us note that Eqs. (3.2) and (3.3) imply the following relation between T_D and T_W:

$$\frac{T_D}{T_W} = \frac{|\log \theta_o|}{2} \quad . \tag{3.4}$$

This equation can be used to define "a posteriori" the value of the effective parameter $|\log \theta_o|$ from the experimental values of T_D/T_W. If the above considerations are correct, this ratio must be practically constant in each experimental situation whatever is the dependence on the density and on the length of the sample. Let us stress that in MF theory the quantity L/L_c is essential not only for the change of the time scale but also in determining the number of lobes N which is ruled by

$$N \simeq L/L_c \quad . \tag{3.5}$$

B. Numerical Results of Reference 2

In Ref. 2 Eqs. (2.1)-(2.3) have been solved numerically assuming a "small angle" θ_o which represents a noise polarization. The numerical data obtained for T_D and T_W are fitted by the following expressions:

$$T_W = 4\tau_R B \quad , \quad B = \frac{1}{4}\left|\log\left(\frac{\theta_o}{2\pi}\right)\right| \quad , \tag{3.6}$$

$$\frac{T_D}{T_W} = B \quad . \tag{3.7}$$

The authors of Ref. 2 claim that these expressions are valid for $\theta_o \ll 1$. Under this assumption they pretend to justify Eqs. (3.6) and (3.7) analytically using the linearized sine-Gordon equation obtained by replacing $\sin\theta$ with θ in Eq. (2.7). However, this procedure seems to us highly unreliable for estimating T_W which is the time spent by θ to go from θ_o $(\ll 1)$ to π. Furthermore, these authors give a formula for the number of lobes which fits their numerical data

$$N \simeq B \quad . \tag{3.8}$$

We stress that these numerical results have been obtained under very peculiar conditions, i.e. the ones which the authors say apply to their experiment

$$\theta_o \simeq 10^{-8} \quad , \quad L/L_c \lesssim 1 \quad . \tag{3.9}$$

Let us note that with such a small value of θ_o the parameter B is 5. This implies large quantitative disagreement with the time scale and the ringing predicted by MF theory. However, as we shall see, these conclusions and in particular Eqs. (3.6)-(3.8) are correct only for $B >> 1$ and $L/L_c \lesssim 1$. On the contrary, if $B \lesssim 1$ or $L/L_c > 1$ the conclusions of Ref. 2 do not apply at all whereas MF theory is always in good qualitative and quantitative agreement with MB equations.

Finally, let us note that in the experiment of Ref. 2 $T_D \simeq 500$ ns, $T_W \simeq 100$ ns and the value of τ_R (as estimated from Eq. (3.1)) is 5 ns. These values are consistent with Eqs. (3.6) and (3.7) if $B = T_D/T_W = 5$. Hence, from (3.8) the number of lobes should be 5. However, experimentally one observes no more than 2 or 3 lobes. Therefore, the authors of Ref. 2 are forced to introduce strong diffraction losses into MB equations. These diffraction losses require a Fresnel number $\mathfrak{F}$ smaller than 1 (e.g. 0.1) whereas the value given in Ref. 2 is nearly 1. We think that this is a serious inconsistency between the experiment and its interpretation which has to be clarified.

C. Present Numerical Results

We have solved numerically Eqs. (2.1)-(2.3) from z = 0 to z = L for different values of θ_o and L/L_c, normalizing the time to $\tau_R = (c/L)\tau_c^2$, as given by (3.1). In all our Figures, the broken line represents the behaviour predicted in Ref. 2 and given by Eqs. (3.6) and (3.7); the dotted one represents the behavior predicted by MF formulas (3.2)-(3.4); the solid one represents the behaviour of our best fit formula (see next Subsection); and the dots are the numerical results.

First we plot in Figs. 1-3 T_D/τ_R as a function of $|\log \theta_o|$ for different values of L/L_c, respectively 1, 2 and 4. Let us recall that MF theory predicts a linear dependence (see Eq. (3.3)) whereas Eqs. (3.6) and (3.7) give a quadratic dependence. As can

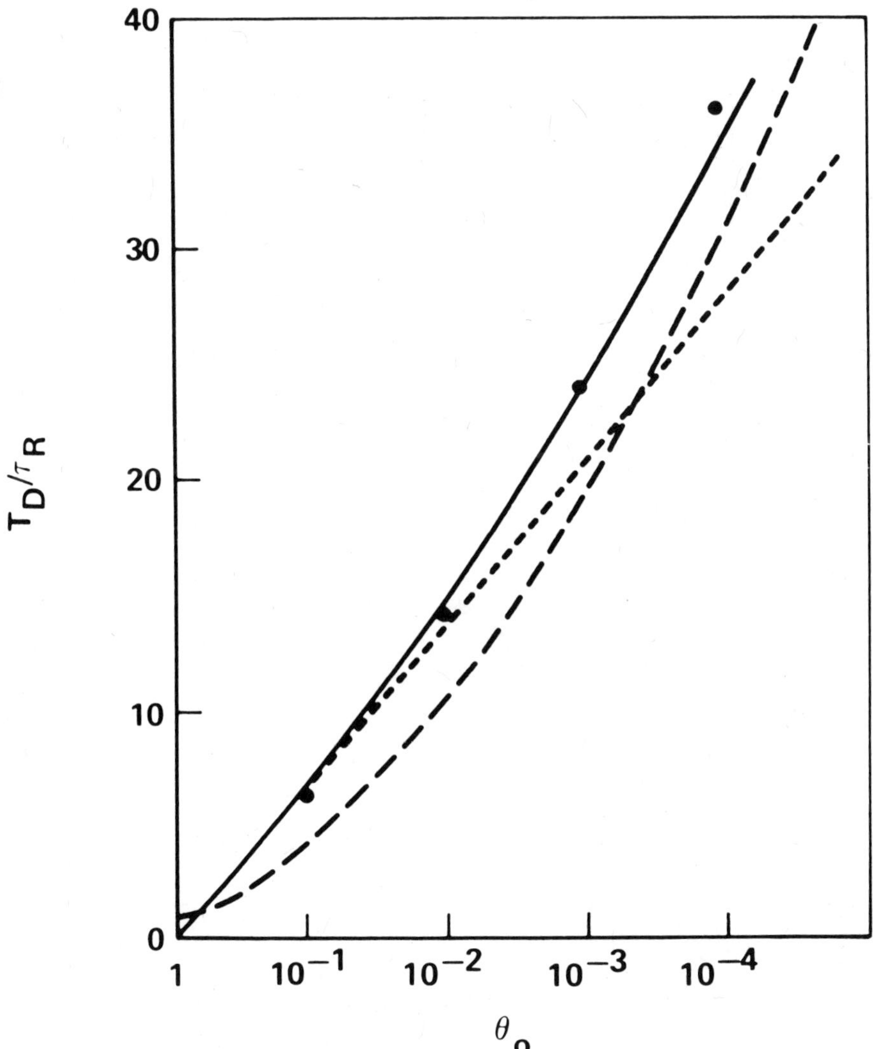

Figure 1. Plot of T_D/τ_R versus θ_o for $L/L_c = 1$. In all figures the broken line represents the predictions of Ref. 2 as described by Eqs. (3.6) and (3.7); the dotted one represents the MF formula (3.3); the solid one represents our best fit formula (3.10); and the dots are the numerical results.

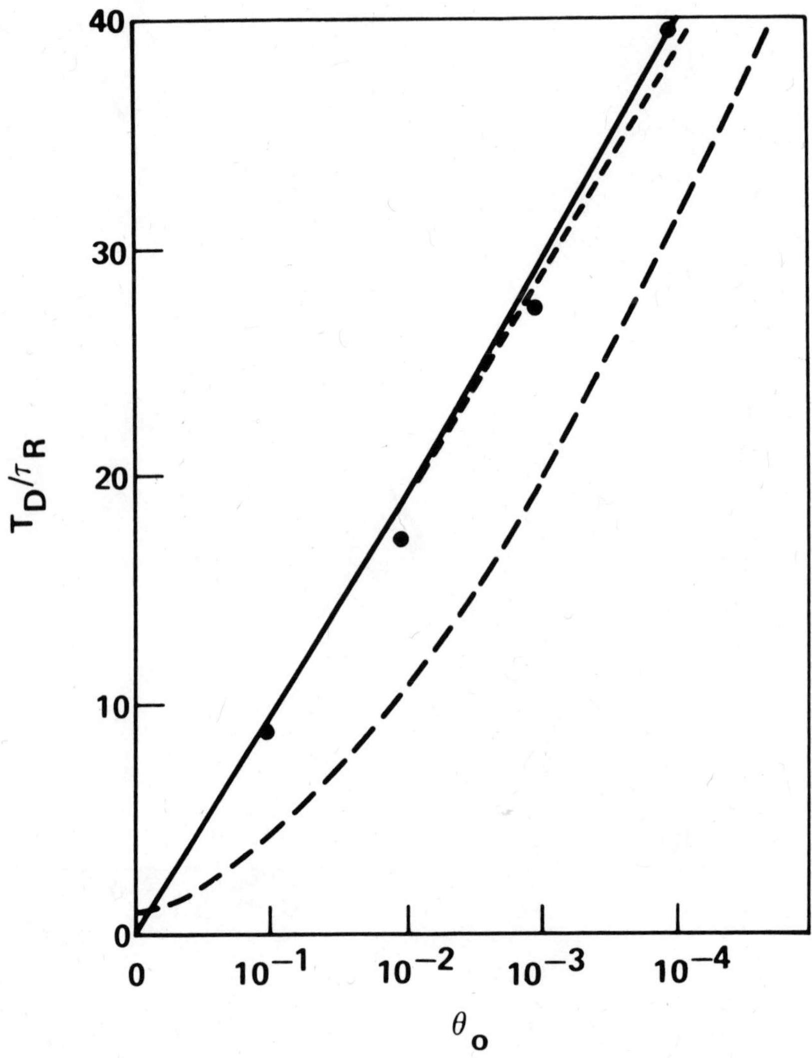

Figure 2. Plot of T_D/τ_R versus θ_o for $L/L_c = 2$.

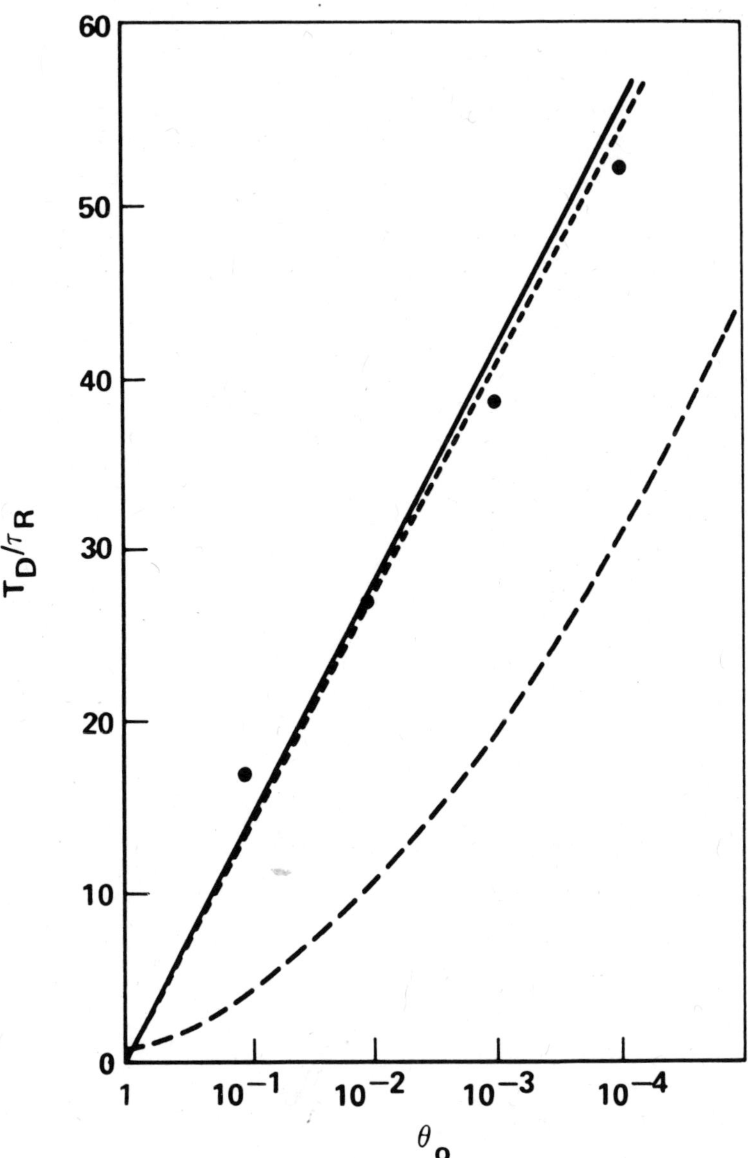

Figure 3. Plot of T_D/τ_R versus θ_o for $L/L_c = 4$.

be seen from these Figures, the agreement of the numerical points with the straight line behavior predicted by MF is already good for $L/L_c = 1$ (unless θ_o is not very small) and becomes quasi perfect for $L/L_c = 2$ and 4. On the contrary, the parabolic behaviour predicted by Ref. 2 is far from the numerical points already for $L/L_c = 2$ and goes even worse for $L/L_c = 4$ (at least for values of θ_o down to 10^{-4}). Hence, MF theory seems to hold unless $L/L_c < 1$ <u>and</u> $\theta_o < 10^{-4}$, which is the case considered in Ref. 2; whereas if $L/L_c > 1$ <u>or</u> θ_o not so small, MF works very well and the behaviour predicted by Eqs. (3.6) and (3.7) does not apply.

In Figs. 4 and 5 we plot T_D/τ_R as a function of L/L_c for $\theta_o = 10^{-2}$ and 10^{-4}. As one can see, the numerical points deviate strongly from the constant broken lines predicted by Eqs. (3.6) and (3.7) for $L/L_c > 1$, approaching the dotted straight lines of MF. For $L/L_c < 1$ the agreement with MF theory becomes worse going from $\theta_o = 10^{-2}$ to $\theta_o = 10^{-4}$.

D. The Best Fit Formula

We worked out a best fit formula for our numerical data which includes as particular cases the behaviours predicted by MF and by Ref. 2. We propose the following one:

$$\frac{T_D}{\tau_R} = \left[\frac{L}{L_c} + 2 + (B - 1)e^{-L/L_c}\right] |\log \theta_o| \quad . \tag{3.10}$$

As can be seen, this formula provides a very good approximation to all numerical data.

Equation (3.10) includes as particular cases Eqs. (3.3) and (3.7) showing their limits of validity. In fact, Eq. $(3.7)^2$ is recovered from (3.10) in the limit $L/L_c \ll 1$ <u>and</u> $B \gg 1$. (We disregard the difference between $|\log \theta_o|$ and $|\log(\theta_o/2\pi)|$!) On the contrary, if θ_o is not so small (e.g. $\theta_o \gtrsim 10^{-4}$) the MF formula (3.3) is a good approximation for all values of L/L_c. If $B \gg 1$ the MF formula is valid only for $L/L_c > 1$. Even in the case $B \gg 1$ and $L/L_c \ll 1$, however, the MF formula needs only a quantitative correction $(B - 1)$ which has to be added to the

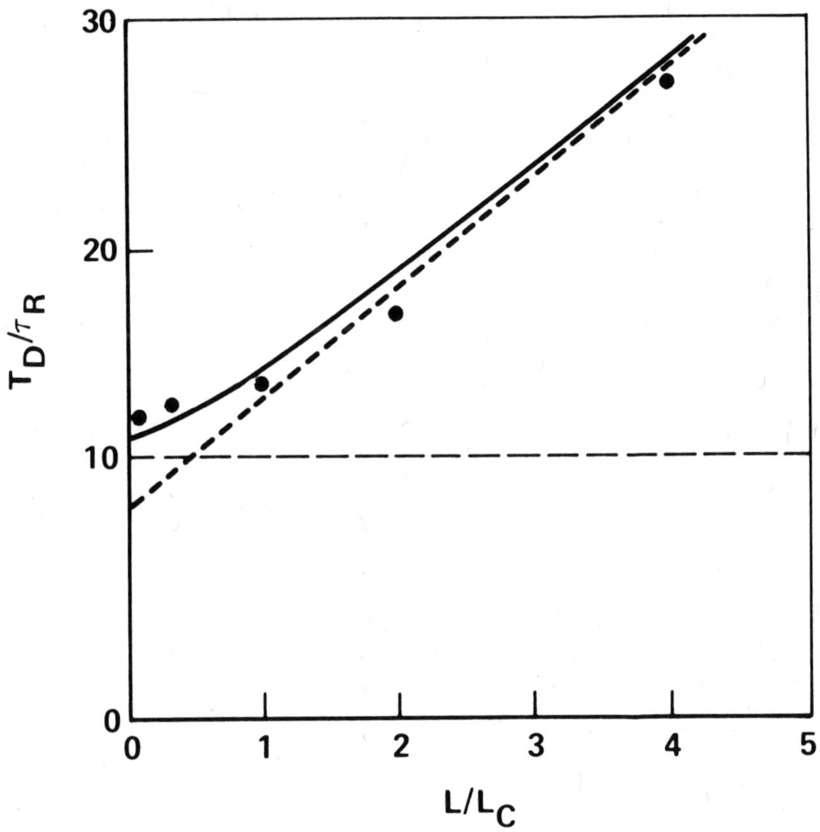

Figure 4. Plot of T_D/τ_R versus L/L_c for $\theta_o = 10^{-2}$.

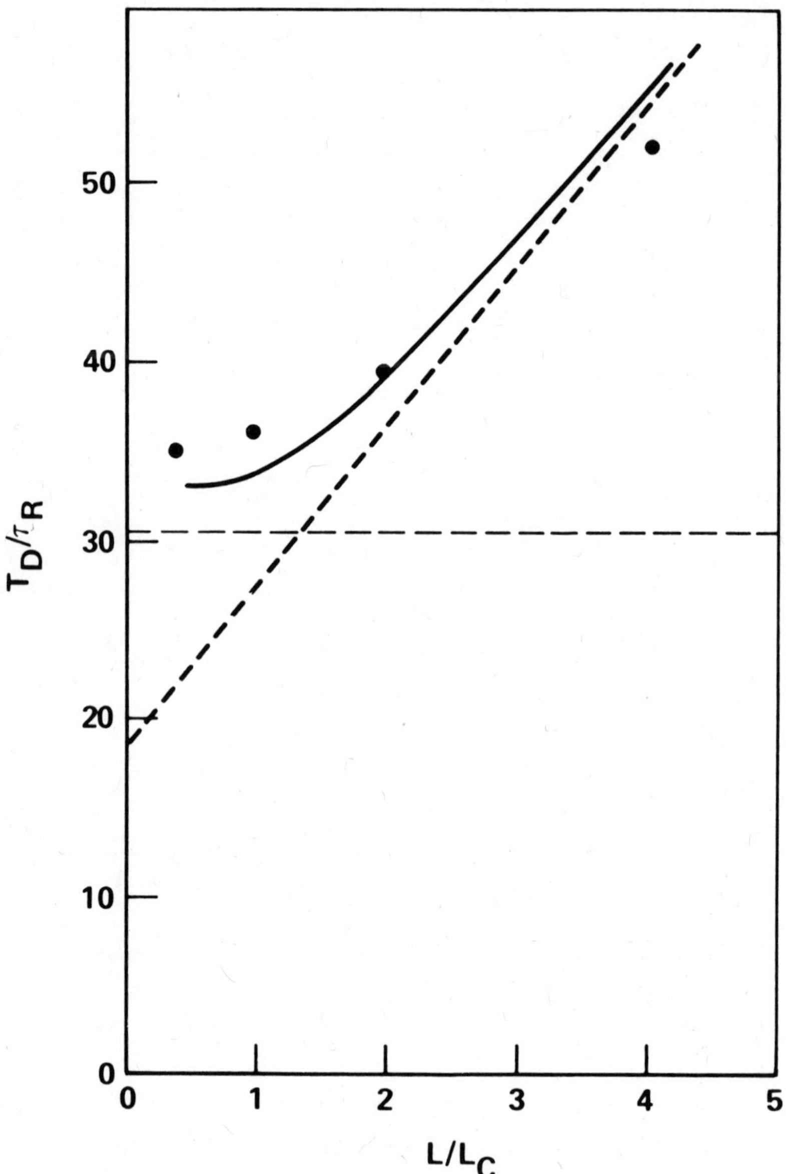

Figure 5. Plot of T_D/τ_R versus L/L_c for $\theta_o = 10^{-4}$.

fundamental part linear in L/L_c. We stress that this part gives the fundamental dependence of the time scale on density and length and the change of time scale from τ_R to τ_c. In other words, MF theory looses only a logarithmic correction to the time scale which is a constant factor in each experiment. When this logarithmic correction becomes quantitatively dominant, one recovers the expression of Ref. 2. Such a correction appears when field inhomogeneity cannot be described by a damping term, as done by MF theory, and this happens when $L/L_c \ll 1$ and $B \gg 1$. T_W seems to be related to T_D through Eq. (3.4), where T_D is given by (3.10), so that

$$\frac{T_W}{\tau_R} = 2\left[\frac{L}{L_c} + 2 + (B - 1)e^{-L/L_c}\right] \quad . \tag{3.11}$$

Here one recovers the MF formula behavior within the logarithmic correction due to propagation effects.

Our numerical analysis gives also indications about the ringing. We find that the number of lobes N is ruled by the quantity

$$N \simeq (B - 1)e^{-L/L_c} + \frac{L}{L_c} \quad . \tag{3.12}$$

Even in this case we recover the expression of Ref. 2, (3.8), in the particular case $L/L_c < 1$ and $B \gg 1$ and the MF expression, (3.5), when $B \lesssim 1$ or $L/L_c > 1$.

As can be seen from Eq. (3.12) there are two contributions to ringing. The L/L_c contribution comes from stimulated emission and reabsorption as correctly described by MF theory. For $L/L_c > 1$ this is the only ringing which survives. The logarithmic contribution comes from the field inhomogeneity inside the active volume and it is the one described in Ref. 2. Since B is a factor which cannot be changed in each experiment, one has two different kinds of experimentally possible situations.

1) $B \gg 1$. In this case one has always ringing due to inhomogeneity for $L/L_c \gtrless 1$ and MF does not describe the propagation ringing when $L/L_c < 1$. Furthermore, the logarithmic correction to the time scale becomes quantitatively relevant. This seems to be the experimental situation of Ref. 2.

2) $B \lesssim 1$. In this case inhomogeneity effects are completely negligible and the predictions of MF are quantitatively and qualitatively correct. Ringing appears or does not appear depending on the value of the ratio L/L_c. This seems to be the experimental situation of Ref. 5.

The extension of the present analysis in order to take into account coherent coupling between opposite propagating waves as well as the comparison with all the experimental results will be presented elsewhere.

References

*Work partially supported by CNR under Contract No. 76.00981.02.

§Work supported by CNR and CISE under Contract No. 73.01435.

(1) R. Bonifacio and L. A. Lugiato, Phys. Rev. A 11, 1507 (1975); 12, 587 (1975).

(2) J. C. MacGillivray and M. S. Feld, Phys. Rev. A 14, 1169 (1976).

(3) R. Saunders, S. S. Hassan and R. K. Bullough, J. Phys. A 9, 1725 (1976).

(4) F. T. Arecchi and E. Courtens, Phys. Rev. A 2, 1730 (1970).

(5) H. M. Gibbs and Q. H. F. Vrehen, this volume.

(6) F. T. Arecchi and R. Bonifacio, IEEE J. Quantum Electronics QE-1, 169 (1965); QE-2, 105 (1966).

(7) R. Bonifacio, P. Schwendimann and F. Haake, Phys. Rev. A 4, 302 (1971); 854 (1971).

(8) This choice of K can be understood as follows. Starting from Equation (2.7) one can deduce the following equation for the space average of the field $\overline{E}(t)$:

$$\frac{d\overline{E}(t)}{dt} + K\overline{E}(t) = \frac{1}{\tau_c^2}\,\overline{P}(t)$$

where $K = C/L\; E(L,t)/\overline{E}(t)$. MF description is obtained neglecting the time variation of K. In particular, the choice $K = 2c/L$ arises assuming a linear variation of the field as a function of space or even an exponential variation as $e^{\gamma z}$ with $\gamma \simeq L/2$.

THEORY OF FIR SUPERFLUORESCENCE

R. Saunders and R.K. Bullough

Department of Mathematics
University of Manchester Institute of Science and Technology, PO Box 88, Manchester M60 1QD, U.K.

Abstract: We show that the second quantised theories of superfluorescence (SF) from rod-like geometries and the semi-classical approach based on the Bloch-Maxwell equations are compatible with each other. From the semi-classical theory we show that two regimes of SF are expected according as τ_{SF}, the superfluorescence time, exceeds or is exceeded by τ_E, the photon escape time: $\tau_{SF} \gtrsim \tau_E$ defines the regime of oscillatory SF characterised by a first pulse intensity $\propto \tau_{SF}^{-2}$ and a small number of subsequent ringing pulses; $\tau_{SF} << \tau_E$ defines a regime of steady oscillation characterised by a first pulse intensity $\propto \tau_{SF}^{-1}$ and a steady train of similar ringing pulses. Diffraction or other losses from a rod-like geometry damps the ringing in the oscillatory SF regime and single pulse emission without subsequent ringing is possible in this regime. Pulse widths τ_W and delays t_D depend on damping: ringing intensities depend on this and very significantly on the initial conditions. The two spatially inhomogeneous fields travelling in opposite directions inside a rod damp each other's ringing. We summarize the difficulties which still prevent the provision of a comprehensive *ab initio* second quantised theory of SF from extended systems.

I. INTRODUCTION

The purpose of this paper is to draw together the apparently incompatible treatments of super-radiance which on the one hand start from the quantised radiation field coupling N_o two-level atoms[1] and on the other hand start from the coupled Bloch-Maxwell equations.[2,3] The main drive of the paper is to establish both the existence of different regimes of fluorescence and the possibility of different ringing behaviours even within these different regimes. A preliminary note in this connection has already been published;[4] but the analysis sketched there will be greatly extended and amplified here.

As we now know from the recent experiments[2,5,6,7] super-radiant emission from a system of inverted atoms or molecules, n cm^{-3}, is characterised by the following: after complete inversion of each atom of the population of density n at time $t = 0$ the system emits short intense spatially directed pulses of radiation of intensity $\propto n^2$ at a time $t_D > 0$ where t_D, the delay time, $\propto n^{-1}$. Bonifacio and Lugiato (BL)[1] introduced a useful distinction between the super-radiant emission from an <u>inverted</u> population and the emission $\propto n^2$ from atomic samples phased by an initial exciting pulse by calling the former "superfluorescence" (SF). This is the terminology we adopt in this paper and it is the regimes of SF we are concerned with within it.

We remark immediately that SF thus defined is a subtle process. The strong directional property it exhibits has to evolve through the co-operation achieved between the atoms in the course of their motion and is governed solely by the macroscopic geometry. An initially inverted population has no direction otherwise associated with it. This already illustrates something of the difficulty of a theory which has been under continuous scrutiny since R.H. Dicke's first imaginative, but in its latter part obscure, paper[8] on co-operative spontaneous emission.

The problem of N_o two-level initially inverted atoms coupled by a quantised interatomic field was solved more or less exactly a few years ago[9,10] in the non-physical case when all N_o atoms occupy a single site. Features induced by a macroscopic geometry do not occur in this connection; nor does any effect of retardation. The extension which includes these effects in a realistic model of a macroscopic system has proved inordinately difficult and is not yet complete. Arecchi and Courtens early pointed out[11] that N_o super-radiating atoms occupying an extended medium can co-operate only if their co-operative super-radiant time τ_R is more than the time required to cover all the atoms by interatomic radiation travelling at speed c. This puts upper bounds, called τ_c^{-1} and N_c, on τ_R^{-1} and N_o. A result we show in this paper is that there is indeed a natural co-operation time τ_c in the theory defined by $\tau_c^{-2} = 2\pi\, p^2\, n\, \omega_s\, \hbar^{-1}$: p is the dipole matrix element for transitions of frequency ω_s between the two non-degenerate states of a model two-level atom and the definition differs from that of Arecchi and Courtens[11] only in a factor of $2\pi/3$. We also find that $\tau_R \lesssim \tau_c$ defines a distinct regime of SF. This is characterised by intensities which go as n rather than n^2. However we are not yet in a position to comment on the further suggestion[11] that initially inverted macroscopic geometries of length $L > \ell_c \equiv c\tau_c$ "spike" into uncorrelated superfluorescent regions of lengths $\ell \sim \ell_c$. We cannot comment on this suggestion because as yet no theory combines an adequate treatment of the spontaneous emission which initiates the SF pulse with a valid description of the co-operative evolution of that pulse.

One direction for a quantitative fully quantised analysis of a macroscopically extended model system has been to assume[12] a rod-like geometry of length L and small cross section A, and to argue that this approximates to a cavity occupied by atoms in which a single resonant mode of the free field is excited in the direction of L. This model proves to be mathematically equivalent to the point N_o-atom model;[12] it can be solved exactly for large N_o [12,13] and[13] its photon statistics can be analysed and a comparison with the corresponding classical limit of the theory established.

It is not easy to establish whether a single mode theory of this type can ever be an applicable theory. We believe it cannot. But Bonifacio and co-workers[1] have now obtained comparable results from the different point of view in which coupled modes of the free field quantised in a large volume contribute to new quantised modes inside the cavity. These modes are quasi-modes damped by the external radiative process. By focussing now on a single such quasi-mode equations mathematically equivalent to the point N_o-atom model have been obtained. This model remains the only exactly soluble second quantised model of an extended system, and this is one of its virtues. We show in this paper nevertheless that it is an approximate theory which does not sufficiently properly treat the ringing process in the pure superfluorescent regime[1] where it would otherwise seem to be most applicable.

For the quantitative comparison with experiment now needed we develop in this paper the semi-classical Bloch-Maxwell theory used already by Feld and co-workers[2,3] and ourselves.[4] For this purpose of comparison with experiment we are able to bridge the gap between the semi-classical approach and that of the quantised theories.[1,12,13] But a substantial theoretical gap, namely the passage from the fully quantised to the semi-classical theory, still remains as we have noted. To see the nature of that gap we consider one feature already apparent in the single mode quantised theory[12,13] when this is compared with the semi-classical theory obtained by replacing all operator products by products of expectation values. Throughout almost all of the motion in the quantised theory the intensity emitted $\propto \frac{1}{4} N_o^2 \operatorname{sech}^2 [\frac{1}{2}N_o I_1(t-t_D)]$ and the inversion to $-\frac{1}{2}N_o \tanh [\frac{1}{2}N_o I_1(t-t_D)]$: $I_1 = (\Delta\Omega/4\pi)\Gamma_o$, Γ_o is the single atom A-coefficient, and $\Delta\Omega$ the small solid angle ($\sim 4\pi^2 c^2 \omega_s^{-2} A^{-1}$) into which the single mode of the theory super-radiates (superfluoresces) as "end fire mode".[14] The same results follow in the semi-classical approximation but there is the fundamental difference that the expectation value of the field amplitude vanishes in the quantised theory and not in the semi-classical theory: the former is therefore incoherent in this sense and the difference will be observable at least in the photon statistics of the two descriptions.

Since the very difficult problem of the classical limit of SF from an extended system of N_o atoms coupled to all the modes of the quantised e.m. field is not yet solved it is not possible to check the semi-classical point of view theoretically. Fortunately incontravertible evidence of SF observed in the FIR is at last now reported in the literature.[2,5] Further Hartmann's recent work on metal vapours[7] suggests that the observations of SF characterized by n^2-dependent intensities will increase rapidly in number from now on. We can therefore directly compare different theoretical approaches with experiment. The precise experiments on Cs at λ $(= \lambda_s) = 2.9\mu$ and 3.1μ already reported by Gibbs and Vrehen[6] in particular show that the quantitative comparison with experiment has become both possible and mandatory. It is a purpose of this paper to indicate through numerical data the measure of agreement between theory and experiment so far achieved.

Before we can go further either with this or our main purpose of drawing together the quantised and semi-classical theories we need to introduce ideas surrounding the "pure super-fluorescent regime" first introduced by BL.[1]

II. THE PURE SUPERFLUORESCENT REGIME

Discussion of the pure SF regime concerns in part discussion of the ringing pulses actually observed in SF emission. The observations on HF[2] at $\lambda_s = 84\mu$ show that the first SF pulse is in general followed by a train of ringing pulses usually of diminishing intensity. The observations on Na[5] at $\lambda_s = 3.41\mu$ and 2.21μ show only a single ringing pulse or no evidence of ringing. The precise observations on Cs at $\lambda_s = 2.9\mu$ and 3.1μ show that SF may occur as a single pulse with the intensity following roughly a sech^2 form; but this is often unsymmetric, displaying a discernible but attenuating tail.

Pure SF exhibiting single sech^2 emission without any ringing pulses is predicted by BL and co-workers[1] in the regime in which the characteristic times T_2^*, T_2, τ_{SF} $(= \frac{1}{2}\tau_R)$, τ_c and τ_E $(\equiv 2Lc^{-1})$ satisfy the inequalities $T_2^*, T_2 >> \tau_{SF} >> \tau_c >> \tau_E$. These times are defined as follows: T_2^* is the inhomogeneous broadening time - we assume it is sensibly infinity in this paper; T_2 is the usual T_2 incorporating the contribution of Γ_o, the atomic A-coefficient and the shortening effect of collisions - except for some discussion of spontaneous emission through Γ_o, T_2 is ignored in this paper; $\tau_{SF} \equiv (4\pi\ p^2\ n\ k_s\ L\ \hbar^{-1})^{-1}$ in terms of the notation already introduced $(k_s \equiv \omega_s c^{-1} \equiv 2\pi\ \lambda_s^{-1})$ and is the superfluorescent time for a rod of atoms- $\tau_{SF} \equiv \frac{1}{2}\tau_R$ where τ_R is the superradiant time used in our previous note; τ_E $(\equiv k^{-1}) = 2Lc^{-1}$ (k is the BL[1] notation) is the escape time;[12] τ_c is the co-operation time introduced above - $\tau_c^{-2} = 2\pi\ n\ p^2\ n\ \omega_s\ \hbar^{-1}$ and $\tau_c^2 = \tau_{SF}\ \tau_E$.

BL[1] start from a quantised field but ultimately obtain c-number equations which have appeared in two forms (for $T_2^* = \infty$)

$$\pm\ c\ \frac{\partial \varepsilon^{(\pm)}}{\partial z} + \frac{\partial \varepsilon^{(\pm)}}{\partial t} + k\ \varepsilon^{(\pm)} = \tau_c^{-2}\ P^{(\pm)} \tag{2.1}$$

together with an associated Bloch equation, and

$$\sigma_{tt} + k\ \sigma_t - \tau_c^{-2} \sin\sigma = 0\ . \tag{2.2}$$

The $(\pm)$ notation in (2.1) describes left $(-)$ and right $(+)$ going wave envelopes travelling inside the rod. Equation (2.2) follows from (2.1) by $\varepsilon^{(+)} = \sigma_t$ and $P^{(+)} = \sin\sigma$ for right going waves in the homogeneous (i.e. spatially independent) limit when the $\varepsilon^{(\pm)}$ fields do not interact with each other. This is so when they do not overlap.

Equation (2.2) reduces to the pendulum equation $\sigma_{tt} = \tau_c^{-2} \sin\sigma$ when L is large so that $k\tau_c << 1$; it becomes a damped pendulum for $k\tau_c \approx 1$; and, using $\tau_c^{-2} = kk^{-1}\ \tau_c^{-2} = k\ \tau_{SF}^{-1}$, is the equation

$$\sigma_t = \tau_{SF}^{-1} \sin\sigma = 2\tau_R^{-1} \sin\sigma \tag{2.3}$$

with solution

$$\sigma = \sin^{-1} \text{sech}\left[2\tau_R^{-1}(t-t_D)\right] \quad , \tag{2.4a}$$

$$\varepsilon^2 = 4\tau_R^{-2} \operatorname{sech}^2[2\tau_R^{-1}(t-t_D)] \qquad (2.4b)$$

when n is small such that $\tau_{SF} >> \tau_c >> \tau_E$ and $k\tau_c$ $(= \tau_c \tau_E^{-1}) >> 1$. This regime is exactly the pure SF regime defined by BL,[1] and the result in this regime is essentially that of the N_o point atom or single mode models as already described in Sec. I. There is therefore no ringing in the pure SF regime defined by $k\tau_c >> 1$. There is ringing of reducing amplitude in the "oscillatory SF regime" defined by $k\tau_c \approx 1$; there is strong ringing in the "regime of steady oscillation" defined by $k\tau_c << 1$. Three regimes are therefore predicted by BL: two have intensities proportional to n^2 but differ in their ringing behaviour.

We can now summarize the main results reported in this paper: from the semi-classical theory we find two regimes defined by $k\tau_c << 1$ and $k\tau_c \gtrsim 1$. The former is indistinguishable from the regime of steady oscillation predicted by BL; the latter covers both the oscillatory and pure SF regimes and is typified by n^2 intensities and at least one ringing pulse even when $k\tau_c >> 1$. However we further show that additional damping processes and the initiating spontaneous emission can reduce the ringing to values below the observable level. We believe this is why single pulses without ringing have been seen in the experiments on Cs.[6]

The understanding summarised by these results emerged in discussion with Rudolfo Bonifacio at the Redstone Arsenal Meeting[15] and we wish to acknowledge this fact very firmly here. The broad measure of agreement on the regime of steady oscillation was clear already in our note.[4] But the status of the pure SF regime remained less certain until we benefitted from the discussions at the Redstone Arsenal Meeting. In this paper we are able to give the analysis for the two regimes and provide numerical data (Figs. 4 and 5 for example) which show that a single ring persists far into the pure SF regime.

The work of the paper is organized in the following way: in Sec. III we trace rapidly through a second quantised theory which has led us to conclude that until much more can be done with that theory the semi-classical approach is simpler and better; in Sec. IV we use that semi-classical approach to derive the equations for the numerical work presented in Sec. V. In Sec. VI we include diffraction loss as additional damping and present further numerical data showing the effect of this. In Sec. VII we analyse the factors governing the numerical simulation of the delay t_D and indicate its connection with the level of ringing. Our conclusions are summarised in Sec. VIII.

III. SECOND QUANTISED THEORY

We start from the usual Hamiltonian for a system of two-level non-degenerate atoms coupled to the quantised e.m. field in dipole approximation.

$$H = H_o + \hbar \sum_{\vec{k},\lambda} \omega_k a^{\dagger}_{\vec{k},\lambda} a_{\vec{k},\lambda} - \sum_{\vec{x}} \vec{e}(\vec{x}, t) \cdot \vec{\mu}(\vec{x}, t) \tag{3.1}$$

$$\vec{e}(\vec{x}, t) = i \sum_{\vec{k},\lambda} \vec{g}_{\vec{k},\lambda} \left\{ a_{\vec{k},\lambda} e^{i\vec{k}\cdot\vec{x}} - a^{\dagger}_{\vec{k},\lambda} e^{-i\vec{k}\cdot\vec{x}} \right\} ; \tag{3.2}$$

$$\omega_k \equiv ck, \quad \vec{g}_{\vec{k},\lambda} \equiv (2\pi \hbar \omega_k V_o^{-1})^{\frac{1}{2}} \hat{\varepsilon}_{\vec{k},\lambda} :$$

"summation" over $\vec{x}$ is integration; $\hat{\varepsilon}_{\vec{k},\lambda}$ is a polarisation direction, λ a polarisation index (1 or 2); V_o is the large region in which the field $\vec{e}$ is quantised. The dipole operator density

$$\vec{\mu}(\vec{x}, t) = \sum_{i=1}^{N_o} \delta(\vec{x} - \vec{x}_i) \vec{\mu}^{(i)}(t) . \tag{3.3}$$

The atomic sites $\vec{x}_i \varepsilon V$ a macroscopic but finite region of definite geometry inside V_o;

$$\vec{\mu}^{(i)}(t) = p \hat{u}[\sigma_+^{(i)}(t) + \sigma_-^{(i)}(t)] ; \tag{3.4}$$

p is the dipole matrix element the same for all the atoms in the system.

$$\begin{aligned} H_o &= \tfrac{1}{2}\hbar \omega_s \sum_{\vec{x}} \sigma_3(\vec{x}, t) \\ &= \tfrac{1}{2}\hbar \omega_s \sum_{\vec{x}} \sum_{i=1}^{N_o} \delta(\vec{x} - \vec{x}_i) \sigma_3^{(i)}(t) ; \end{aligned} \tag{3.5}$$

$\sigma_\pm^{(i)}$, $\sigma_3^{(i)}$ constitute spin operators (an SU_2 lie algebra) for the i th atom.

By using reaction field theory[16] it is possible to derive master equations which in different ways[10] can be reduced to those obtained by others.[1,9] An alternative route is this: From the Hamiltonian (3.1) with the definitions (3.2) - (3.5) we find operator equations of motion for the field driven by the dipole density (3.3). The field splits into two parts - a reaction field which properly describes[16] both spontaneous emission and radiative level shifts, and a co-operative (inter-atomic) field $\vec{e}_{SR}$. This field operator satisfies

$$\nabla^2 \vec{e}_{SR}(\vec{x}, t) - c^{-2}\partial^2/\partial t^2 \, \vec{e}_{SR}(\vec{x}, t) = 4\pi \, c^{-2}\partial^2/\partial t^2 \vec{\mu}(\vec{x}, t) . \tag{3.6}$$

(This field is transverse and the Hamiltonian (3.1) omits longitudinal (Coulomb) field contributions relevant to the chirp[17] discussed below.)

We introduce ensemble averaged matter and field operators by

$$\begin{aligned} R_\pm(\vec{x}, t) &= n^{-1} \langle\sigma_\pm(\vec{x}, t)\rangle_{Avr.} \; e^{\mp i\omega_s t} , \\ R_3(\vec{x}, t) &= n^{-1} \langle\sigma_3(\vec{x}, t)\rangle_{Avr.} , \\ E^{(\pm)}(\vec{x}, t) &= \langle e^\pm(\vec{x}, t)\rangle_{Avr.} \; e^{\pm i\omega_s t} , \\ n &\equiv N_o/V . \end{aligned} \tag{3.7}$$

The fields $e^\pm$ are positive and negative frequency parts.[10,16] The step (3.7) is a coarse graining step which seems unavoidable. We do not wish to know where the atoms are inside the region V, but we do wish to know they are inside and not outside V since it is here the macroscopic geometry enters. An alternative is to assume definite sites $\vec{x}_i$, for example a 3-dimensional lattice[1] filling the interior of V. We prefer to assume a smooth distribution inside V. A consequence is that we do not know the commutation relations for the operators (3.7). So as is usual in the transition from a fine grained theory we assume the commutation relations for the operators (3.7) are the same as those for the corresponding fine grained operators.

From Heisenberg's equations in r.w.a. we now find the operator Bloch equations

$$\dot{R}_\pm(\vec{x}, t) = -\tfrac{1}{2}\Gamma_o \, R_\pm(\vec{x}, t) \pm i E^{(\pm)}(\vec{x}, t) \, R_3(\vec{x}, t) , \tag{3.8a}$$

$$\dot{R}_3(\vec{x}, t) = -\Gamma_o(1 + R_3(\vec{x}, t)) + i\left\{E^{(-)}(\vec{x}, t) \, R_-(\vec{x}, t) + R_+(\vec{x}, t) \, E^{(+)}(\vec{x}, t)\right\} . \tag{3.8b}$$

In (3.8a) $E^{(\pm)}$ and R_3 are ordered:[18] $E^{(+)}$ should appear to the right of R_3, $E^{(-)}$ to the left. We have included the reaction fields which introduce the spontaneous emission through the A-coefficient Γ_o: we ignore radiative shifts although these are easily included.

For $t > Lc^{-1}$ we find from (3.6) now extended to include longitudinal contributions[19] that

$$E^{(+)}(x, t) = np^2\hbar^{-1}\int_V \hat{u}\cdot\overleftrightarrow{F}(\vec{x}, \vec{x}'; \omega_s) \cdot \hat{u}\, R_-(\vec{x}', t)\, d\vec{x}' ,$$

$$E^{(-)}(\vec{x}, t) = \left\{E^{(+)}(\vec{x}, t)\right\}^* \qquad , \tag{3.9}$$

where

$$\overleftrightarrow{F}(\vec{x}, \vec{x}'; \omega) \equiv (\nabla\nabla + \omega^2 c^{-2}\overleftrightarrow{U})(\exp i\omega r c^{-1})r^{-1}$$

and $r = |\vec{x}-\vec{x}'|$; $\overleftrightarrow{U}$ is the unit tensor.

We introduce internal modes $\vec{k}$ by the Fourier transform (an exact transform)

$$R_\pm(\vec{x}, t) = \sum_{\vec{k}} R_\pm(\vec{x}, t)\, e^{\pm i\vec{k}\cdot\vec{x}} \quad . \tag{3.10}$$

The operator theory which includes all these operator Fourier components is intractable but a look at the total radiation rate is immediately instructive.

This total radiation rate is determined by $nV = N_o$ times

$$-\tfrac{1}{2}\dot{R}_3(t) = (2V)^{-1}\int_V \dot{R}_3(\vec{x}, t)\, d\vec{x} \quad . \tag{3.11}$$

We find

$$\dot{R}_3(t) = -\Gamma_o(1 + R_3(t)) - N_o \sum_{\vec{k}, \vec{k}'} I(\vec{k}, \vec{k}')\, R_+(\vec{k}, t)\, R_-(\vec{k}, t) \quad . \tag{3.12}$$

The key quantities are

$$I(\vec{k}, \vec{k}') \equiv 2p^2\hbar^{-1}V^{-2}\int_V d\vec{x}\int_{V'} d\vec{x}'\, \hat{u}\cdot\overleftrightarrow{F}(\vec{x}, \vec{x}'; \omega_s) \cdot \hat{u}\, \exp i(\vec{k}\cdot\vec{x} - \vec{k}'\cdot\vec{x}') \tag{3.13}$$

(where for (3.12) the real part must be taken). We have been able to evaluate this c-number quantity analytically for only one geometry, namely a parallel sided slab of width 2d. We find it is <u>maximised</u> when $\vec{k} = \vec{k}'$, $|\vec{k}| = k_s$ $(\equiv \omega_s c^{-1})$ and the direction of k lies along the longest available length. For the slab this direction is in the plane of the slab. However we read this as a general result and deduce that in the case of the rod (which as a problem in its own right we have only partially solved) $I(\vec{k}, \vec{k}')$ is maximised with $\vec{k}$ along the axis of the rod.

These results show that the radiation process is dominated by resonant modes which have their direction along the axis of the rod. They suggest the two-mode ansatz

$$R_{\pm}(\vec{x}, t) = R_{\pm}^{-}(t)\, e^{\pm ik_s z} + R_{\pm}^{+}(t)\, e^{\mp ik_s z} \quad , \tag{3.14}$$

where upper $\mp$ mean left going and right going modes respectively. From this it follows that the dipole operator density takes the form

$$P(\vec{x}, t) = n\Big\{P^{+}(t) \sin \Phi^{+}(z, t) + Q^{+}(t) \cos \Phi^{+}(z, t) + P^{-}(t) \sin \Phi^{-}(z, t) + Q^{-}(t) \cos \Phi^{-}(z, t)\Big\} , \tag{3.15}$$

where

$$\Phi^{\pm}(z, t) = \pm k_s z - \omega_s t \quad ,$$
$$P^{\pm}(t) = i^{-1}\Big\{R_{+}^{\pm}(t) - R_{-}^{\pm}(t)\Big\} \quad ,$$
$$Q^{\pm}(t) = R_{+}^{\pm}(t) + R_{-}^{\pm}(t) \quad . \tag{3.16}$$

<u>Consistency with the Bloch equations (3.8) now demands $E^{\pm}(\vec{x}, t)$ take similar form</u>, that is

$$\langle e_{SR}(\vec{x}, t)\rangle_{Avr.} \equiv E^{+}(\vec{x}, t)\, e^{-i\omega_s t} + E^{-}(\vec{x}, t)\, e^{+i\omega_s t}$$
$$= \Big\{\varepsilon^{+}(t) + \varepsilon^{+}(t)^{*}\Big\} \cos \Phi^{+} + i^{-1}\Big\{\varepsilon^{+}(t) - \varepsilon^{+}(t)^{*}\Big\} \sin \Phi^{+}$$
$$+ \Big\{\varepsilon^{-}(t) + \varepsilon^{-}(t)^{*}\Big\} \cos \Phi^{-} + i^{-1}\Big\{\varepsilon^{-}(t) - \varepsilon^{-}(t)^{*}\Big\} \sin \Phi^{-} \quad . \tag{3.17}$$

Although we are assuming the geometry to be that of the rod we are obliged to calculate for the slab. If we assume (as we do) that the rod is modelled by the slab with the modes $\vec{k}$ parallel to the <u>axis</u> of the slab we find that

$$E^{+}(\vec{x}, t) = \frac{4\pi}{3} n\, p^2 \hbar^{-1} (e^{ik_s z} R_{-}^{-}(t) + e^{-ik_s z} R_{-}^{+}(t))$$
$$+ 2\pi\, n\, k_s\, p^2 \hbar^{-1} i\Big[(z+d)\, e^{-ik_s z} R_{-}^{+}(t) + (-z+d)\, e^{+ik_s z} R_{-}^{-}(t)\Big] \quad ,$$

$$E^{-}(\vec{x}, t) = \left\{E^{+}(\vec{x}, t)\right\}^{*} . \qquad (3.18)$$

Because of the dependence on the spatial co-ordinate z it is not possible to put the fields into the required form (3.17). The situation is roughly equivalent to a spatially dependent chirp with frequency shift

$$\sim (2\pi\, n\, p^{2}\, \hbar^{-1}\, k_{s}z) \tan k_{s}z\, R_{3}(t) + \frac{4\pi}{3}\, n\, p^{2}\, \hbar^{-1}. \qquad (3.19)$$

The Coulomb contribution is the second term and for the densities n of interest is in fact totally negligible: the term is a Lorentz field correction however and is the contribution of a smooth distribution n of surrounding atoms: local density correlations change this.[20] The first term has singularities and cannot be negligible. Thus (3.18) and (3.19) show that the two mode ansatz (3.14) is internally inconsistent and unacceptable.

It is interesting to note that consistency can be achieved by averaging the equations over z (that is by working with integrated quantities like (3.11)): both consistency and closure of the equations is also achieved by working with the two operator densities

$$R_{3}(\vec{x}, t)$$

and

$$R_{+}^{+}(\vec{x}, t)\, R_{-}^{+}(\vec{x}, t) + R_{+}^{-}(\vec{x}, t)\, R_{-}^{-}(\vec{x}, t),$$

that is by working with the inversion and the intensity. It is therefore at quite a subtle level, that of the chirp, at which inconsistency appears. For this reason (and see below) experimental investigation of both the SF spectrum and the statistics of the radiation seem desirable.

The failure of the two-mode ansatz (3.14) leads us to introduce a spread of modes $\vec{k}$. This spread is presumably axial and peaked about the resonant modes of the two-mode ansatz by extension of the argument leading from (3.11). Thus it is convenient to use envelope operators depending on the co-ordinate z as well as on t and modulating harmonic carrier waves with wave vectors $\pm \vec{k}_{s}$ parallel to the rod axis (This remains the axis of the slab in our actual model). We try the formal ansatz

$$\sigma_{+}(\vec{x}, t) = n\Big\{R_{+}^{+}(z, t) \exp[-i(k_{s}z - \omega_{s}t + \phi^{+}(z, t))] + R_{+}^{-}(z, t) \exp[-i(-k_{s}z - \omega_{s}t + \phi^{-}(z, t))]\Big\} ,$$

$$\sigma_{-}(\vec{x}, t) = \left\{\sigma_{+}(\vec{x}, t)\right\}^{*} , \qquad (3.20)$$

in which $R_{\pm}^{\pm}$, $\phi_{\pm}^{\pm}$ vary slowly. An operator theory on these lines has so far proved intractable and we therefore leap, without further justification here, straight to the semi-classical theory which is obtained by simply replacing all operators of the theory by c-numbers.

IV. THE SEMI-CLASSICAL EQUATIONS

To the dipole operator density

$$\sigma_x(\vec{x}, t) \equiv \sigma_+(\vec{x}, t) + \sigma_-(\vec{x}, t) \qquad (= p^{-1}\, \mu(\vec{x}, t))$$

corresponds

$$n\Big\{P^+(z, t) \sin \Phi^+ + Q^+(z, t) \cos \Phi^+ + P^-(z, t) \sin \Phi^- + Q^-(z, t) \cos \Phi^-\Big\} \tag{4.1a}$$

where

$$\Phi^{\pm}(z, t) = \pm k_s z - \omega_s t + \phi^{\pm}(z, t) ,$$

$$P^{\pm}(z, t) = i^{-1}\Big\{R_+{}^{\pm}(z, t) - R_-{}^{\pm}(z, t)\Big\} ,$$

$$Q^{\pm}(z, t) = R_+{}^{\pm}(z, t) + R_-{}^{\pm}(z, t) , \tag{4.1b}$$

and E is scaled to a frequency:

$$p\hbar^{-1} e_{SR}(\vec{x}, t) \rightarrow E(z, t) = \varepsilon^+(z, t) \cos \Phi^+(z, t) + \varepsilon^-(z, t) \cos \Phi^-(z, t) . \tag{4.2}$$

For comparison with previous work on self-induced transparency[21] we also define an inversion density per atom by $\sigma_3(\vec{x}, t) \rightarrow n\, N(z, t)$. In terms of these several quantities the semi-classical equations in slowly varying envelope and phase approximation prove then to be

$$c^{-1} \varepsilon_t{}^{\pm} \pm \varepsilon_z{}^{\pm} = \alpha P^{\pm} ,$$

$$\varepsilon^{\pm}\Big\{c^{-1} \phi_t{}^{\pm} \pm \phi_z{}^{\pm}\Big\} = \alpha Q^{\pm} ,$$

$$P_t{}^{\pm} = \phi_t{}^{\pm} Q^{\pm} + \varepsilon^{\pm} N^{\pm} ,$$

$$Q_t{}^{\pm} = - \phi_t{}^{\pm} P^{\pm} ,$$

$$N_t = - \Big\{\varepsilon^+ P^+ + \varepsilon^- P^-\Big\} . \tag{4.3}$$

We use the notation $\varepsilon_t \equiv \partial\varepsilon/\partial t$, etc., and we define the coupling constant α (not the usual α!) by $\alpha = 2\pi n k_s p^2 \hbar^{-1}$ (where $k_s \equiv \omega_s c^{-1}$).

The argument easily extends to include homogeneous and inhomogeneous broadening. We drop radiative shifts

$$\Delta = \tfrac{1}{2}\Gamma_o \left\{ \frac{2}{\pi} \ln \left| \frac{\omega_c}{\omega_s} \right| \right\}$$

(ω_c is the Compton cut off frequency $m_e c^2 \hbar^{-1}$ and Δ is half the Bethe shift because of r.w.a.): these shifts simply add to $\phi_t^{\pm}$ in the equations for P_t and Q_t. We define a frequency offset $\Delta\omega \equiv \omega - \omega_s$ for the group of atoms with Doppler shifted resonance frequency ω: we define

$$\langle F(\Delta\omega)\rangle \equiv \int_{-\infty}^{\infty} g(\Delta\omega)\, F(\Delta\omega) d\, \Delta\omega$$

for any function F of $\Delta\omega$. Then

$$\begin{aligned}
c^{-1} \varepsilon_t^{\pm} \pm \varepsilon_z^{\pm} &= \alpha \langle P^{\pm}(z, t; \Delta\omega)\rangle , \\
\varepsilon^{\pm}(c^{-1}\phi_t^{\pm} \pm \phi_z^{\pm}) &= \alpha \langle Q^{\pm}(z, t; \Delta\omega)\rangle , \\
P_t^{\pm}(z, t; \Delta\omega) &= (\Delta\omega + \phi_t^{\pm})\, Q^{\pm} - \tfrac{1}{2}\Gamma_o P^{\pm} + \varepsilon^{\pm} N , \\
Q_t^{\pm}(z, t; \Delta\omega) &= -(\Delta\omega + \phi_t^{\pm})\, P^{\pm} - \tfrac{1}{2}\Gamma_o Q^{\pm} , \\
N_t(z, t; \Delta\omega) &= -\Gamma_o(1 + N) - (\varepsilon^+ P^+ + \varepsilon^- P^-) \qquad (4.4)
\end{aligned}$$

(for convenience the arguments $(z, t; \Delta\omega)$ of P, Q, N are omitted on the right sides; ε, ϕ depend only on z, t).

These equations are incomplete in one respect and are actually totally inadequate in another. They are incomplete in that they omit "standing wave" contributions from the coupling of oppositely directed harmonics of the plane wave modes adopted in (4.1). We comment on this below. They are totally inadequate in that they do not properly describe the motion in the small time region close to the start of the SF process at $t = 0$.

From the definition of SF in Sec. I $N = +1$ everywhere inside V at $t = 0$. There are no fields inside V and $P^{\pm}$, $Q^{\pm}$, $\varepsilon^{\pm}$ vanish everywhere there. Under this initial condition a solution of (4.4) is $N(z, t) = 2e^{-\Gamma_o t} - 1$ at all points z inside V and no fields develop. This is a measure of the failure of the de-correlation procedure which has resulted in (4.4) and points to a very interesting problem. For present purposes we meet it as others have done however[1,2,3] and set at $t = 0$[22]

$$N(z, 0) = 1 - \delta \qquad (0 < \delta << 1)$$

$$P^+(z, 0) = P^-(z, 0) = \left\{\tfrac{1}{2}\left[1 - (1-\delta)^2\right]\right\}^{\frac{1}{2}} . \qquad (4.5)$$

The argument for (4.5) relies in part on the results of the analysis[13] for the single mode theory. After the short time during which spontaneous emission starts the motion, that motion is governed by the evolution of the co-operative intensity. In the semi-classical theory there can be intensity only if there are fields $\varepsilon^\pm$. We therefore introduce δ and discard Γ_o. We shall find nevertheless that the precise choice for the value of δ has a significant effect on the motion and δ is an important parameter of the theory.

The model we adopt henceforth includes δ as a parameter to be chosen in a way which is as consistent as possible with the physics underlying it. It sets $\Gamma_o \equiv 0$, $T_2{}^* = \infty$ and $\phi_t = \phi_z = 0$. It seems doubtful whether $T_2{}^* = \infty$ will be good enough for the data[6] on Cs but we are primarily concerned here to establish the most prominent features of the theory. The choice $\phi_t = \phi_z = 0$ eliminates any chirp and seems drastic in view of the analysis of Sec. III. Further work both theoretical and experimental is needed on the chirp but no inconsistency has arisen in the numerical work through this assumption.

V. NUMERICAL INTEGRATION OF THE SEMI-CLASSICAL EQUATIONS

The equations we integrate are

$$c^{-1}\,\varepsilon_t{}^\pm \pm \varepsilon_z{}^\pm = \alpha\, P^\pm ,$$

$$P^\pm{}_t = \varepsilon^\pm N ,$$

$$N_t = -(\varepsilon^+ P^+ + \varepsilon^- P^-) , \qquad (5.1a)$$

with

$$N(z, 0) = 1 - \delta ,$$

$$P^\pm(z, 0) = \left\{\tfrac{1}{2}\left[1 - (1-\delta)^2\right]\right\}^{\frac{1}{2}} ,$$

$$\varepsilon^\pm(z, 0) = 0 , \qquad (5.1b)$$

for all points z in $0 < z < L$; $\varepsilon^\pm(z, t)$ is continuous across $z = 0$ and $z = L$ and we follow the fields both inside and outside $0 < z < L$ and compute the intensities outside. The theory is now a one-dimensional theory and effectively refers to the plane wave problem and a slab-like geometry in three space dimensions. We apply the theory to the rod later by taking a rough account of the sideways losses which must occur in this case.

To fix ideas for the later analysis we note that, in the case where fields ε^+ and ε^- emerge in two well separated regions inside $0 < z < L$, the inversion $N(z, t)$ must break up into two time evolving non-overlapping parts, say

$$N(z, t) = N^+(z, t) + N^-(z, t) \quad , \qquad (5.2)$$

such that equations (5.1a) now break up into two uncoupled sets describing the evolution of the right going (+) and left going (-) fields separately. For the + case (dropping that label)

$$\begin{aligned} c^{-1}\varepsilon_t + \varepsilon_z &= \alpha P \quad , \\ P_t &= \varepsilon N \quad , \\ N_t &= -\varepsilon P \quad , \end{aligned} \qquad (5.3)$$

the familiar sharpline limit of the SIT equations.[21] Thus with $\varepsilon = \sigma_t$ and for $N = 1-\delta$ at $t = 0$, $N = \cos\sigma$, $P = \sin\sigma$ and

$$c^{-1}\sigma_{tt} + \sigma_{zt} = \alpha \sin\sigma \quad . \qquad (5.4)$$

This reduces to the usual Lorentz invariant form of the sine-Gordon equation for the amplifier if

$$\xi = \alpha^{\frac{1}{2}} c^{-\frac{1}{2}}(2z-ct), \qquad \tau = \alpha^{\frac{1}{2}} c^{+\frac{1}{2}} t \quad ,$$

so that $\sigma_{\xi\xi} - \sigma_{\tau\tau} = -\sin\sigma$. However the region L is finite with $\sigma_t = (\alpha c)^{\frac{1}{2}}(\sigma_\tau - \sigma_\xi)$ continuous across $z = 0$, L - whilst for $L \to \infty$ the conditions require $\sigma = 0$, $t \to -\infty$; $\sigma = \pi, t \to +\infty$. Recent analytical results for the sine-Gordon equation scarcely extend to these conditions.[23]

Figs. 1 and 2 (all the Figs. appear together at the back of this paper) show numerical results for equations (5.1) (both ways going fields) for the data for HF gas[2] ($L = 100$ cm, $\lambda = 84\mu$m, $p = 6.7 \times 10^{-19}$ c.g.s. units): δ is chosen to have the large value $\delta = 10^{-3}$. For $\tau_R = 2\tau_{SF} \sim 5$ n sec (Fig. 1) ($n = 9.4 \times 10^{11}$ molecules cm^{-3}), intensities go $\propto \tau_R^{-2}$ ($\propto n^2$) and delays $t_D \sim 3\tau_R|\log_{10}\delta|$ (both results are obtained empirically by varying the parameters n and δ). The delays are too short for agreement with the observations[2] by a factor ~ 12 (but compare Fig. 9). The fields shown in the left hand trace are manifestly inhomogeneous; ringing (seen in the output intensities in the right hand trace) certainly occurs but the (±) fields overlap only in the later ringing pulses. Under these conditions the approximation (5.2)(also used by Feld and co-workers[2,3]) is acceptable at least for the leading radiated pulses.

For $\tau_R \sim 5 \times 10^{-10} - 5 \times 10^{-11}$ sec $(n \sim 10^{13} - 10^{14}$ molecules cm^{-3}, typified by Fig. 2) we find intensities $\propto \tau_R^{-1}$ $(\propto n)$, delays and pulse widths $\propto \tau_R^{\frac{1}{2}}$ (for δ-fixed) and the $\pm$ fields interfere throughout the motion. Ringing is pronounced and the fields are inhomogeneous. In Fig. 3 we compare the results for (5.3) (one way going) against (5.1) (both ways going) for τ_R = 10 n sec. There is some reduction of ringing in the both ways case.

The Fig. 1 $(\tau_R\ (= 2\tau_{SF}) = 5$ n sec, $\tau_c = 4.1$ n sec, $\tau_E = 6.7$ n sec) is in the regime of oscillatory SF. The Fig. 2 $(\tau_R = 0.27$ n sec, $\tau_c = 0.30$ n sec, $\tau_E = 6.7$ n sec) approaches the regime of steady oscillation. The Fig. 4 $(\tau_R = 100$ n sec, $\tau_c = 18.2$ n sec, $\tau_E = 6.7$ n sec) is well into the regime of pure SF. Ringing persists and the general profile is very little different from the case of $\tau_R = 50$ n sec (Fig. 5) or indeed from Fig. 1. This numerical work leads us to conclude that the semi-classical equations (5.1) describe two and only two distinct regimes. To see analytically that this is so we can conveniently move for the moment to the one-way going system (5.3) and the result (5.4) derived from it.

The numerical results (Figs. 1 – 5) have shown that field gradients $\partial^2\sigma/\partial t\partial z = \partial\varepsilon/\partial z$ are of <u>order</u> ε/L. In terms of the "stretched variable" $\zeta \equiv z/2L$ equation (5.4) is

$$\sigma_{tt} + k\,\sigma_{t\zeta} - \tau_c^{-2}\sin\sigma = 0 \quad . \tag{5.5}$$

The co-operation time τ_c emerges by noting that αc (recall α is the coupling constant $2\pi np^2\hbar^{-1}k_s$) is identically τ_c^{-2} as we have defined this quantity.

We first introduce the "stretched" time $\tau \equiv t\tau_c^{-1}$: this scales (5.5) to

$$\sigma_{\tau\tau} + k\,\tau_c\,\sigma_{\tau\zeta} - \sin\sigma = 0 \quad . \tag{5.6}$$

Since $\sigma_{\zeta\tau}$ is the size of σ_τ from the numerical work, it follows that for $k\tau_c << 1$, the regime of steady oscillation, (5.6) is governed by just the pendulum equation $\sigma_{\tau\tau} - \sin\sigma = 0$. Since $\varepsilon = \sigma_t = \tau_c^{-1}\sigma_\tau$, $\varepsilon \propto n^{\frac{1}{2}}$ and $\varepsilon^2 \propto n$. In the limit $k = c/2L \to 0$ this regime is indistinguishable from the regime of steady oscillation predicted by BL.[1] The output is a train of close to squared hyperbolic secant intensities (BL[1]) with widths on the scale of τ_c (The Fig. 2 which is both ways going shows that this limit is still being approached but is not yet reached).

For the opposite regime $k\tau_c >> 1$ we use $\tau_c^{-2} = k\tau_{SF}^{-1}$ in (5.4) and introduce the stretched time $t\tau_{SF}^{-1} \equiv \eta$ to see that

$$(k\tau_c)^{-2}\,\sigma_{\eta\eta} + \sigma_{\eta\zeta} - \sin\sigma = 0 . \qquad (5.7)$$

Since σ_η is $O(1)$ in this regime, the equation reduces to the sine-Gordon equation

$$\sigma_{\eta\zeta} - \sin\sigma = 0 . \qquad (5.8)$$

The fields are $\tau_{SF}^{-1}\sigma_\eta$ and intensities are proportional to n^2. For the initial and boundary conditions (5.1b) the analytical problem is complicated. For the related conditions ($\sigma = -\sqrt{2\delta}$, $\sigma_\eta = 0$, $\eta = 0$; $\sigma \to -\pi$, $\eta \to \infty$), we can take the similarity solution[24]

$$\sigma = \phi(2\sqrt{\eta\zeta}) = \phi(Y)$$

for which (5.8) is

$$\phi'' + 2Y^{-1}\phi' - \sin\phi = 0 \quad , \qquad (5.9)$$

with $\phi = -\sqrt{2\delta}$, $\phi' = 0$, at $Y = 0$. The solution is $\phi \sim -\sqrt{2\phi}\, I_o(Y)$ for small Y where ϕ is small ($I_o(Y) = J_o(iY)$) and $J_o(X)$ is the Bessel function of the first kind). Both this small amplitude solution and the exact solution, which applies for all Y (for given δ), falls from $\phi = -\sqrt{2\delta}$ towards $\phi = -\pi$: the exact (non-linear) solution crosses $\phi = -\pi$ and oscillates about this value with decreasing amplitude.[25] The solution for $\tau_{SF}^{-1}\sigma_\eta = \sigma_t$ ($= \varepsilon$) therefore rings: the delay, determined by the first zero of σ_{tt} at $z = L$, is roughly compatible with the formula $\tau_D = 4\tau_R|\ln\sqrt{2\delta}|$[24,25] (see Sec. VII). None but oscillatory solutions are known to the writers for the assumed boundary conditions. There is no sech solution, and this is consistent with the numerical results based on (5.1).

It is important to notice that the equation (2.3) which determines the pure SF regime according to BL[1] is obtained from (5.8) by the rather cavalier approximation $\sigma_z = \sigma/2L$ (ie. $\sigma_\zeta = \sigma$) or $\sigma_{tz} = \varepsilon/2L$. The numerical results show that σ_{tz} has this order but it is not correct to change the functional form. From this point of view the results for the pure SF regime from the second quantised theories are more approximate than the results from the semi-classical theory.[26]

Rudolfo Bonifacio illustrated this connection between the two different approaches to the theory at the Redstone Arsenal Meeting. With this connection established we think it is now possible to achieve a united understanding of the phenomenon of superfluorescence. Our view is that the semi-classical theory is easier to use for accurate analysis of intensities and pulse shapes at this stage. But a comprehensive theory must ultimately return to the second quantised arguments.

Our conclusion from the work of this Sec. V is that semi-classical theory predicts two and only two regimes. However Gibbs and Vrehen[6] reported evidence of a fairly sharp drop in the ringing observed from Cs for τ_{SF} $(=\frac{1}{2}\tau_R) \sim \frac{1}{2}\tau_E$. For $\tau_{SF} \sim \frac{1}{2}\tau_E$ pulse profiles show ringing typical of the regime of oscillatory SF: for $\tau_{SF} \sim 5/4\tau_E$ a single pulse is emitted with an asymmetric tail but without discernible ringing (see the data for the atomic beam, $L = 2.0$ cm $n = 7.6 \times 10^{10}$ cm^{-3}, $\tau_R = 3.5 \times 10^{-10}$ sec, $\tau_E = 1.3 \times 10^{-10}$ sec from Gibbs and Vrehen[6]). These results suggest that the regime of oscillatory SF occurs for $\tau_{SF} < \tau_E$ and that a ringing free regime occurs for $\tau_{SF} \gtrsim \tau_E$. We attempt to achieve this situation empirically in the work of the next section. We certainly show that ringing can be either reduced or eliminated. But we are not able to show that this occurs as the sharp transition into a third distinct superfluorescent regime where τ_{SF} exceeds τ_E. Nor do we simulate any asymmetrical pulse tail - which however may be associated with finite $T_2{}^*$ or perhaps with strictly dynamical transverse effects.

VI. INCLUSION OF DIFFRACTION LOSS

The theory (5.1) is a plain wave theory modelled on the slab and applied to the rod. We correct for diffraction loss (Fresnel number $F \equiv A/\lambda L < \infty$), and perhaps other losses, by introducing a term $\kappa\sigma_t$ on the ℓ.h.s. in (5.4). MacGillivray and Feld[3] quote the divergent Gaussian beam formula $L = \frac{1}{2}\ln(1+F^{-2})$ from which, for $F \simeq 0.77$, $\kappa L = 0.5$. For this choice equation (5.4) is

$$\sigma_{tt} + c\sigma_{tz} + k\sigma_t = \tau_c{}^{-2} \sin\sigma . \tag{6.1}$$

More generally if $\kappa L = \frac{1}{2}\nu$, k is replaced by νk.

It is clear that the argument for the regime of steady oscillation $(k\tau_c \ll 1)$ is unchanged: equation (6.1) reduces to $\sigma_{\tau\tau} - \sin\sigma = 0$. There is also a transition to oscillatory SF for $k\tau_c \approx 1$ where the n dependent intensities of the regime of steady oscillation again move over to an n^2 dependence; but this is now governed by the damped sine-Gordon equation

$$\sigma_{\eta\zeta} + \nu\sigma_\eta - \sin\sigma = 0 . \tag{6.2}$$

For large ν (large diffraction loss) the equation (6.2) describes pure SF behaviour since, for $\sigma_\eta \sim \nu^{-1}$,

$$\sigma_\eta = \nu^{-1} \sin \sigma \quad , \tag{6.3a}$$

with solution

$$\sigma = \sin^{-1} \operatorname{sech}\left[(t-t_D)/\nu\tau_{SF}\right]$$

and

$$\varepsilon^2 = \nu^{-2} \tau_{SF}^{-2} \operatorname{sech}^2\left[(t-t_D)/\nu\tau_{SF}\right] \quad . \tag{6.3b}$$

For reasonable values of ν ($\lesssim 7$ certainly for $\kappa L \lesssim 3.5$) the behaviour is less obviously pure SF. The Figs. 6a-e calculated from (5.1a), for both ways going fields with a term $\kappa\varepsilon^\pm$ added to the Maxwell equations for $\varepsilon^\pm$, show that ringing is reduced as L increases from 0.0 to 0.5 to 1.5 to 2.5 to 3.5. The data are for HF with $\tau_R = 5$ n sec ($\tau_{SF} = 2.5$ n sec) and Fig. 6a ($\kappa L = 0.0$) repeats the Fig. 1 with however a different scaling.

The behaviour changes little, well inside the pure SF regime: Figs. 7a, b are for HF with τ_R ($= 2\tau_{SF}$) $= 50$ n sec and $\kappa L = 0.5$ and 1.5 (compare Fig. 5 where $\kappa L = 0$). Figs. 8a, b, c have $\tau_R = 50$ n sec and $\kappa L = 0.0$, 0.5 and 1.5; but now the fields are one way going rather than both ways. In both cases κL damps the ringing and increases the delays and widths. Approximate numbers taken from the data in Figs. 5, 7 and 8 appear in the Table 1.

Table 1

	κL	Intensity ratio	t_D	τ_W	$\tau_W/3.52\tau_R$
OW	0	24%	472	279	1.56
	0.5	16%	528	288	1.64
	1.5	9%	623	334	1.90
BW	0	16%	531	282	1.60
	0.5	12%	574	315	1.75
	1.5	0%	703	347	1.97

In the Table 1 we quote the approximate ratio (in %) between the intensity of the first pulse emitted and the intensity of the first ringing pulse following it in the two cases one way going (OW) and both ways going (BW): $\delta = 10^{-3}$, and the delays are indicated. The value $3.52\tau_R$ is the pulse width τ_W expected for a pure sech^2 pulse like (2.4) when $\nu = 0$ ($\kappa L = 0.0$): thus all widths in the Table are too big to be consistent with simple sech^2 emission. This feature is consistent with the single pulses observed from Cs[6] where ratios $\tau_W/3.52\tau_R \sim 2.5$ are observed.

However the Figs. 5, 7 and 8 exhibit no smoothly attenuating tails, and discernible asymmetry in the outputs is introduced only through the ringing pulses.

Further, we find no evidence that weak damping suppresses the ringing more efficiently for τ_{SF} large: compare for example the Figs. 6 ($\tau_R = 5$ n sec) with the Figs. 7 or 8 ($\tau_R = 50$ n sec). Diffraction loss as included here does not introduce a regime of single pulse emission characterised by $k\tau_c >> 1$; and to achieve a passage into such a regime from the regime of oscillatory SF, ν must depend on $k\tau_c$, increasing with it.

We should stress, however, that by introducing "diffraction loss" in the fashion of this Sec. VI, we treat this aspect of the theory as cavalierly as in the approximation $\sigma_{tc} \approx \sigma_t/2L$ relating the quantised and semi-classical theory as discussed in Sec. V. Conceptually this loss is a secondary aspect of the theory and it merits a rough approximation on this ground: but its effects are of first order significance if the loss is large enough. A careful dynamical treatment of sideways loss from a rod-like geometry is therefore now very desirable.

With the qualification that such a treatment is not yet available we predict from the results of this section three regimes of SF, steady oscillation, oscillatory SF, and pure SF, in agreement with BL.[1] However both the physics and character of the pure SF regime are changed. The physics is ascribed to transverse or other loss mechanisms; and the character is simply that of single pulse emission - not that of sech^2 intensities given by (2.4). The boundary of this pure SF regime is not sharp: for $\tau_c k << 1$ the regime of steady oscillation is unchanged; with damping losses included intensities change from n to n^2 dependence for $\tau_{SF}k \sim 1$; but these damping losses suppress ringing for $\tau_{SF}k \approx 1$ if they suppress ringing for $\tau_{SF}k >> 1$, and the distinction between the regime of oscillatory SF and pure SF must now be drawn somewhere where $\tau_{SF}k \lesssim 1$.[27]

We turn finally to the analysis of the delay.

VII. THE DELAY

We are concerned here with the details of the delay t_D in the regimes of pure and oscillatory SF, $\tau_{SF} \gtrsim \tau_E$. It is obvious that the delay scales as $\tau_c \propto n^{-\frac{1}{2}}$ in the regime of steady oscillation. For $\tau_{SF} \gtrsim \frac{1}{4}\tau_E$ an empirical formula for the delay t_D based on the numerical data in Sec. V is $t_D \sim 3\tau_R|\log_{10}\delta| = 3\tau_R|\ln\delta|/(2.303)$. Thus $t_D \sim \tau_R|\ln\delta| = 2\tau_R|\ln\delta^{\frac{1}{2}}|$. In the one way going approximation $N = \cos\sigma$, so that $N = 1-\delta$ at $t = 0$

is $\frac{1}{2}\sigma_o^2 = \delta$ (where $\sigma_o \equiv \sigma(z, 0)$, the initial tipping angle). Then in terms of tipping angle $t_D \sim 2\tau_R \,\ell n\{\sigma_o/\sqrt{2}\}$. These empirical results assume that $10^{-3} \gtrsim \delta \gtrsim 10^{-6}$.

Expressions of similar type have been obtained analytically;[1,12] and δ can be related to the total number of atoms N_o taking part in the superfluorescence essentially in the following way: from the two mode ansatz for (3.11) we find

$$N_o \dot{R}_3(t) = -\Gamma_o N_o(1 + R_3(t)) - 2N_o^2 I(\vec{k}_s, \vec{k}_s) R_+(t) R_-(t) . \quad (7.1)$$

Since $R_\pm(t)$ are normalised to a single atom, we expect that (at $t = 0$) the expectation value $<|R_+(0) R_-(0)|> = N_o^{-2} N_o = N_o^{-1}$. Also $N_o I = 4\pi n p^2 \hbar^{-1} k_s L = \tau_{SF}^{-1}$. Thus at $t = 0$ the total radiation rate is

$$\tfrac{1}{2} N_o \dot{R}_3(0) = -\Gamma_o N_o - \tau_{SF}^{-1}. \quad (7.2)$$

The parameter δ was introduced in Sec. VI by dropping Γ_o and using δ to act to replace it. If we simply discard Γ_o in (7.2) the initial total radiation rate is precisely τ_{SF}^{-1}. The de-correlated solution for the two-mode ansatz is

$$R_+ R_- = \tfrac{1}{4} \operatorname{sech}^2 \tau_R^{-1}(t - t_D) ,$$

$$R_3 = -\tanh \tau_R^{-1}(t - t_D) . \quad (7.3)$$

Thus

$$\dot{R}_3(0) = -\tau_R^{-1} \operatorname{sech}^2 (t_D\tau_R^{-1})$$

$$= -2N_o^{-1}\tau_{SF}^{-1} = -4N_o^{-1}\tau_R^{-1} , \quad (7.4)$$

and

$$t_D = \tau_R \,\ell n\, N_o^{\frac{1}{2}} = 2\tau_{SF} \,\ell n\, N_o^{\frac{1}{2}} \quad (7.5)$$

(We use $\ell n \equiv \log_e$ throughout this paper). From $R_3 = (1-N_o^{-1})/(1+N_o^{-1}) \approx 1-2N_o^{-1}$ at $t = 0$, we find $\delta = 2N_o^{-1}$ and $t_D = 2\tau_{SF} \,\ell n(2/\delta)^{\frac{1}{2}} = \tau_{SF}|\ell n \tfrac{1}{2}\delta|$. This compares with the empirical result $t_D \sim \tau_R|\ell n\, \delta|$ and only agrees with it at best up to a factor of one half.[28]

For HF $N_o \sim 10^{12}$ and $\delta \sim 2 \times 10^{-12}$ if $\delta = 2N_o^{-1}$: this compares with the choice $\delta = 10^{-3}$ used in the numerical work so far. Even from the formula $t_D \sim \tau_R|\ell n\, \delta|$ the delays are too short by a factor ~ 4 therefore. However this result scarcely agrees with the published data[2] where e.g. for $\tau_R = 5$ n sec, $t_D \sim 600$ n sec. Figs. 9a, b nevertheless show numerical results for $\tau_R = 5$ n sec ($\tau_{SF} = 2.5$ n sec) with $\delta = 10^{-12}$ and $\kappa L = 0$ (Fig. 9a) and $L = 3.5$ (Fig. 9b). Fields are going both ways. The delays and widths prove to be: $\kappa L = 0 - t_D = 399$ n sec, $\tau_W = 81$ n sec; $\kappa L = 3.5 - t_D =$ 543 n sec, $\tau_W = 89$ n sec. We note first that the dramatic change in δ changes t_D from $t_D \sim \frac{3}{2}\tau_R|\ell n\ \delta|$ to about $3\tau_R|\ell n\ \delta|$ for $\kappa L = 0$. This is then further substantially increased by changing κL to 3.5. The profile Fig. 9b ($\kappa L = 3.5$) and the delays and width now agree fairly well with the observations.[29]

A conclusion from this is that $t_D = \tau_R\ \ell n\ N_o^{\frac{1}{2}}$ (equation (7.5)) is a rather poor formula for the delay. The formula $t_D \sim 2\tau_R\ \ell n\ N_o^{\frac{1}{2}}$ is good (by construction) for $10^{-3} \gtrsim \delta \gtrsim 10^{-6}$ ($10^3 < N_o < 10^6$) but becomes more like $t_D \sim 6\tau_R\ \ell n\ N_o^{\frac{1}{2}}$ for $N_o \sim 10^{12}$. On this evidence no very simple connection exists between δ and t_D. But $\delta \approx N_o^{-1}$ is apparently a very proper choice for the numerical work.

The most striking feature observable in Figs. 9 is the enhancement of ringing for smaller values of δ. (Compare Figs. 9 with Fig. 1 or Figs. 6, and note in particular that $\kappa L = 3.5$ previously suppressed all ringing). The same general behaviour persists into the pure SF regime (ie. for $\tau_{SF} >> \tau_E$) Figs. 10 show the two cases $\tau_R = 25$ n sec ($\tau_{SF} = 12.5$ n sec) and $\delta = 10^{-3}$ (Fig. 10a) and $\delta = 10^{-12}$ (Fig. 10b) for $\kappa L = 0$. Ringing is strong for $\delta = 10^{-12}$ and pulse widths are increased as before ($t_D = 268$ n sec, $\tau_W = 155$ n sec (10a); $t_D = 1940$ n sec, $\tau_W = 393$ n sec (10b)).

From these results it is impossible to doubt that the choice of δ has a very significant effect on t_D, τ_W, the level of ringing, and the onset of any regime such as that of pure SF defined by the level of that ringing. The general feature is that reducing δ greatly increases t_D; but the "spring" once released after the longer delay rings more violently on a longer, but only slightly longer, time scale.

On the other hand we should not expect to find simple formulae, like that of (7.5) for the delay, in terms of δ. The analysis leading to (7.5) relies, strictly speaking, on a number of particular features: spontaneous emission is ignored, the two-mode ansatz (or an equivalent assumption) is implied; and that ansatz is used semi-classically (the operators are de-correlated). The two-mode ansatz in particular leads to pure SF and sech^2 intensity emission. But even in the pure SF regime the results of Secs. V and VI show that the radiated intensity follows other than sech^2 behaviour. Further in Sec. VI and this Sec. VII we found that t_D was influenced by the additional damping through κL. These points and the discrepancies between (7.5) and the empirical results (as well as the dramatic changes induced in these by $\delta = 10^{-3} \to \delta = 10^{-12}$) suggest that (7.5) is by no means of fundamental significance.

One further remark seems worth making: the weakest point in (7.5) could well be that it ignores spontaneous emission. Equation (7.2) shows already that for fixed τ_{SF}^{-1} the total radiation rate is dominated by spontaneous emission for large enough N_o (compare the analysis[11] leading to the suggestion of "spiking" referred to in Sec. I). In fact, however, perturbation theory[30] shows that

$$\tfrac{1}{2} N_o \dot{R}_3(0) = - N_o \Gamma_o \tag{7.6}$$

identically at $t = 0$, and the co-operative contribution governed by τ_{SF}^{-1} emerges only during the early part of the delay time. It is correct to retain τ_{SF}^{-1} to determine t_D since the delay is clearly co-operative. Even so, if we accept (7.2) instead of (7.6) we might wish to retain the larger, simply, of the two rates $N\Gamma_o$ or τ_{SF}^{-1} at $t = 0$.

If $N_o\Gamma_o$ is retained instead of τ_{SF}^{-1}, it is easy to see that the analysis leading to (7.5) could go through with $\delta = 2N_o^{-1}$ replaced by $\delta = 2N_o^{-1}[N_o \tau_{SF} \tau_{SP}^{-1}] = 2N_o^{-1} \times (4\pi A/3\lambda_s^2)$, where $\tau_{SP}^{-1} = \Gamma_o$, (the A-coefficient), and A is again the cross-section of the rod of length L containing N_o superfluorescing atoms.

We do not find that this choice for δ improves agreement with the observations: the factors on $2N_o^{-1}$ are large ($\sim 10^4$ for HF, $\sim 10^5$ for Cs) and act to reduce the effective value of N_o. For Cs in particular we find our empirical results $t_D \sim 2\tau_R|\ln \delta^{\frac{1}{2}}| = 2\tau_R \ln (\tfrac{1}{2}N_o)^{\frac{1}{2}}$ (or $2\tau_R \ln N_o^{\frac{1}{2}}$) are in good agreement with observed t_D (e.g.[31] $2\tau_R = 1.4$, $N_o = 4.5 \times 10^7$, t_D (calc) = 12.3 n sec, t_D (obs) = 12.5 n sec). The main point of this analysis therefore is to show that the theory of the delay is markedly incomplete. Our conclusion is then that δ is a parameter which should be roughly of order N_o^{-1}, but should be specifically chosen to achieve the best fit with the given data.

HF is a high ringing system with a very small Γ_o. For systems with larger Γ_o and smaller observed ringing (Na[5] and Cs[6]) it may be appropriate to reduce ringing by reducing the effective value of N_o providing, of course, that this is compatible with the observed delays. Other factors influencing both ringing and delays are (cf. Table 1) both ways going fields and (cf. Sec. VI) diffraction loss damping. In each case a reduction of ringing increases the delay. We have not yet examined the standing wave effects referred to in Sec. IV in sufficient detail to determine their consequences. We expect that the inclusion of these additional terms for two-way going fields reduces ringing. We do not yet know how they influence the delay.

VIII. CONCLUSION

One conclusion from this work is that there is certainly no incompatibility between the second quantised theories of SF and the semi-classical theories.[2,3,4] Results agree with those of BL[1] in predicting three regimes of SF. However, except in the regime of steady oscillation where the semi-classical theory coincides with the results obtained by BL[1] (for $T_2^* = \infty$), it is clear that propagational effects are important. The spatial derivatives $\varepsilon^{(\pm)}_z$ introduced and analysed in Secs. IV and V have an important effect on the SF emission. They must certainly be retained in derivative form and may be easier to handle for that reason in terms of the semi-classical theory.

We have also concluded that semi-classical theory (here given as a plane wave theory adapted for the rod but presented for the slab) actually predicts only two regimes - steady oscillation for $\tau_{SF} \ll \tau_E$ and oscillatory SF for $\tau_{SF} \gg \tau_E$. The numerical data (Figs. 4 and 5) seem conclusive in this respect and the analysis (Sec. V) already predicted the same finding.

However a regime of pure SF characterised by single pulses and negligible ringing, but by no means pure sech^2 intensities, is made possible by introducing "diffraction loss" damping to account phenomenologically for sideways losses in the rod. A consequence of this relatively crude description is that ringing can be substantially reduced for $\tau_{SF} \sim \frac{1}{4}\tau_E$ $(\tau_R \sim \frac{1}{2}\tau_E)$, and the passage from the regime of oscillatory SF to this regime of pure SF is pushed back (from $\tau_{SF} \approx \tau_E$) to somewhere in this region.

The regimes are not sharp however; this is in part because both delays, widths and ringing depend on the choice of an initial tipping angle $\sigma_o = \sqrt{2}\,\delta^{\frac{1}{2}}$. The parameter δ should be of order N_o^{-1}; but its theoretical status is such that this is at best an order of magnitude estimate only, and δ can be treated as a parameter, not wholly free, and used to combine best fits with both the observed delays and the observed ringing levels. There seems to be no relatively simple <u>analytical</u> expression which connects δ with the calculated delay t_D.

We have deliberately adopted a "low ringing model" for HF by choosing $\delta = 10^{-3}$ for most of the numerical work reported in this paper: this has shown rather convincingly that low ringing can be a feature of semi-classical theory and is one reason for concluding that semi-classical theories and quantised theories are not in disagreement. On the other hand by choosing $\delta = 10^{-12}$ (Figs. 9 and 10), the ringing becomes high and the measure of agreement with the observations on HF[2] very satisfactory. We believe that the rather dramatic effect of δ on the level of ringing has not been appreciated and that this is one reason for a wider misunderstanding of the relationship between the semi-classical and quantised theories.

As far as the work reported in this paper goes we need now to include T_2^*, $T_2 < \infty$, whilst the need for a dynamical study of three (rather than one)-dimensional emission is abundantly clear. The most intriguing theoretical problem remaining is that of providing a treatment of spontaneous emission in the region close to $t = 0$ which goes smoothly over to the co-operative regime of larger t studied in this paper. This all amounts to requiring an exact second quantised treatment of a long rod-like geometry and an adequate investigation of its semi-classical limit.

Even so we believe that the significant processes in superfluorescence are now well understood. It is too early to say how well an exact agreement with the observations[6] on Cs can be achieved within the present relatively simple semi-classical theory or how far further refinements are still needed. Certainly we are in a position to find out. It is clear that in the study of quantum beats,[6] cascading SF,[5] and the effects of degeneracy, interesting multi-level extensions of the theory will be needed. We have already acknowledged the fundamental gap which still lies between the second quantised theory and the semi-classical theory of SF which has been presented in this paper. This is perhaps the most tantalising problem still remaining in the whole theory.

References

(1) G. Banfi and R. Bonifacio, Phys. Rev. Lett. 33, 1259 (1974); G. Banfi and R. Bonifacio, Phys. Rev. A 12, 2068 (1975); R. Bonifacio and L.A. Lugiato, Phys. Rev. A 11, 1507 (BL), 12, 587 (1975); R.J. Glauber and F. Haake in "Co-operative Phenomena", H. Haken ed. (North Holland, Amsterdam, 1974) p.71 (these authors discuss a single mode theory both in the semi-classical and quantised forms); R. Bonifacio, ibid, p.97; N. Rehler and J.H. Eberly, Phys. Rev. A 3, 1735 (1971) (Refs. 12 and 13 below also come within the general category of the quantised theories).

(2) N. Skribanowitz, I.P. Herman, J.C. MacGillivray and

M.S. Feld, Phys. Rev. Lett. 30, 309 (1973).

(3) J.C. MacGillivray and M.S. Feld, Phys. Rev. A 14, 1169 (1976).

(4) R. Saunders, R.K. Bullough and S.S. Hassan, J. Phys. A: Math. Gen. 9, 1725 (1976).

(5) M. Gross, C. Fabre, P. Pillet and S. Haroche, Phys. Rev. Lett. 36, 1035 (1976).

(6) H.M. Gibbs "Quantum Beat Superfluorescence in Cs" and Q.H.F. Vrehen "Single Pulse Fluorescence in Cs". Papers given at the Co-operative Effects Meeting, Redstone Arsenal, Alabama, Dec. 1-2, 1976.

(7) S.R. Hartmann "Co-operative Effects in Metal Vapours". Paper given at the Co-operative Effects Meeting, Redstone Arsenal, Alabama, Dec. 1-2, 1976.

(8) R.H. Dicke, Phys. Rev. 93, 99 (1954).

(9) G.S. Agarwal, Phys. Rev. A 2, 2038 (1970); A 3, 1783, A 4, 1791 (1971); "Proceedings of the Third Conference on Coherence and Quantum Optics, Rochester", L. Mandel and E. Wolf eds. (Plenum, New York, 1973) p.157; R. Bonifacio, ibid, p.465; R.K. Bullough, ibid, p.121; M. Dillard and H.R. Robl, Phys. Rev. 184, 312 (1969); D.F. Walls and R. Barakat, Phys. Rev. A 1, 446 (1970) (Also see Refs. 12 and 13 below).

(10) R. Saunders, Ph.D. Thesis, U. of Manchester, 1973.

(11) F.T. Arecchi and E. Courtens, Phys. Rev. A 2, 1730 (1970).

(12) R. Bonifacio, P. Schwendimann and F. Haake, Phys. Rev. A 4, 302, 854 (1971).

(13) F. Haake and R. Glauber, Phys. Rev. A 5, 1457 (1972). R.J. Glauber and F. Haake, Phys. Rev. A 13, 357 (1976). (See also R.J. Glauber and F. Haake, Ref. 1.)

(14) R.H. Dicke in "Proceedings of the Third International Conference on Quantum Optics, Paris, 1963" (Columbia U.P., New York, 1964) p.35.

(15) Meeting on "Co-operative Effects" (Co-operative Effects Meeting[6,7]), Redstone Arsenal, Alabama, Dec. 1-2, 1976. The proceedings of this meeting are reported in this volume.

(16) R. Saunders, R.K. Bullough and F. Ahmad, J. Phys. A: Math. Gen. 8, 579 (1975).

(17) R. Friedberg and S.R. Hartmann, Optics Comm. 10, 298 (1974); Phys. Rev. A 10, 1728 (1974).

(18) The ordering of the field operators is discussed in Ref. 10, Chapter 6 (§6.2), particularly that of the +ve and -ve frequency parts of the inter-atomic fields (The conclusion is that the inter-atomic fields need not be ordered for a Hamiltonian like (3.1)).

(19) The tensor propagator $\overleftrightarrow{F}$ is a Green's function for $\nabla\nabla. - \nabla^2 + c^{-2}\,\partial^2/\partial t^2$. It has longitudinal part $\nabla\nabla r^{-1}$ and transverse part $\overleftrightarrow{F} - \nabla\nabla r^{-1}$. We use * for adjoint in (3.9).

(20) R. Saunders and R.K. Bullough, J. Phys. A: Math. Nucl. Gen. 6, 1360 (1973).

(21) J.C. Eilbeck, J.D. Gibbon, P.J. Caudrey and R.K. Bullough, J. Phys A: Math. Nucl. Gen. 6, 1337 (1973).

(22) $P^{+2} + P^{-2} + Q^{+2} + Q^{-2} + N^2$ is almost a constant of the motion in (4.4).

(23) When $L \to \infty$ the dielectric is the half space $z > 0$: we need left going not right going waves for emission across $z = 0$. This analytical problem is not solved. Mostly analytical solutions of the sine-Gordon equation are for the real line $-\infty < z < +\infty$. For example, the inverse scattering method (e.g. M. Ablowitz, D.J. Kaup, A.C. Newell and H. Segur, Phys. Rev. Lett. 30, 1262 (1973); Studies in Appl. Math. 53, 249 (1974)) solves the initial value problem for the sine-Gordon equation in the form (5.8) which is $\sigma = f(\zeta)$ at $t = 0$, with boundary conditions $\sigma \to 0$ (mod 2π), σ_ζ etc. $\to 0$ as $|\zeta| \to \infty$. Solutions of (5.3) and (5.4) for an amplifier ($P = -\sin\sigma$, $N = -\cos\sigma$ and $\sigma = \pi$ (mod 2π), $\sigma_z \to 0$, etc. at $t = 0$ with $\sigma_t \neq 0$ at $z = 0$) are given by Lamb (G.L. Lamb, Phys. Rev. A 12, 2052 (1975)) who also justifies a π-pulse similarity solution of (5.8) like that of (5.9) below.[24] Lamb previously considered this solution (G.L. Lamb, Rev. Mod. Phys. 43, 99 (1971)) and (and see below also) infers from the theorem that no solution σ of (5.8) remains within $0 < \sigma < \pi$ that all solutions ring. (This is the point we make here below (5.9)). Solutions of the sine-Gordon equation are in principal available for periodic boundary conditions (cf. the article by S.P. Novikov in "Solitons" R.K. Bullough and

P.J. Caudrey, eds. (Springer Verlag, Heidelberg. To be published, 1977).

(24) The boundary condition $\phi = -\sqrt{2\delta}$, $\phi' = 0$, at $Y = 0$ on equation (5.9) means $\phi = -\sqrt{2\delta}$, $\phi' = 0$ for all $\eta > 0$ at $\zeta = 0$. The solution is a half-space solution which shows some of the behaviour required of it for $1 > \zeta > 0$. MacGillivray and Feld[3] use this solution differently to reach their delay formula $t_D \approx \frac{1}{4}\tau_R|\ell n\ \sigma_o/2\pi|^2$. ($\sigma_o$ is a tipping angle - see Sec. VII.) Note that (5.8) is in laboratory co-ordinates, i.e. $\eta = t\ \tau_{SF}^{-1}$ is not a retarded time.

(25) S.L. McCall, Ph.D. Thesis, U. of California at Berkeley, 1969.

(26) Applied to (2.1) the approximation introduces a second σ_η term.

(27) The point is this: for $\nu \lesssim 7.0$ the regime of steady oscillation is unchanged. As the n dependence of the intensity in this regime changes to the n^2 dependence of the oscillatory SF regime ν becomes effective in damping the pulses. If ν is large enough there must be a smooth transition from n dependence and strong ringing through to n^2 dependence and no ringing.

(28) At the time of going to press we have not tracked down a possibly discrepant factor of two (Compare e.g. (2.4) against (7.3)). There is no general agreement on the definition of τ_R in the expressions for t_D: our definition assumes non-degenerate two-level transitions with fixed polarisation orthogonal to z, the axis of the rod: this must be averaged for a real gas. Note that the discussion throughout uses our definitions of τ_R and τ_{SF}. BL[1] use $2\tau_R$ for our τ_R and e.g. the regimes described in Sec. II are changed accordingly.

(29) Compare Fig. 2c in Ref. 2: $\tau_W \sim 120$ n sec there however.

(30) R. Saunders and R.K. Bullough, J. Phys. A: Math. Nucl. Gen. 6, 1348 (1973).

(31) We are grateful to Hyatt Gibbs and Quirin Vrehen for this particular data on Cs: agreement is with the empirical $t_D \sim 2\tau_R\ \ell n\ N_o^{\frac{1}{2}}$ obtained from HF data; and we have yet to calculate the delay from this data for Cs.

THEORY OF FIR SUPERFLUORESCENCE:

THE FIGURES

The following Figures are referred to in the main text: Figs. 1(V), 2(V), 3(V), 4(V), 5(V); 6a,b,c,d,e(VI); 7a,b(VI); 8a,b,c(VI); 9a,b(VII); 10a,b(VII). The parentheses refer to the Section where reference to the particular Figure is first made. The Figures themselves are grouped together and follow in order after these remarks. All the data are strictly speaking for HF (L = 100 cm, λ = 84 μm, p = 6.7 x 10^{-19} e.g.s. units) but have quite general application for the given values of τ_R, τ_E, etc.

Typically each Figure is labelled as on the Fig.1, namely:

Fig.1 Data for HF

τ_R = 5 n sec

τ_E = 6.7 n sec

τ_C = 4.1 n sec

κL = 0.0

δ = 10^{-3}

τ_W = 31.1 n sec

t_D = 54.4 n sec

Two ways.

In the right hand trace is the output intensity from one end of the 'rod' as a function of time (the symmetry about either end of the rod has been checked). In the left hand trace are the fields inside the rod: successive frames are at equal intervals at times which can be obtained by reference to the time scale for the output intensity on the right. The output fields and intensities are in arbitrary units which may vary between the differently numbered Figs. but not between the differently lettered Figs. (e.g. Figs.6a,b,c,d,e are all scaled to the same arbitrary units). Figs.5 and 7a,b are scaled to the same unit and can be compared one against the other. Output intensities are always formally scaled against n^2 except in the Fig.2 which is scaled against n: the scaling is formal because the units are arbitrary, but the scaling against n^2 or n indicates the n dependence. Note Figs.4 and 5 (for τ_R = 100 n sec and 50 n sec) are apparently identical: they are scaled to the same arbitrary intensity unit and the numerical outputs show that they differ only in the third figures. "Two ways" and "One way" refer as in the text to fields going both ways (two ways) or one way. Fig.3 is exceptional in comparing the both ways going (solid line) and one way going (broken line) cases.

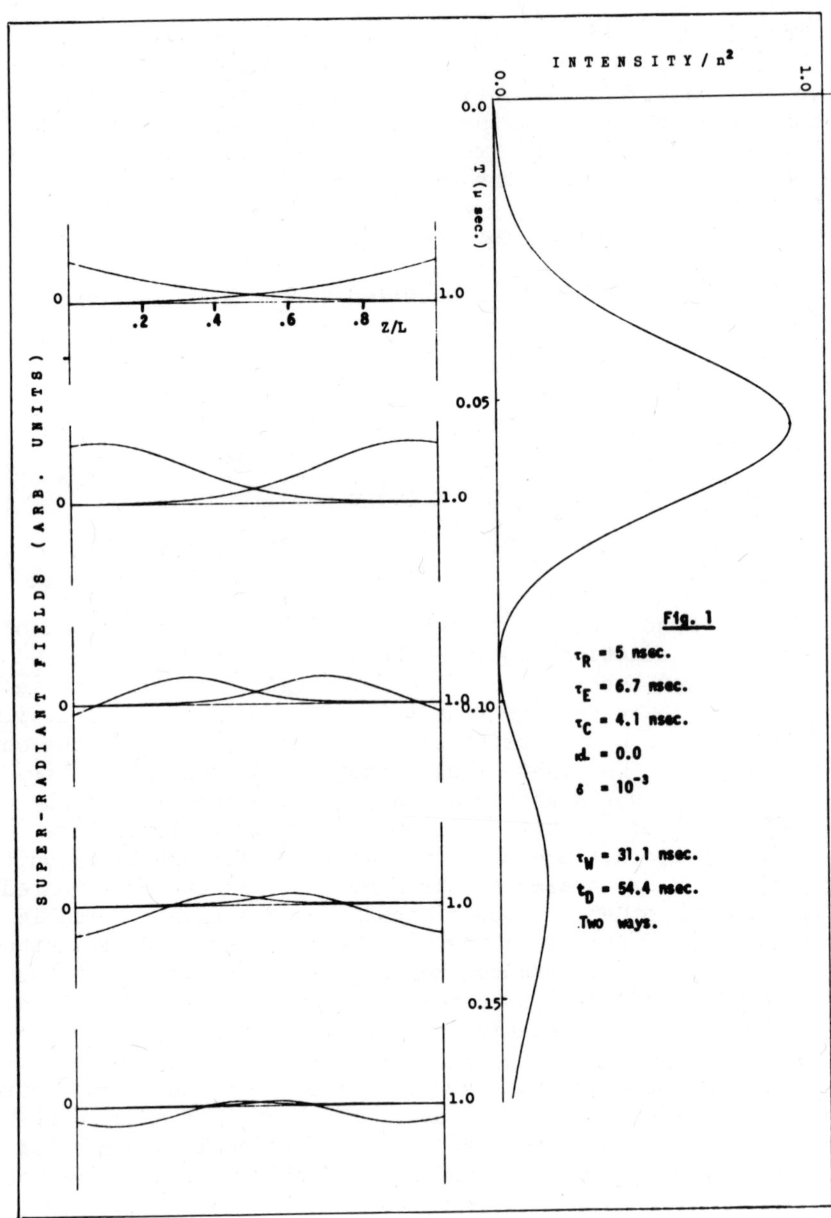
INTENSITY / n²
0.0
1.0
T (μ sec.)
0.05
0.10
0.15
SUPER-RADIANT FIELDS (ARB. UNITS)
.2 .4 .6 .8 Z/L
τ_R = 5 nsec.
τ_E = 6.7 nsec.
τ_C = 4.1 nsec.
κL = 0.0
δ = 10^{-3}
τ_W = 31.1 nsec.
t_D = 54.4 nsec.
Two ways.

Fig. 1

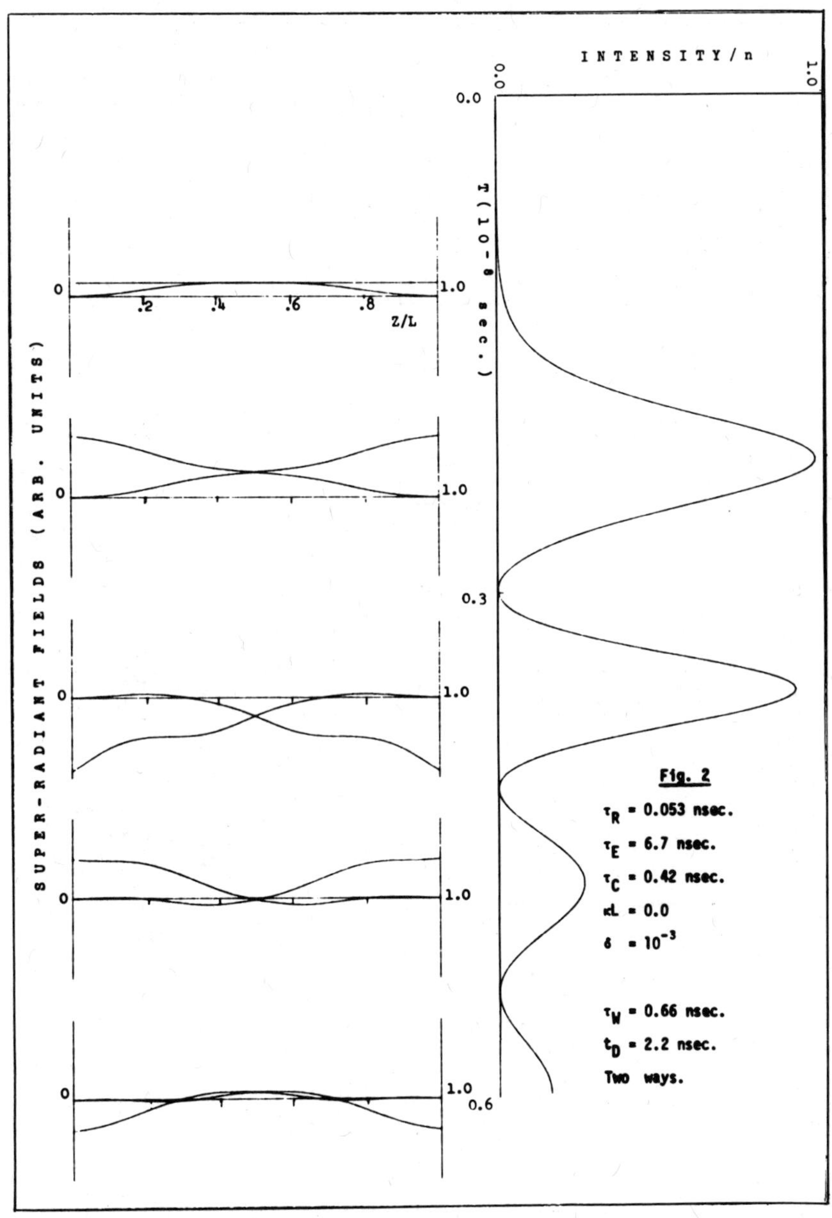
INTENSITY/n
0.0
1.0
T(10⁻⁸ sec.)
0.0
0.3
0.6
SUPER-RADIANT FIELDS (ARB. UNITS)
0
.2
.4
.6
.8
1.0
Z/L
τ_R = 0.053 nsec.
τ_E = 6.7 nsec.
τ_C = 0.42 nsec.
κL = 0.0
δ = 10^{-3}
τ_W = 0.66 nsec.
t_D = 2.2 nsec.
Two ways.

Fig. 2

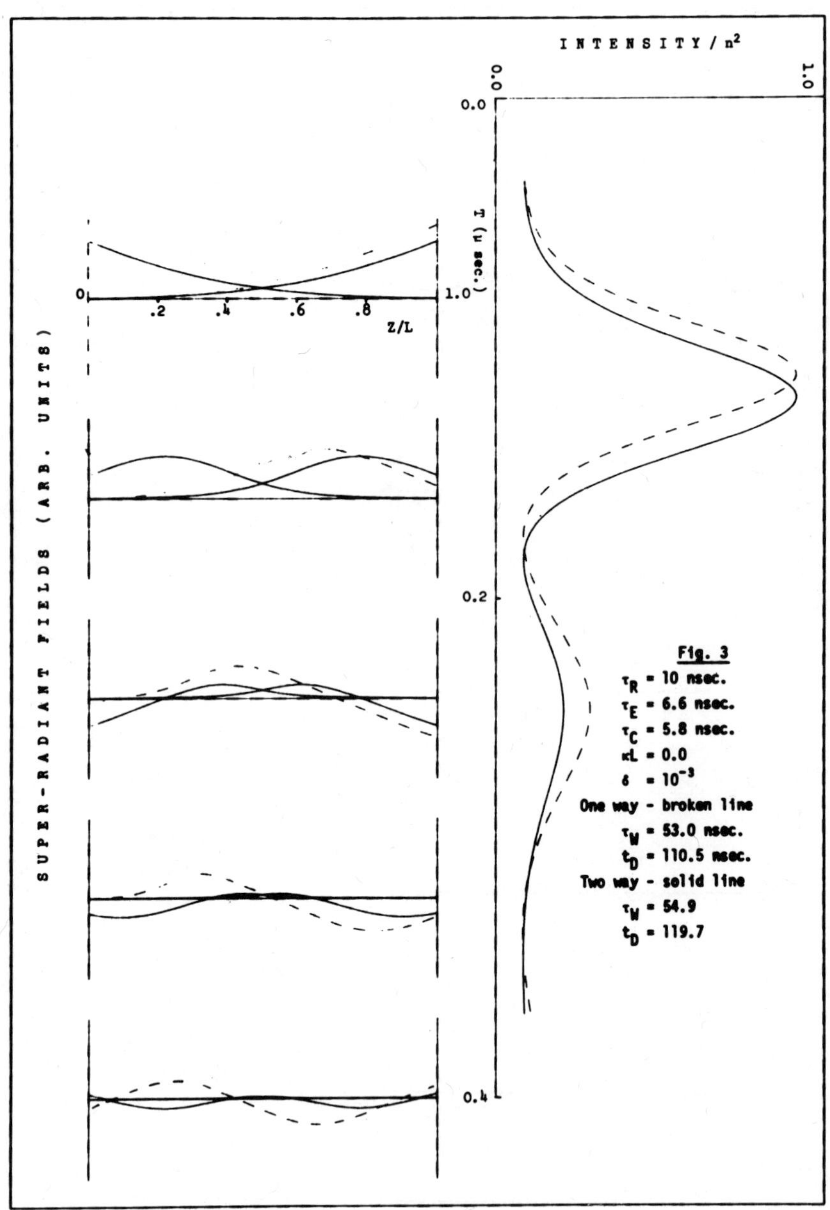
INTENSITY / n²
0.0
1.0
0.0
0.2
0.4
T (μ sec.)
0
.2
.4
.6
.8
1.0
Z/L
SUPER-RADIANT FIELDS (ARB. UNITS)
Fig. 3
τR = 10 nsec.
τE = 6.6 nsec.
τC = 5.8 nsec.
κL = 0.0
δ = 10⁻³
One way - broken line
τW = 53.0 nsec.
tD = 110.5 nsec.
Two way - solid line
τW = 54.9
tD = 119.7

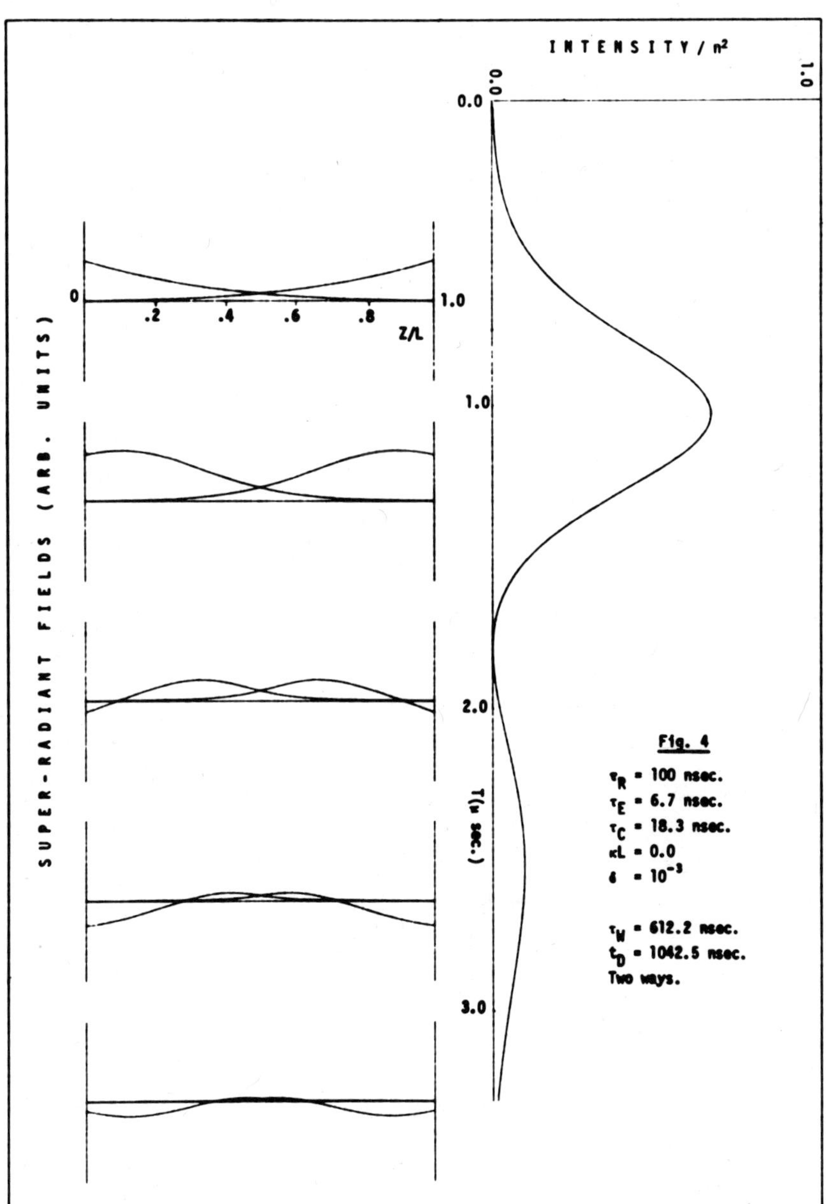
INTENSITY / n^2
0.0
1.0
0.0
1.0
1.0
2.0
3.0
T(μ sec.)
0
.2
.4
.6
.8
1.0
Z/L
SUPER-RADIANT FIELDS (ARB. UNITS)
τ_R = 100 nsec.
τ_E = 6.7 nsec.
τ_C = 18.3 nsec.
κL = 0.0
δ = 10^{-3}
τ_W = 612.2 nsec.
t_D = 1042.5 nsec.
Two ways.

Fig. 4

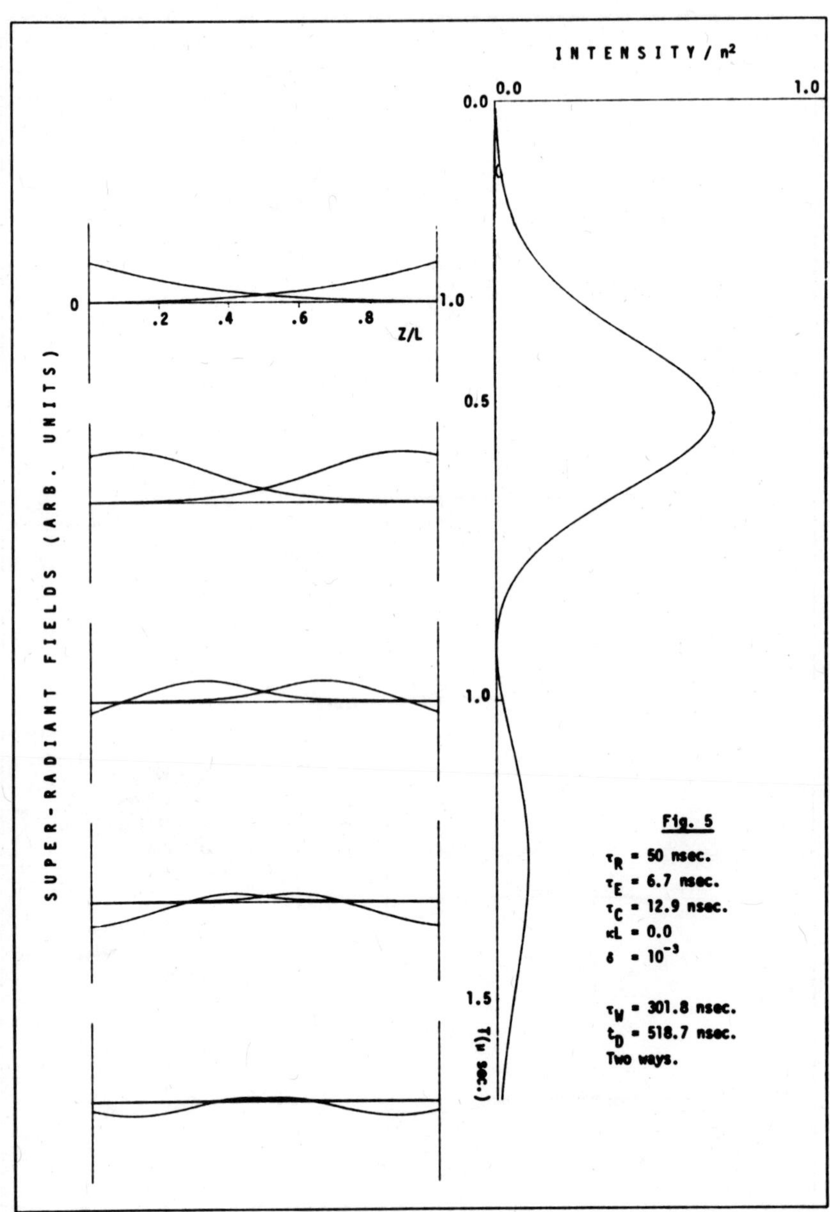
INTENSITY / n^2
0.0
1.0
0.0
0.5
1.0
1.5
T(μ sec.)
0
.2
.4
.6
.8
1.0
Z/L
SUPER-RADIANT FIELDS (ARB. UNITS)
τ_R = 50 nsec.
τ_E = 6.7 nsec.
τ_C = 12.9 nsec.
κL = 0.0
δ = 10^{-3}
τ_W = 301.8 nsec.
t_D = 518.7 nsec.
Two ways.

Fig. 5

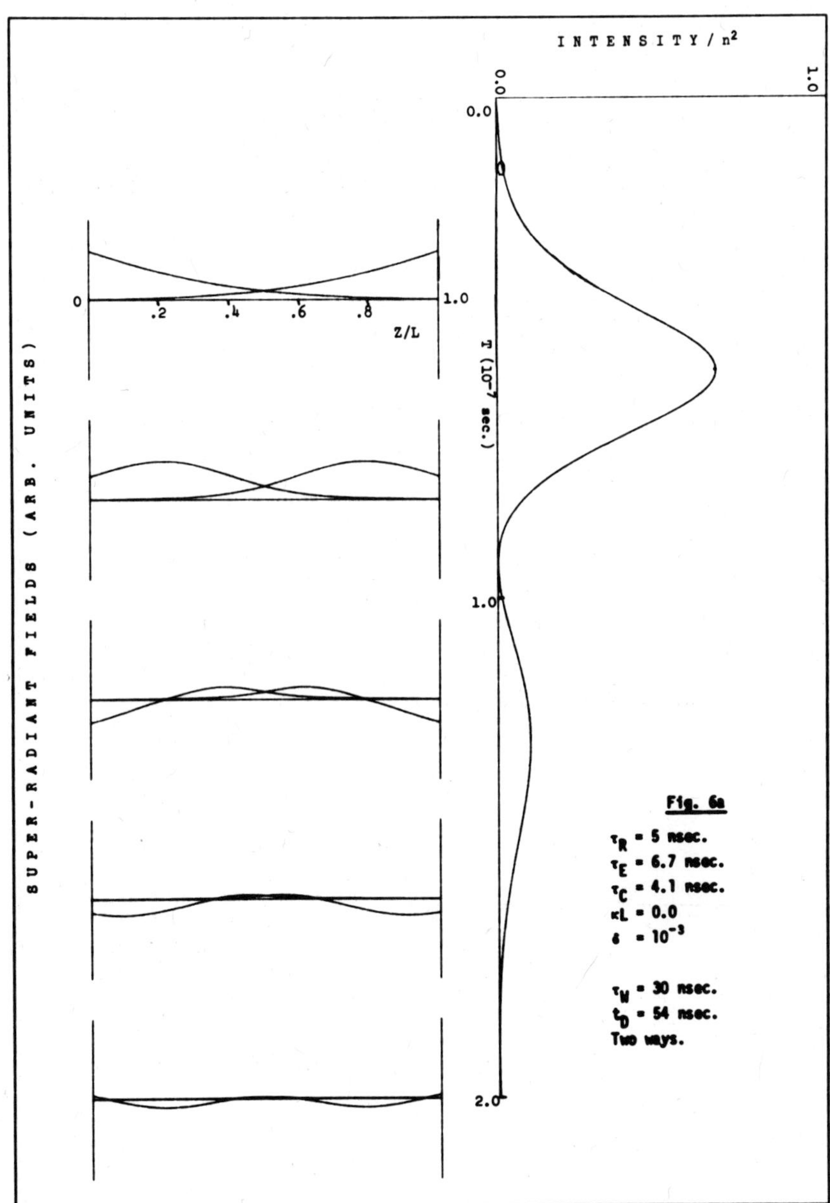
INTENSITY / n²
0.0
1.0
0.0
1.0
2.0
T (10⁻⁷ sec.)
0
.2
.4
.6
.8
1.0
Z/L
SUPER-RADIANT FIELDS (ARB. UNITS)
τR = 5 nsec.
τE = 6.7 nsec.
τC = 4.1 nsec.
κL = 0.0
δ = 10⁻³
τW = 30 nsec.
tD = 54 nsec.
Two ways.

Fig. 6a

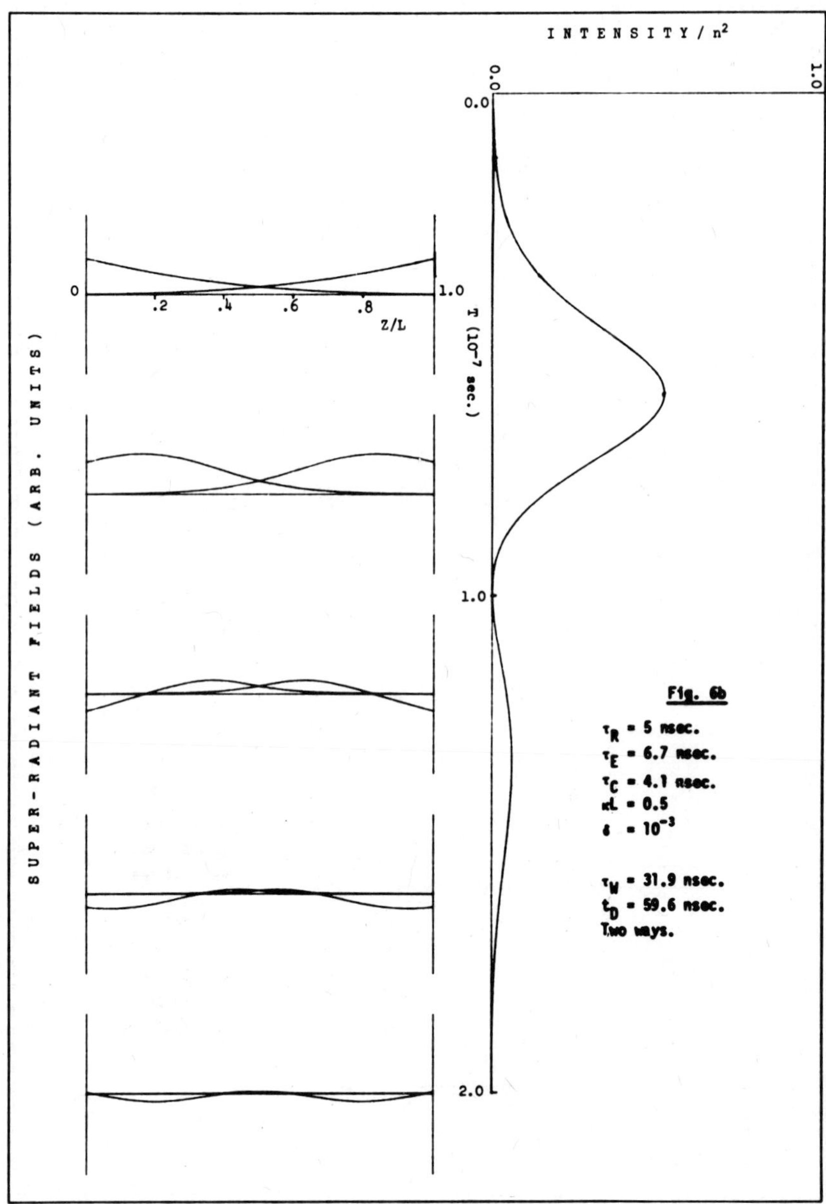
INTENSITY / n^2
0.0
1.0
0.0
1.0
2.0
T (10^{-7} sec.)
0
.2
.4
.6
.8
1.0
Z/L
SUPER-RADIANT FIELDS (ARB. UNITS)
τ_R = 5 nsec.
τ_E = 6.7 nsec.
τ_C = 4.1 nsec.
κL = 0.5
δ = 10^{-3}
τ_W = 31.9 nsec.
t_D = 59.6 nsec.
Two ways.

Fig. 6b

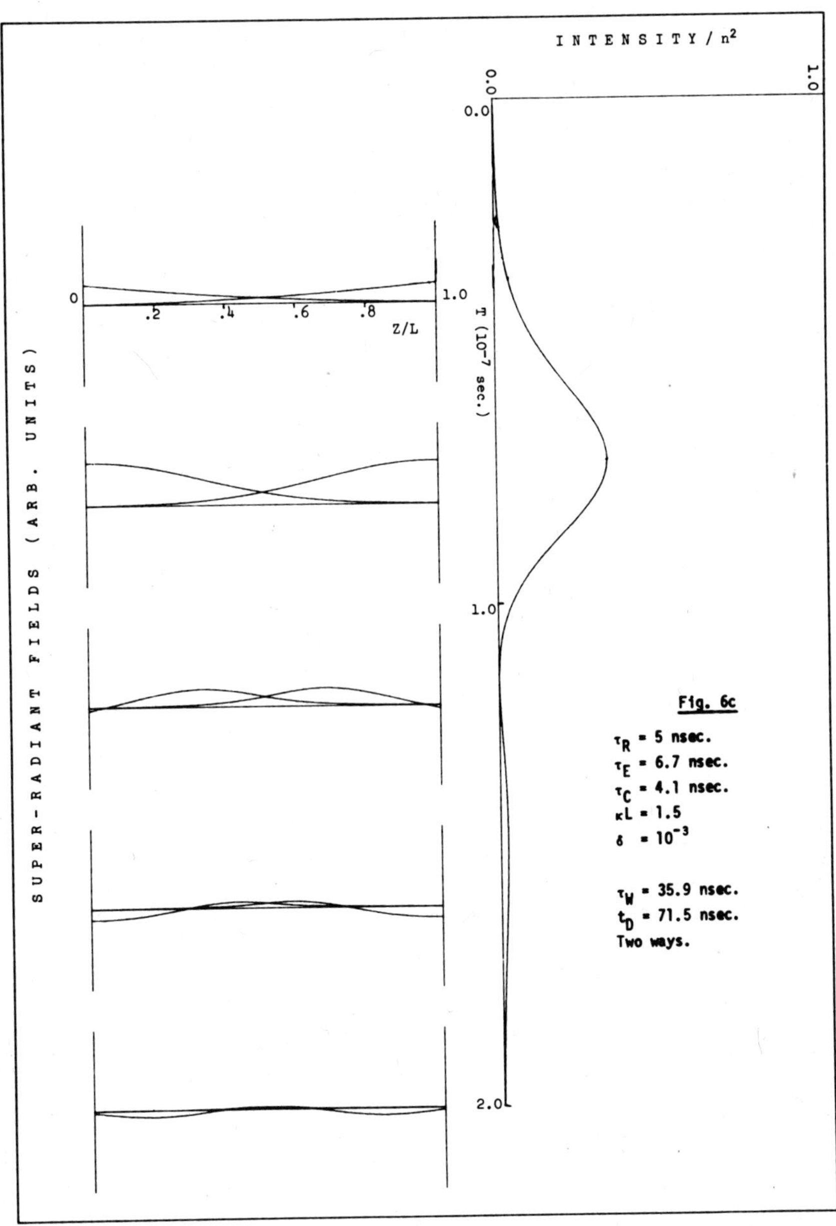
INTENSITY / n^2
0.0
1.0
0.0
1.0
2.0
T (10^{-7} sec.)
0
.2
.4
.6
.8
1.0
Z/L
SUPER-RADIANT FIELDS (ARB. UNITS)
Fig. 6c
τ_R = 5 nsec.
τ_E = 6.7 nsec.
τ_C = 4.1 nsec.
κL = 1.5
δ = 10^{-3}
τ_W = 35.9 nsec.
t_D = 71.5 nsec.
Two ways.

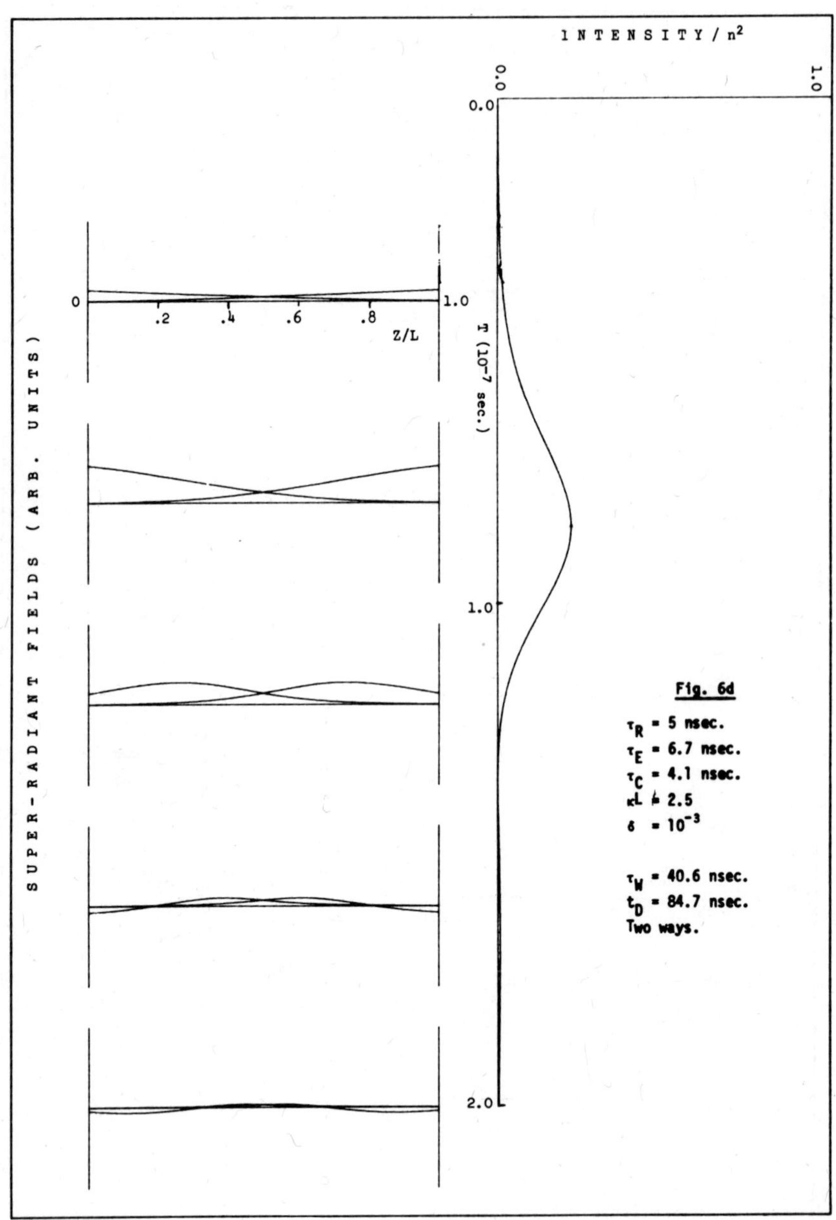

Fig. 6d

τ_R = 5 nsec.
τ_E = 6.7 nsec.
τ_C = 4.1 nsec.
κL = 2.5
δ = 10^{-3}

τ_W = 40.6 nsec.
t_D = 84.7 nsec.
Two ways.

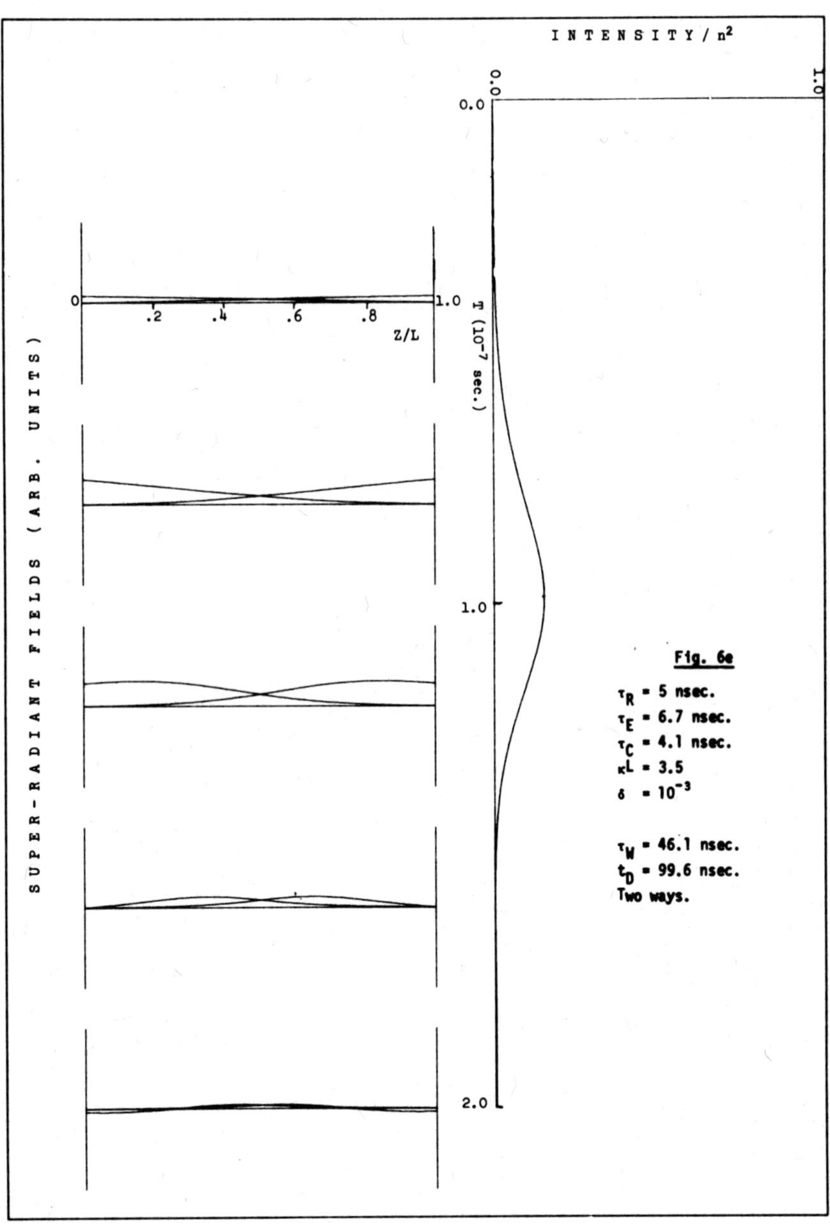
INTENSITY / n^2
0.0
1.0
0.0
1.0
2.0
T (10^{-7} sec.)
0
.2
.4
.6
.8
1.0
Z/L
SUPER-RADIANT FIELDS (ARB. UNITS)
τ_R = 5 nsec.
τ_E = 6.7 nsec.
τ_C = 4.1 nsec.
κL = 3.5
δ = 10^{-3}
τ_W = 46.1 nsec.
t_D = 99.6 nsec.
Two ways.

Fig. 6e

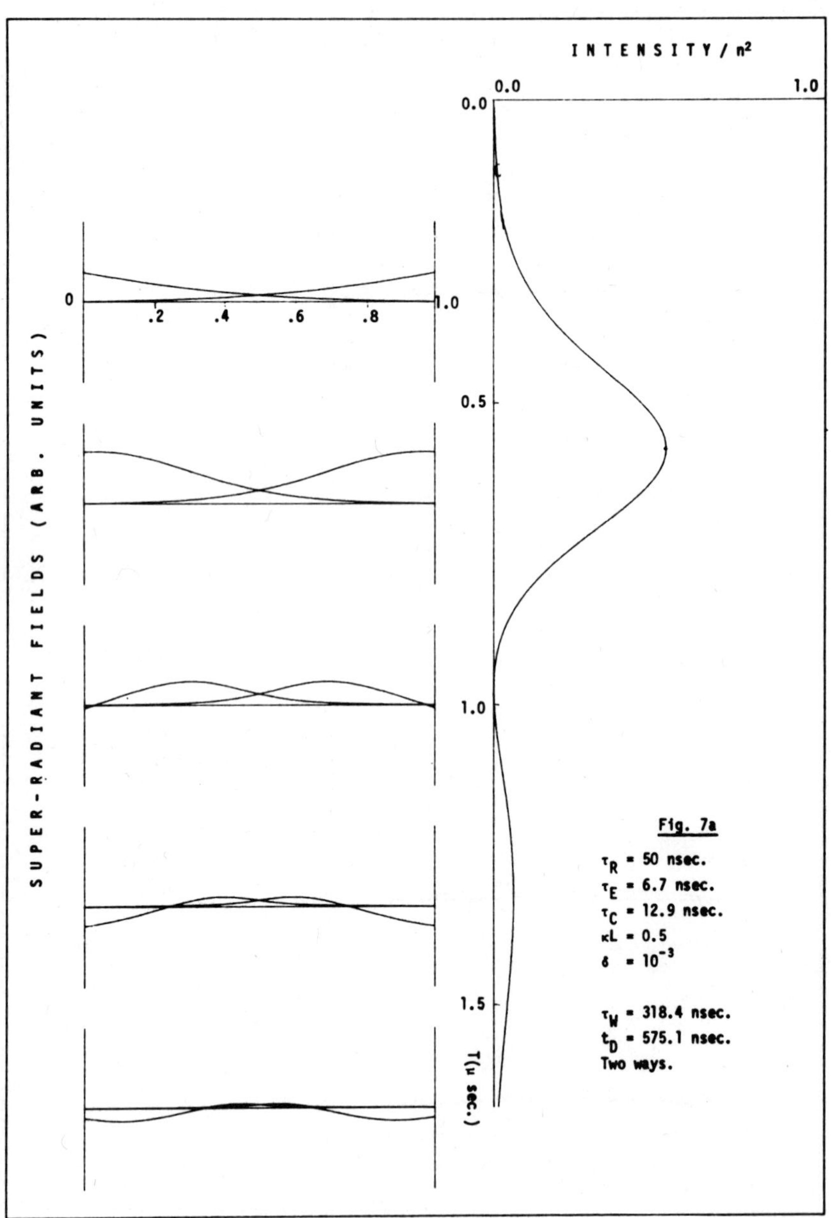
INTENSITY / n^2
0.0
1.0
0.0
0.5
1.0
1.5
T(μ sec.)
SUPER-RADIANT FIELDS (ARB. UNITS)
0
.2
.4
.6
.8
1.0
Fig. 7a
τ_R = 50 nsec.
τ_E = 6.7 nsec.
τ_C = 12.9 nsec.
κL = 0.5
δ = 10^{-3}
τ_W = 318.4 nsec.
t_D = 575.1 nsec.
Two ways.

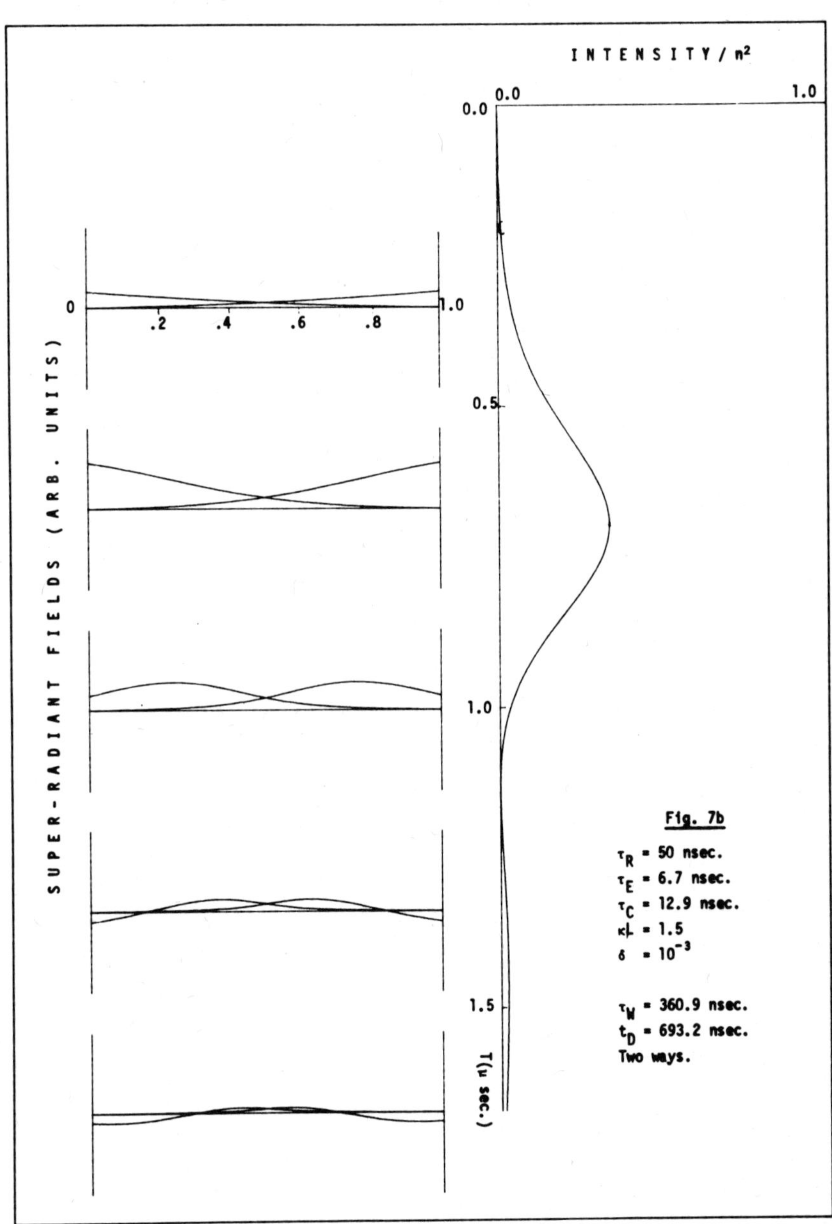
INTENSITY / n²
0.0
1.0
0.0
0.5
1.0
1.5
T(μ sec.)
0
.2
.4
.6
.8
1.0
SUPER-RADIANT FIELDS (ARB. UNITS)
Fig. 7b
τR = 50 nsec.
τE = 6.7 nsec.
τC = 12.9 nsec.
κL = 1.5
δ = 10⁻³
τW = 360.9 nsec.
tD = 693.2 nsec.
Two ways.

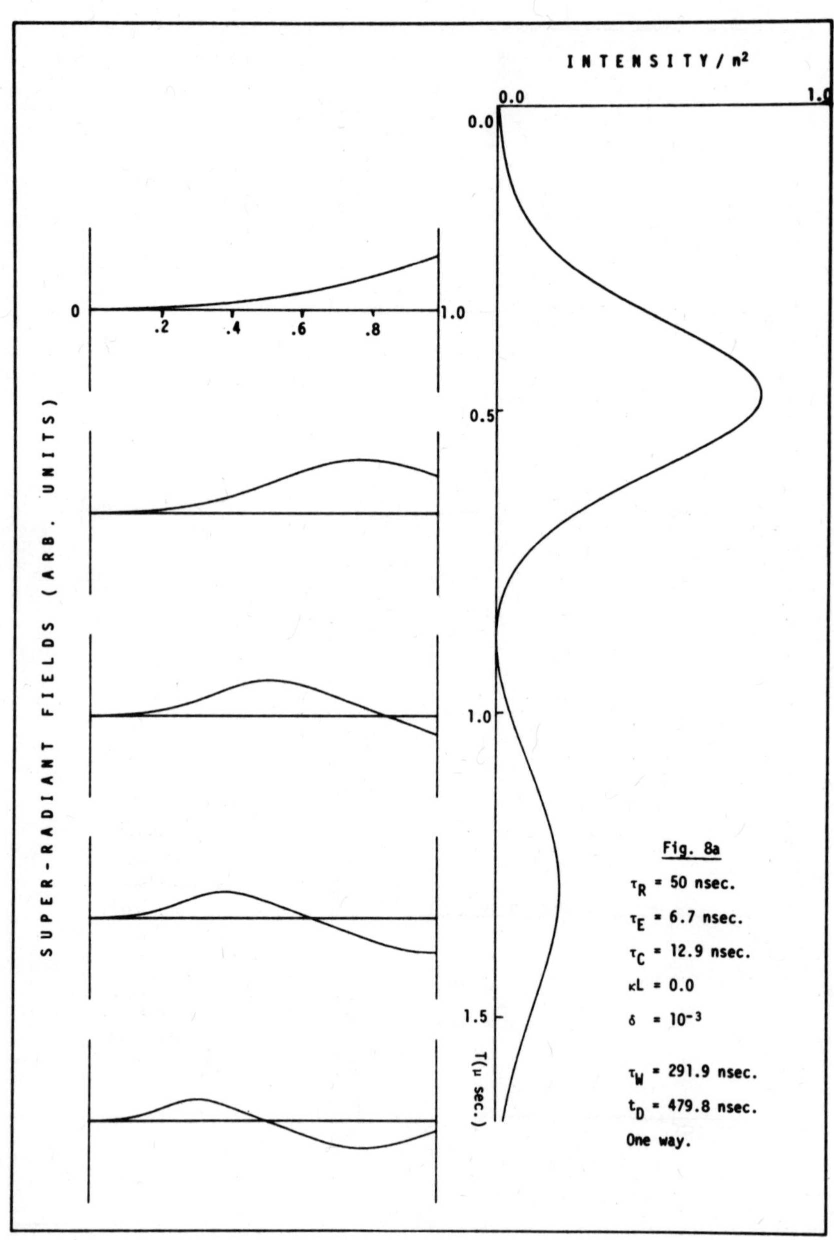
INTENSITY / n^2
0.0
1.0
0.0
0.5
1.0
1.5
T(μ sec.)
0
.2
.4
.6
.8
1.0
SUPER-RADIANT FIELDS (ARB. UNITS)
Fig. 8a
τ_R = 50 nsec.
τ_E = 6.7 nsec.
τ_C = 12.9 nsec.
κL = 0.0
δ = 10^{-3}
τ_W = 291.9 nsec.
t_D = 479.8 nsec.
One way.

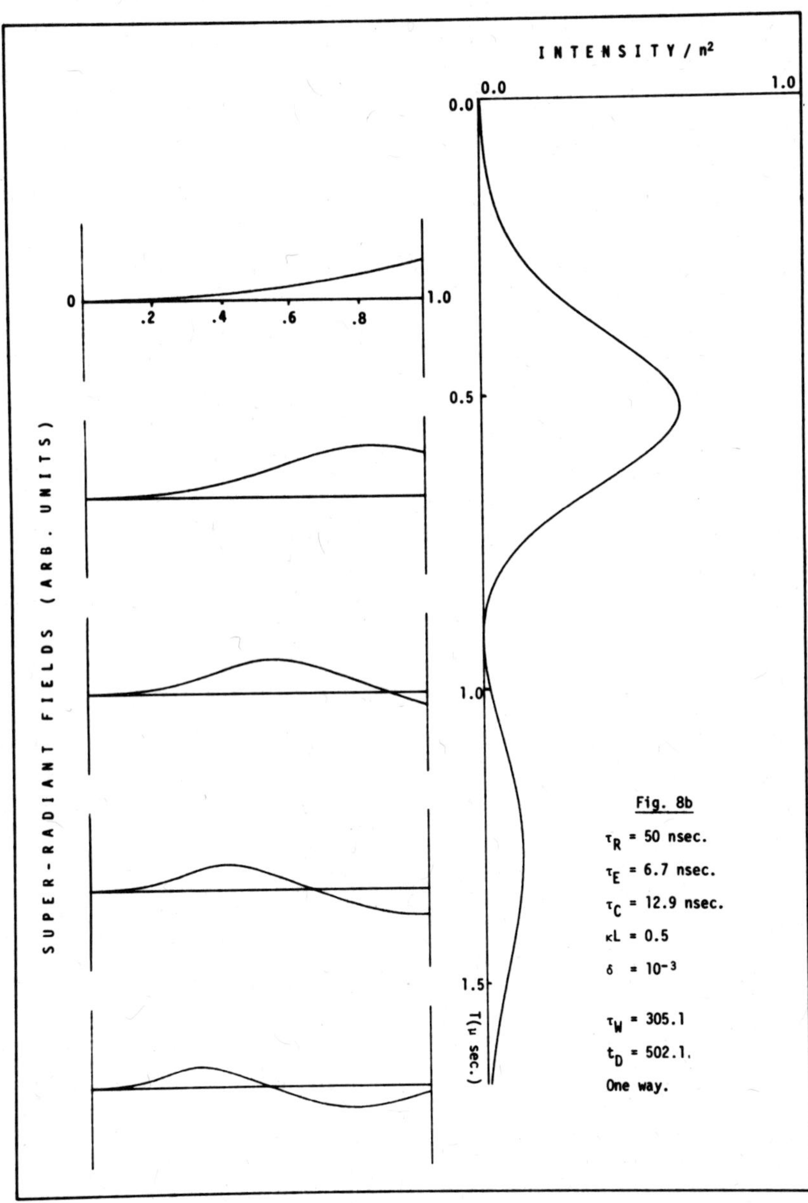

Fig. 8b

τ_R = 50 nsec.

τ_E = 6.7 nsec.

τ_C = 12.9 nsec.

κL = 0.5

δ = 10^{-3}

τ_W = 305.1

t_D = 502.1.

One way.

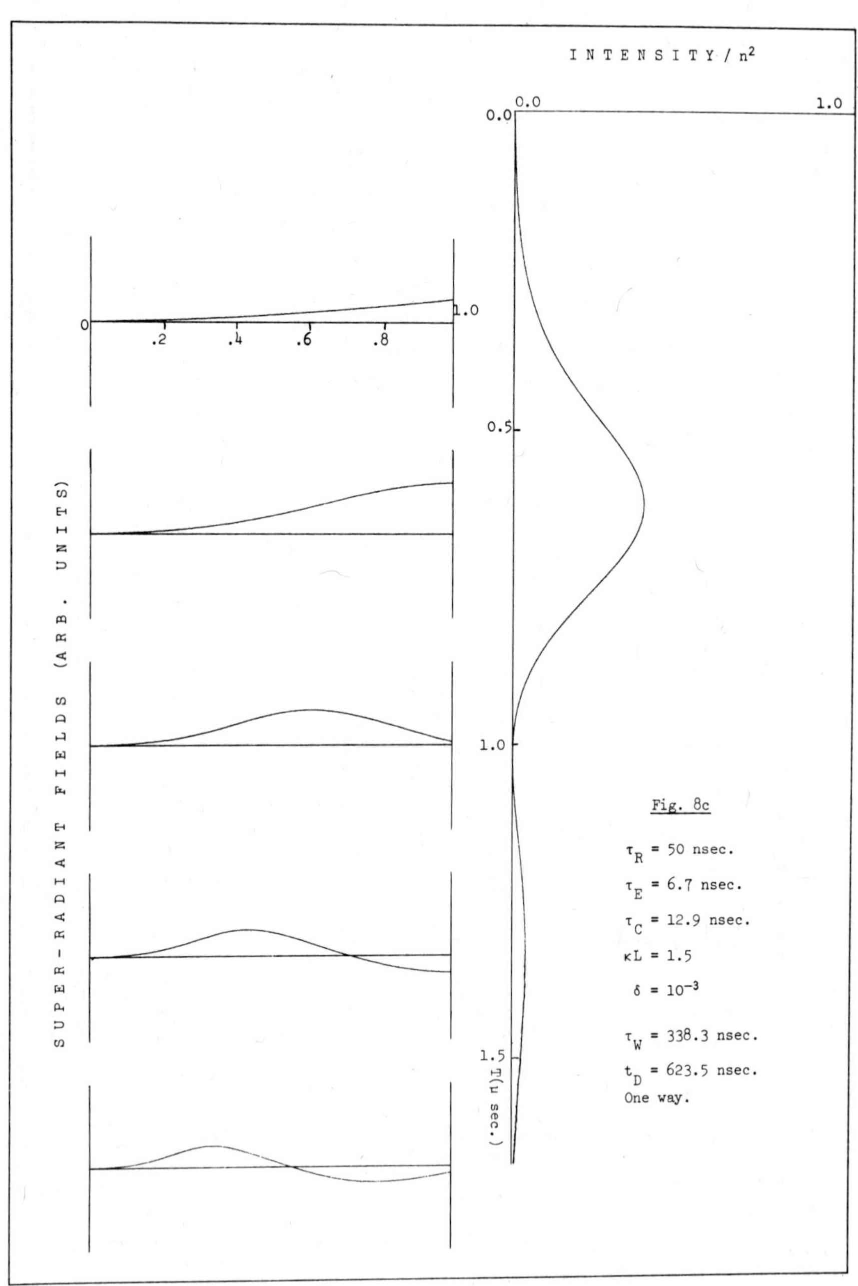
INTENSITY / n^2
0.0
1.0
0.0
0.5
1.0
1.5
T(μ sec.)
0
.2
.4
.6
.8
1.0
SUPER-RADIANT FIELDS (ARB. UNITS)
τ_R = 50 nsec.
τ_E = 6.7 nsec.
τ_C = 12.9 nsec.
κL = 1.5
$\delta = 10^{-3}$
τ_W = 338.3 nsec.
t_D = 623.5 nsec.
One way.

Fig. 8c

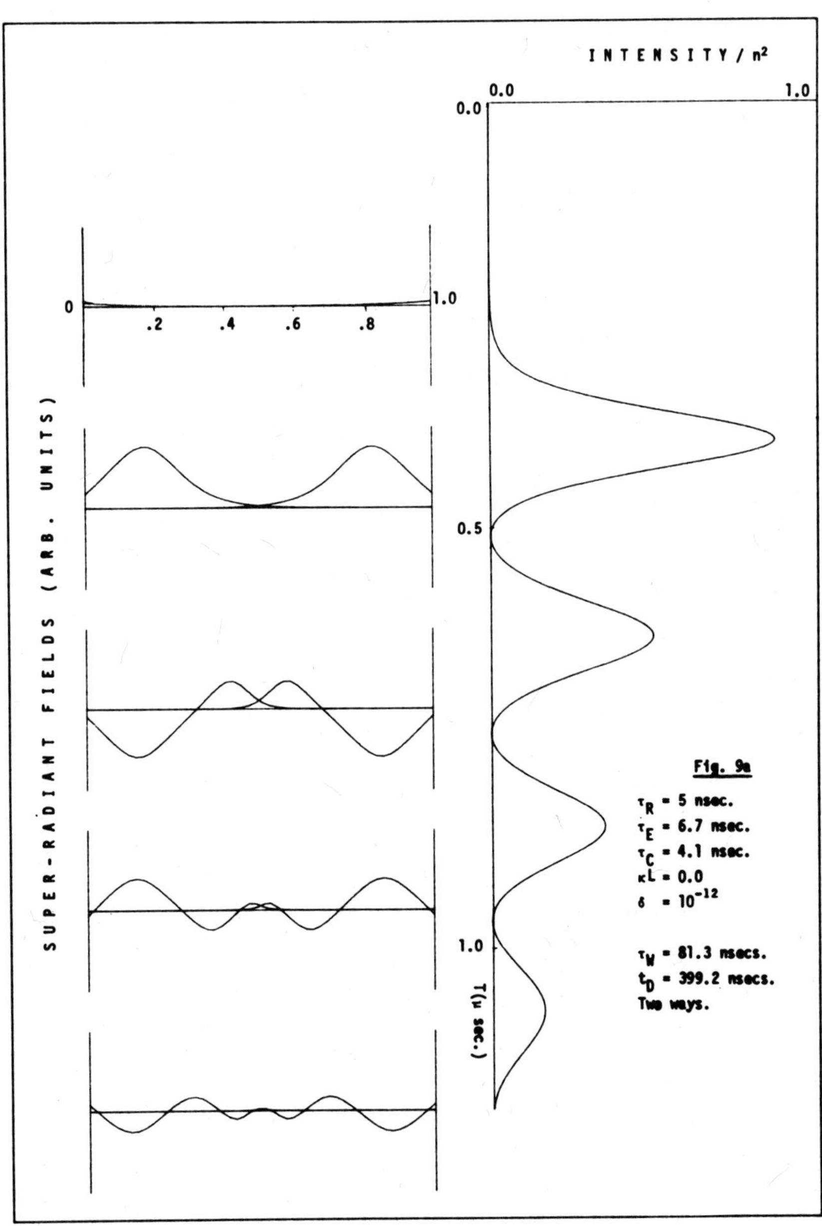
INTENSITY / n^2
0.0
1.0
0.0
0.5
1.0
T(μ sec.)
0
.2
.4
.6
.8
1.0
SUPER-RADIANT FIELDS (ARB. UNITS)
Fig. 9a
τ_R = 5 nsec.
τ_E = 6.7 nsec.
τ_C = 4.1 nsec.
κL = 0.0
δ = 10^{-12}
τ_W = 81.3 nsecs.
t_D = 399.2 nsecs.
Two ways.

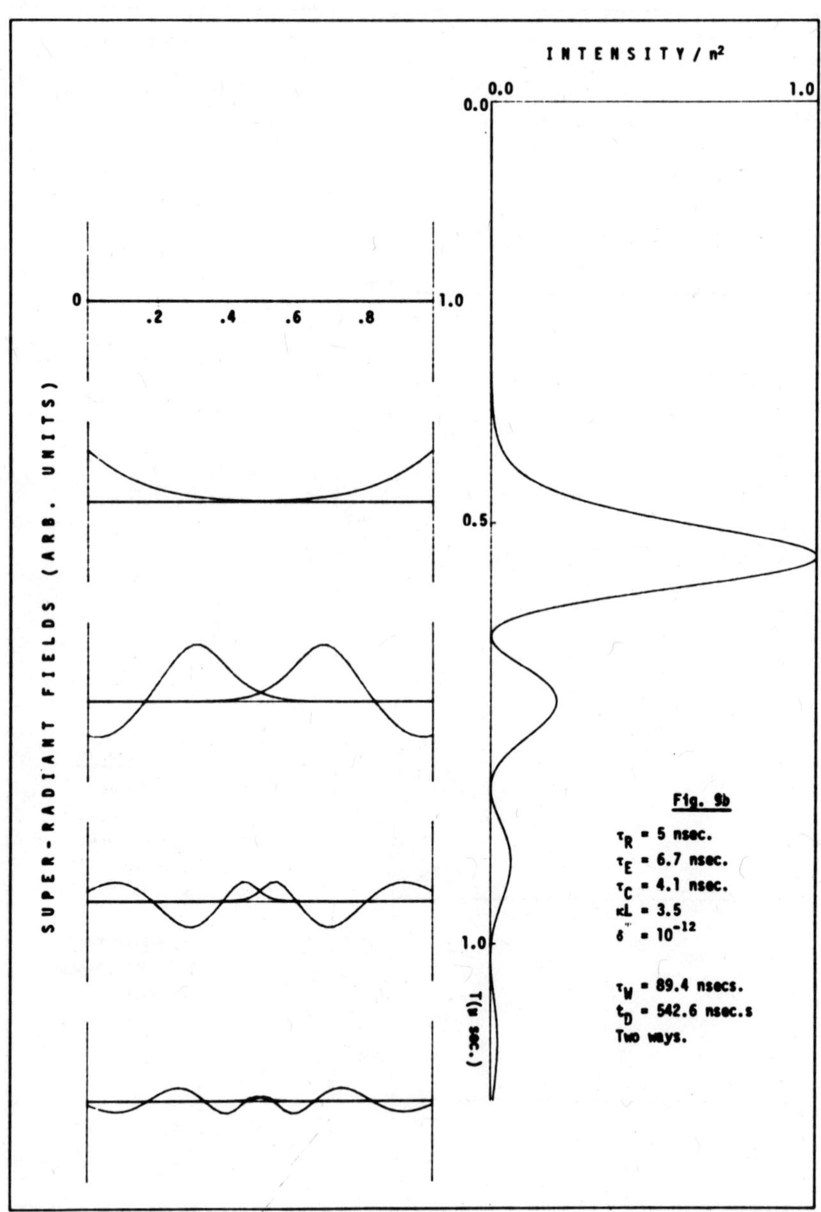

INTENSITY / n^2
0.0
1.0
0.0
0.5
1.0
T(μ sec.)
0
.2
.4
.6
.8
1.0
SUPER-RADIANT FIELDS (ARB. UNITS)
Fig. 9b
τ_R = 5 nsec.
τ_E = 6.7 nsec.
τ_C = 4.1 nsec.
κL = 3.5
δ = 10^{-12}
τ_W = 89.4 nsecs.
t_D = 542.6 nsec.s
Two ways.

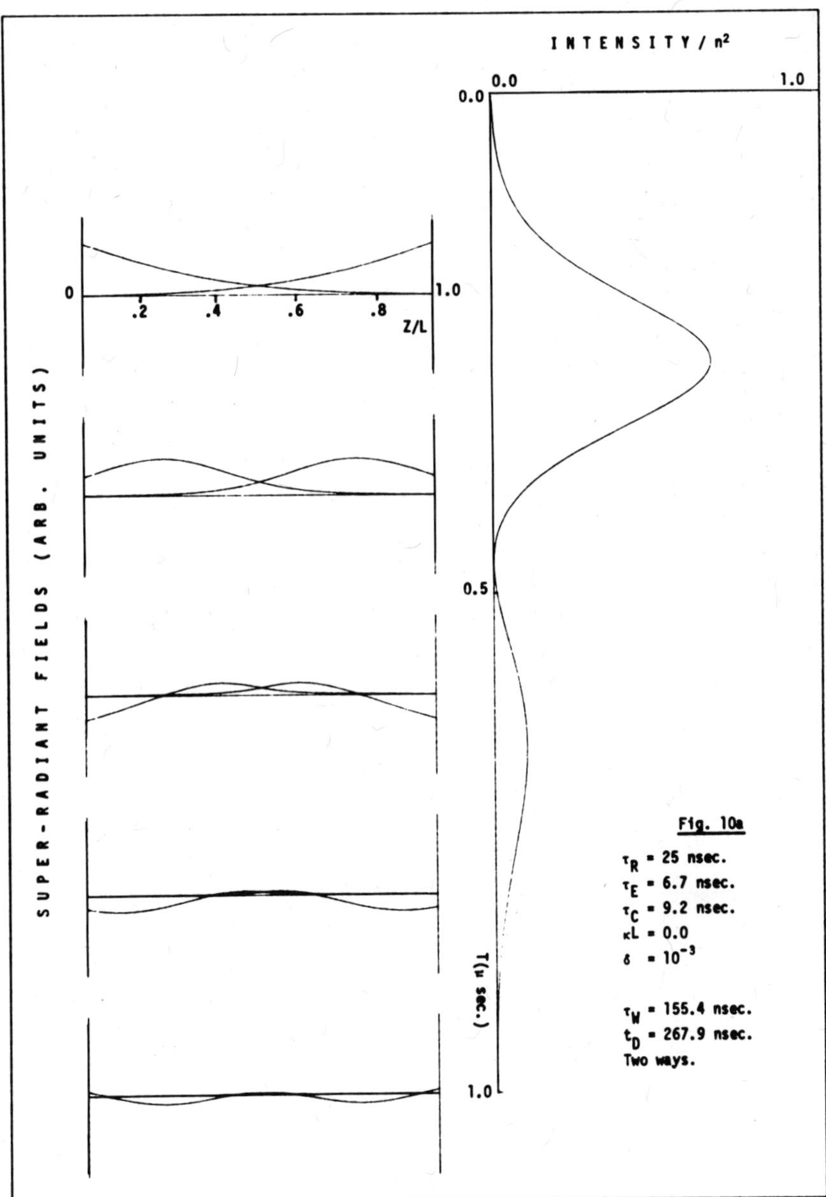
INTENSITY / n²
0.0
1.0
0.0
0.5
1.0
T(μ sec.)
0
.2
.4
.6
.8
1.0
Z/L
SUPER-RADIANT FIELDS (ARB. UNITS)
τR = 25 nsec.
τE = 6.7 nsec.
τC = 9.2 nsec.
κL = 0.0
δ = 10⁻³
τW = 155.4 nsec.
tD = 267.9 nsec.
Two ways.

Fig. 10a

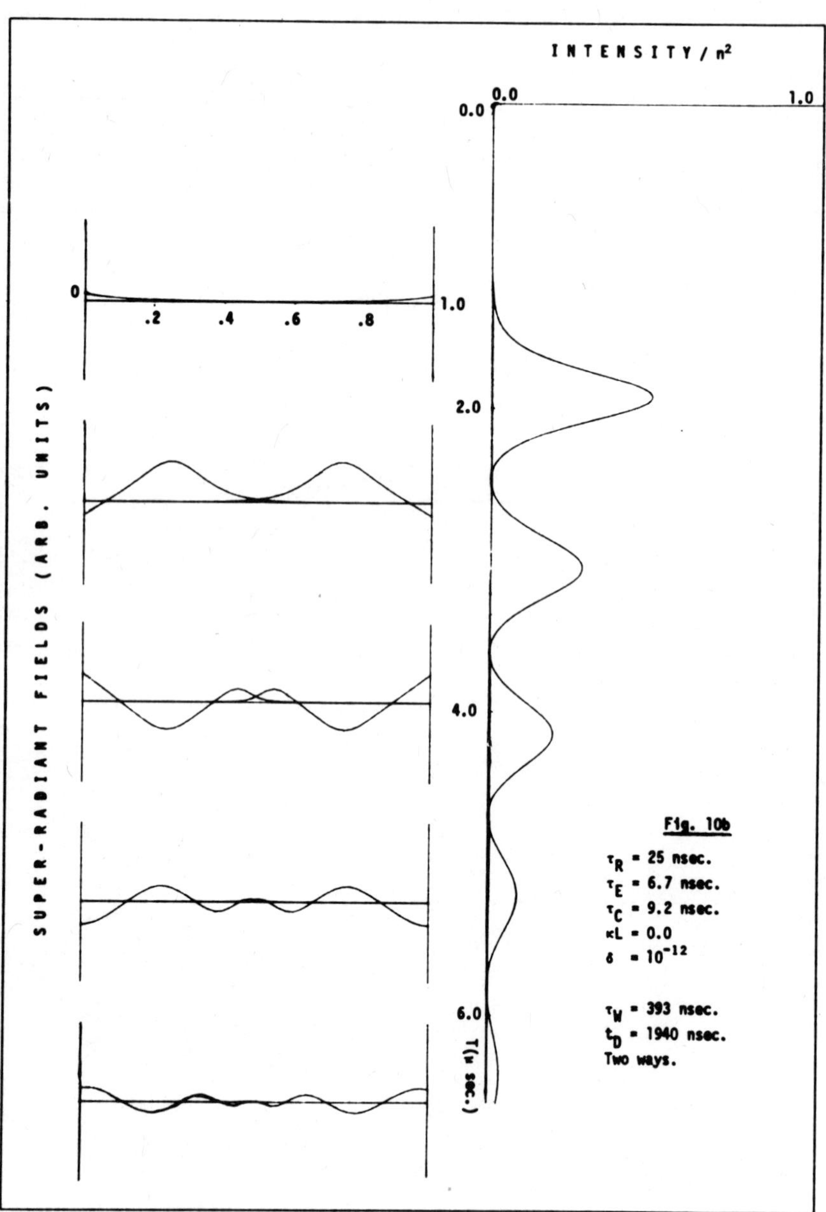
INTENSITY / n^2
0.0
1.0
0.0
1.0
2.0
4.0
6.0
T(μ sec.)
0
.2
.4
.6
.8
SUPER-RADIANT FIELDS (ARB. UNITS)
τ_R = 25 nsec.
τ_E = 6.7 nsec.
τ_C = 9.2 nsec.
κL = 0.0
δ = 10^{-12}
τ_W = 393 nsec.
t_D = 1940 nsec.
Two ways.

Fig. 10b

A MODEL OF A DEGENERATE TWO-PHOTON AMPLIFIER**

L. M. Narducci*, L. G. Johnson*, E. J. Seibert*,

W. W. Eidson*, and P. S. Furcinitti†

*Physics Department, Drexel University, Philadelphia, Pennsylvania 19104

†Department of Biochemistry, Pennsylvania State University, University Park, Pennsylvania 16802

Abstract: This paper is concerned with the propagation of an electromagnetic pulse through a two-photon amplifier medium prepared in a state of inversion between two levels of the same parity, the carrier frequency of the incident pulse being approximately one half of the atomic transition frequency. Upon neglecting competing effects which can cause emission at a frequency other than that of the incident pulse, we describe the coupled evolution of the atom-field system in terms of the usual self-consistent approach. In the coherent limit, i.e., when atomic relaxation effects are negligible, we derive an area equation for the total pulse energy, characterize the threshold condition for power amplification, and classify the multiple steady state solutions of the area equation.

After introducing incoherent atomic relaxation effects into the model, we describe the propagation of the input pulse through the amplifier with the help of a hybrid computer simulation. We analyze the transition between coherent and rate equation

**Work partially supported by the Army Research Office Grant number DAAG29-76-G-0075 and by the Office of Naval Research contract number N0014-76-C-1082.

propagation by varying the ratio T_2/τ_p between the transverse atomic relaxation time and the pulse duration. As expected, the pulse envelope modulation and pulse break-up, which are typical manifestations of coherent interaction, gradually disappear as the rate equation limit is approached.

I. INTRODUCTION

Laser amplification by two-photon stimulated decay was first considered in the early sixties[1,2] as a means of producing high peak-power radiation pulses. Since then, considerable progress has been made in understanding the dynamics of two-photon processes, especially in the coherent regime[3]. (The term coherent is used to signify that the atomic relaxation times are much longer than the duration of the propagating pulse.) Recent investigations have revealed numerous qualitative similarities between coherent two-photon absorption processes and their single-photon counterparts[4]. Little attention, however, has been paid to the physics of two-photon amplification[5,6], and especially to the effects of atomic relaxation on the dynamics of the process.

Here we discuss the feasibility of producing laser amplification by two-photon stimulated emission. The active medium is modeled as a collection of identical atoms, initially prepared in a state of inversion between homogeneously broadened levels of the same parity, separated by a frequency difference ω_{ba}. We are especially concerned with the possibility of inducing a non-linear polarization which oscillates at the same frequency, $\omega \approx 1/2\ \omega_{ba}$, as the incident pulse. The induced polarization is regarded as the source of a local field which adds coherently to the input wave to produce additional atomic polarization deeper in the amplifying medium. Our proposed scheme is based on the well known single-photon amplifier theory of Arecchi and Bonifacio[7]. In fact, similarities and differences between their results and ours will be repeatedly pointed out during our discussion.

There are, however, differences that are worth summarizing at the outset. For convenience we may distinguish between two different dynamical regimes. The first, called coherent propagation, occurs when the incident pulse is much shorter than the characteristic atomic relaxation times.

In the coherent propagation limit we find that an area equation, similar in some respects to the Arecchi-Bonifacio area equation, can be derived to predict the evolution of the pulse energy. Our area equation allows different steady state solutions for the pulse energy, unlike the Arecchi-Bonifacio equation which allows a unique steady state solution.

In addition, the threshold condition for power amplification imposes constraints on both the gain constant of the active medium and on the incident pulse energy. More precisely, we find that the two-photon amplifier has no small signal gain so that, even if the gain constant is large enough to allow power amplification, a weak incident pulse will not be amplified. On the contrary, if the incident pulse energy exceeds a certain threshold value, power amplification will occur with the subsequent formation of single or multiple-pulses in an analytically predictable fashion. Unlike the case of the Arecchi-Bonifacio amplifier, no envelope steady state is possible for a two-photon amplifier because there is no power saturation mechanism in the model to limit the growth of the propagating pulse above threshold.

The rate equation regime offers additional interesting variations relative to the corresponding behavior of a single-photon amplifier. In the case of the Arecchi-Bonifacio theory, it is known that peak amplification occurs when the carrier frequency of the incident pulse coincides with the transition frequency of the two levels of interest. In our case we find that the frequency response of the system is not symmetric with respect to the detuning parameter $2\omega - \omega_{ba}$. In fact, for fixed values of the parameters characterizing the active medium, we find that power amplification is enhanced for positive values of the detuning parameter and depressed for negative values of $2\omega - \omega_{ba}$. This will be easily explained on the basis of the non-linear interaction mechanism proposed for the system.

Our discussion develops along the following main lines. In Section II we describe the basic features of the model, and derive the coupled propagation equations that govern the evolution of the light pulse. In Section III we discuss the atomic relaxation mechanism, and compare the limiting coherent and rate equation regimes. Section IV contains the results of a hybrid computer simulation designed to describe the propagation process for arbitrary values of the pulse duration and of the transverse atomic relaxation time. In Section V we discuss the details of the rate equation regime and the behavior of the amplification process for different values of the detuning parameter.

II. DESCRIPTION OF THE MODEL AND EQUATIONS OF MOTION

Our formulation of the problem closely parallels the self-consistent field approximation theory of a single-photon laser amplifier advanced by Arecchi and Bonifacio[7]. In the same spirit, we visualize the amplification process as resulting from the following steps: an incident quasi-monochromatic light pulse of frequency ω impinges on the medium, it stimulates a macroscopic polarization at the same frequency, and interferes with the re-radiated field from the macroscopic polarization. The new total field repeats the cycle as it propagates through the medium. The system will behave as an amplifier, or as an absorber, depending on the relative phase between the incident and the re-radiated light.

Our system is assumed to be initially prepared in a state of inversion between two levels of the same parity. From a practical point of view, in order to maintain the inversion for a reasonable length of time, it is important that no dipole-allowed intermediate levels exist between the levels of interest (Fig. 1). It will be assumed here that the laser levels are the ground and first excited states of the active atoms. The levels are separated by an energy difference ω_{ba} and are coupled by dipole-allowed transitions to higher lying energy levels.

The incident electric field is assumed to be a quasi-monochromatic plane wave

$$\vec{\varepsilon}(x,t) = \vec{\varepsilon}_o(x,t) \cos(\omega t - kx + \phi(x,t)) \quad , \tag{2.1}$$

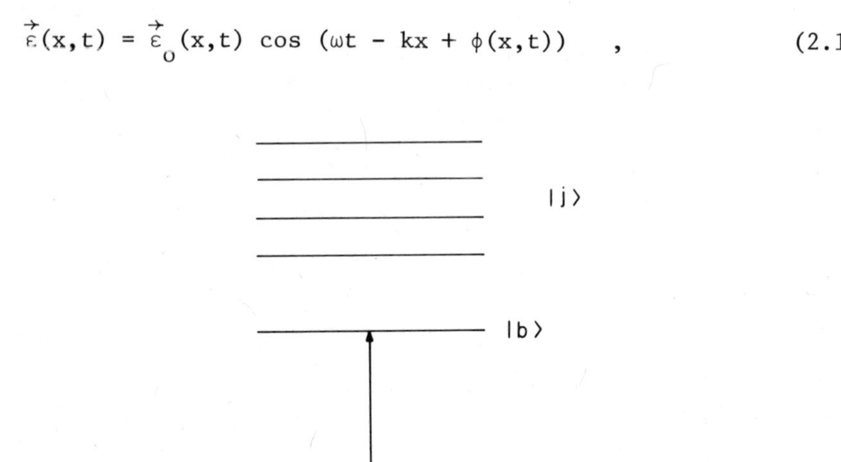

Figure 1. Schematic energy level diagram for an active atom. The energy separation between the states $|a\rangle$ and $|b\rangle$ is approximately twice the energy of an incident photon. The symbol $|j\rangle$ collectively represents all the intermediate states.

where $\vec{\varepsilon}_o(x,t)$ and $\phi(x,t)$ are slowly varying functions of space and time, ω is the carrier frequency, and k the propagation vector directed along the x axis. We choose a frequency ω such that $2\omega \approx \omega_{ba}$. The condition $2\omega = \omega_{ba}$ will be called two-photon resonance.

Let

$$|\psi(t)\rangle = \sum_j C_j(t)e^{-i\omega_j t}|j\rangle + C_a(t)e^{-i\omega_a t}|a\rangle + C_b(t)e^{-i\omega_b t}|b\rangle \tag{2.2}$$

be the arbitrary state vector of an active atom, with $C_j(t)$, $C_a(t)$ and $C_b(t)$ slowly varying amplitudes. The evolution of the state vector (2.2) is induced by the Hamiltonian

$$H = E_a|a\rangle\langle a| + E_b|b\rangle\langle b| + \sum_j E_j|j\rangle\langle j| - \vec{p}\cdot\vec{\varepsilon}(x,t) \quad , \tag{2.3}$$

where the atomic polarization operator $\vec{p}$ is assumed to take the form

$$\vec{p} = \sum_j |a\rangle\langle j|\vec{p}_{aj} + \sum_j |b\rangle\langle j|\vec{p}_{bj} + \text{hermitian adjoint.} \tag{2.4}$$

In Eq. (2.4), we have neglected terms involving dipole matrix elements of the type $\vec{p}_{jj'}$, on the ground that the intermediate levels are never populated for a significant length of time, and interference effects between intermediate state amplitudes are negligible. The Schrödinger equation for the state vector (2.2) with the Hamiltonian (2.3) reduces to the set of coupled linear equations

$$i\hbar\dot{C}_a(t) = -\sum_j \mu_{aj}\varepsilon(x,t)C_j(t)e^{-i(\omega_j-\omega_a)t} \quad , \tag{2.5a}$$

$$i\hbar\dot{C}_b(t) = -\sum_j \mu_{bj}\varepsilon(x,t)C_j(t)e^{-i(\omega_j-\omega_b)t} \quad , \tag{2.5b}$$

$$i\hbar\dot{C}_j(t) = -\mu_{ja}\varepsilon(x,t)C_a(t)e^{-i(\omega_j-\omega_a)t}$$

$$- \mu_{jb}\varepsilon(x,t)C_b(t)e^{i(\omega_j-\omega_b)t} \quad , \tag{2.5c}$$

where μ_{aj}, μ_{bj} are the projections of $\vec{p}_{aj}$ and $\vec{p}_{bj}$ along the direction of polarization of the field. Since we intend to focus our attention on the evolution of the amplitudes C_a and C_b, we eliminate the intermediate amplitudes by formally integrating Eq. (2.5c) and substituting the result into Eqs. (2.5a) and (2.5b). The new exact form of the equation of motion for C_a is

$$\dot{C}_a(t) = - \sum_j \frac{\mu_{aj}}{\hbar} \varepsilon(x,t)\, e^{-i\omega_{ja}t} \int_o^t dt'$$

$$\left(\frac{\mu_{ja}}{\hbar} \varepsilon(x,t')C_a(t')e^{i\omega_{ja}t'} + \frac{\mu_{jb}}{\hbar} \varepsilon(x,t')C_b(t')e^{i\omega_{jb}t'} \right). \tag{2.6}$$

A similar equation holds for the amplitude C_b.

At this point we make the slowly varying envelope approximation. This amounts to replacing $\varepsilon_o(x,t')$, $C_a(t')$, and $C_b(t')$ inside the integrals with their values at the upper limit of integration, and carrying out the exact integration of the remaining exponential factors.

After retaining the slowly varying terms, we arrive at the following equations of motion for the ground and excited state amplitudes C_a and C_b

$$\dot{C}_a = \frac{i}{\hbar} (k_{aa}|E_o|^2 C_a(t) + k_{ab}E_o^2 C_b e^{i(2\omega-\omega_{ba})t}) \quad , \tag{2.7}$$

$$\dot{C}_b = \frac{i}{\hbar} (k_{ab}E_o^{*2} C_a e^{-i(2\omega-\omega_{ba})t} + k_{bb}|E_o|^2 C_b) \quad . \tag{2.8}$$

In Eqs. (2.7) and (2.8) we have introduced the new field amplitude E_o defined by

$$\varepsilon(x,t) = \varepsilon_o(x,t) \cos (\omega t - kx + \phi)$$

$$\equiv E_o(x,t)\, e^{i\omega t} + E_o^*(x,t)\, e^{-i\omega t} \quad , \tag{2.9}$$

and the atomic parameters

$$k_{aa} = \frac{2}{\hbar} \sum_j \mu_{ja}^2 \frac{\omega_{ja}}{\omega_{ja}^2 - \omega^2} \quad ,$$

$$k_{bb} = \frac{2}{\hbar} \sum_j \mu_{jb}^2 \frac{\omega_{jb}}{\omega_{jb}^2 - \omega^2} \quad ,$$

$$k_{ab} = \frac{i}{\hbar} \sum_j \frac{\mu_{ja}\,\mu_{jb}}{\omega_{ja} + \omega} \quad . \tag{2.10}$$

It is worth pointing out that in the identification of the slowly varying terms leading to Eqs. (2.7) and (2.8), we have excluded the possibility of accidental resonances of the type $|\omega_{ja}| \approx \omega$, and $|\omega_{jb}| \approx \omega$.

As a check of consistency on the adiabatic elimination of the intermediate amplitudes, we observe that the probability conservation statement

$$\frac{d}{dt} (|c_a|^2 + |c_b|^2) = 0 \tag{2.11}$$

follows directly from the equations of motion (2.7) and (2.8). At this point we are dealing, in effect, with a fictitious two-level system. The presence of the intermediate states is reflected indirectly in the quadratic field dependence exhibited in Eqs. (2.7) and (2.8), a feature which is in sharp contrast with the results of the one-photon amplifier theory.

Guided by past experience with one-photon processes, one can cast Eqs. (2.7) and (2.8) in the form of a Bloch set of equations[8]. The identification of the appropriate Bloch variables is aided by the calculation of the total polarization source.

By definition, if N is the number of atoms per unit volume, the atomic polarization is given by

$$P = N\langle p\rangle = N\langle\psi(t)|p|\psi(t)\rangle \quad , \tag{2.12}$$

where $|\psi(t)\rangle$ is the state vector given by Eq. (2.2). Upon elimination of the intermediate amplitudes $C_j(t)$, and after performing the slowly varying amplitude approximation as done in the derivation of Eqs. (2.7) and (2.8), the total polarization takes the form

$$P = N\left\{k_{aa}|C_a|^2 + k_{bb}|C_b|^2 + k_{ab}(C_aC_b^*e^{-i\alpha} + C_a^*C_be^{i\alpha})\right\}$$
$$\times\ \varepsilon_o(x,t)\cos(\omega t - kx + \phi)$$
$$+ N\,k_{ab}\,i(C_aC_b^*e^{-i\alpha} - C_a^*C_be^{i\alpha})\,\varepsilon_o\sin(\omega t - kx + \phi) \quad , \tag{2.13}$$

where

$$\alpha = (2\omega - \omega_{ba})t - 2kx + 2\phi \quad .$$

In close analogy with the one-photon amplifier theory, the atomic polarization contains one component which is in phase and one in quadrature with the driving electromagnetic field. In this case, however, the polarization components are explicitly proportional to the electric field envelope. Furthermore, the in-phase component, which is responsible for dispersion effects, depends also on the atomic population through $|C_a|^2$ and $|C_b|^2$.

The polarization (2.13) plays the role of the source term in Maxwell's wave equation

$$\frac{\partial^2\varepsilon}{\partial x^2} - \frac{1}{c^2}\frac{\partial^2\varepsilon}{\partial t^2} = \frac{1}{c^2\varepsilon_o}\frac{\partial^2 P}{\partial t^2} \quad . \tag{2.14}$$

Upon setting

$$P = P_s(x,t)\sin(\omega t - kx + \phi) + P_c(x,t)\cos(\omega t - kx + \phi) \quad , \tag{2.15}$$

where P_c and P_s are the slowly varying in phase and in quadrature components of P(x,t), Eq. (2.14) reduces to the pair of transport equations

$$\varepsilon_o\left(c\frac{\partial\phi}{\partial x}+\frac{\partial\phi}{\partial t}\right)=-\frac{\omega}{2\kappa_o}P_c \quad , \quad \kappa_o = \text{dielectric constant} \quad , \tag{2.16}$$

$$c\frac{\partial\varepsilon_o}{\partial x}+\frac{\partial\varepsilon_o}{\partial t}=-\frac{\omega}{2\kappa_o}P_s \quad . \tag{2.17}$$

As usual, the field equations (2.16) and (2.17) are valid in the slowly varying amplitude and phase approximation.

The formal analogy between Eqs. (2.7) and (2.8) and those describing single-photon transitions has been recognized by previous workers[3,4]. Such analogy is made especially transparent by introducing the new atomic variables

$$R_1 = i(C_a^*C_b e^{i\alpha} - C_aC_b^* e^{-i\alpha}) \quad ,$$

$$R_2 = -(C_aC_b^* e^{-i\alpha} + C_a^*C_b e^{i\alpha}) \quad ,$$

$$R_3 = |C_b|^2 - |C_a|^2 \quad . \tag{2.18}$$

In terms of these variables the atomic equations (2.7) and (2.8) take the form

$$\dot{R}_1 = \left[\frac{k_{bb}-k_{aa}}{4\hbar}\varepsilon_o^2 + \left(2\omega - \omega_{ba} + 2\frac{\partial\phi}{\partial t}\right)\right]R_2 + \frac{k_{ab}}{2\hbar}\varepsilon_o^2 R_3 \quad ,$$

$$\dot{R}_2 = -\left[\frac{k_{bb}-k_{aa}}{4\hbar}\varepsilon_o^2 + \left(2\omega - \omega_{ba} + 2\frac{\partial\phi}{\partial t}\right)\right]R_1 \quad ,$$

$$\dot{R}_3 = -\frac{k_{ab}}{2\hbar}\varepsilon_o^2 R_1 \quad . \tag{2.19}$$

Equations (2.19) are formally identical to the Euler equation for a precessing top

$$\frac{d}{dt}\vec{R} = \vec{\Lambda}\times\vec{R} \tag{2.20}$$

with $\vec{R} \equiv (R_1, R_2, R_3)$ and

$$\Lambda_1 = 0 \quad ,$$

$$\Lambda_2 = \frac{k_{ab}}{2\hbar} \varepsilon_o^2 \quad ,$$

$$\Lambda_3 = -\left[\frac{k_{bb} - k_{aa}}{4\hbar} \varepsilon_o^2 + \left(2\omega - \omega_{ba} + 2 \frac{\partial \phi}{\partial t}\right)\right] \quad . \tag{2.21}$$

Clearly, the length of the Bloch vector is conserved. The atomic polarization (2.13) and the field equations (2.16) and (2.17) can also be expressed in terms of the new atomic variables as follows:

$$P = -N\, k_{ab}\, \varepsilon_o \left[R_2 - \frac{k_{bb} - k_{aa}}{2\, k_{ab}} \left(R_3 + \frac{k_{aa} + k_{bb}}{k_{bb} - k_{aa}}\right)\right] \cos\, (\omega t - kx + \phi)$$

$$- N\, k_{ab}\, \varepsilon_o\, R_1 \sin\, (\omega t - kx + \phi) \quad , \tag{2.22}$$

and

$$\left(c \frac{\partial}{\partial x} + \frac{\partial}{\partial t}\right)\left(2\omega - \omega_{ba} + 2 \frac{\partial \phi}{\partial t}\right) = \frac{\omega N k_{ab}}{\kappa_o} \left(\dot{R}_2 - \frac{k_{bb} - k_{aa}}{2\, k_{ab}} \dot{R}_3\right) \tag{2.23a}$$

$$\left(c \frac{\partial}{\partial x} + \frac{\partial}{\partial t}\right) \varepsilon_o^2 = \frac{\omega N k_{ab}}{\kappa_o} R_1\, \varepsilon_o^2 \quad . \tag{2.23b}$$

Finally, after a few minor changes of notation, we arrive at the entire set of coupled propagation equations for the amplifier

$$\frac{\partial R_1}{\partial \tau} = \left(\frac{\gamma}{\sqrt{1 + \gamma^2}} \omega_R + \Omega\right) R_2 + \frac{\omega_R}{\sqrt{1 + \gamma^2}} R_3 \quad ,$$

$$\frac{\partial R_2}{\partial \tau} = -\left(\frac{\gamma}{\sqrt{1 + \gamma^2}} \omega_R + \Omega\right) R_1 \quad ,$$

$$\frac{\partial R_3}{\partial \tau} = -\frac{\omega_R'}{\sqrt{1 + \gamma^2}} R_1 \quad ,$$

$$\frac{\partial \omega_R}{\partial \eta} = g\, \omega_R\, R_1 - \ell \omega_R \quad ,$$

$$\frac{\partial \Omega}{\partial \eta} = g \left(\frac{\partial R_2}{\partial \tau} - \gamma \frac{\partial R_3}{\partial \tau} \right) \quad . \tag{2.24}$$

The atomic and field variables evolve in the local reference system

$$\eta = \frac{x}{c} \quad ,$$

$$\tau = t - \frac{x}{c} \quad , \tag{2.25}$$

which is traveling in the direction of propagation of the pulse with the speed of light in the inert background. The field Rabi frequency ω_R and the detuning parameter Ω are defined by

$$\omega_R = \sqrt{1 + \gamma^2}\, \frac{k_{ab}}{2\hbar}\, \varepsilon_o^2 \quad ,$$

$$\Omega = (2\omega - \omega_{ba}) + 2 \frac{\partial \phi}{\partial \tau} \quad . \tag{2.26}$$

The parameters γ and g are given by

$$\gamma = \frac{k_{bb} - k_{aa}}{2\, k_{ab}} \quad , \quad g = \frac{\omega N k_{ab}}{\kappa_o} \quad . \tag{2.27}$$

Finally, ℓ describes all other non-resonant loss mechanisms (scattering, diffraction losses, etc.) which the incoming pulse suffers through the propagation.

We observe that ω_R depends quadratically on the field amplitude rather than linearly. We have decided to name ω_R the Rabi frequency of our problem because it plays an analogous role to the usual Rabi frequency introduced in the discussion of one-photon processes[8].

Certain differences between the two-photon amplifier equations and those derived by Arecchi and Bonifacio for the one-photon amplifier can be discussed at once. The effective detuning parameter $\gamma/\sqrt{1+\gamma^2}\ \omega_R + \Omega$ is explicitly field-intensity dependent; the term proportional to ω_R plays the role of a dynamic Stark shift. The field equation for ω_R displays a field-intensity dependent gain coefficient. This indicates that the amplifier will not exhibit small signal gain, and that it will have an entirely different asymptotic (large η) behavior for the propagating pulse. Thus, while one of the main features of the Arecchi-Bonifacio amplifier theory was the prediction of a steady state pulse with an amplitude proportional to the instantaneous induced polarization, no such steady state pulse can be found for a two-photon amplifier. Finally, in the coherent regime (no atomic relaxation), the frequency detuning and the field intensity ω_R evolve, coupled to one another by a conservation law which is an exclusive feature of the two-photon model. This can be seen from the last two equations in (2.24). In fact, the transport equations

$$\frac{\partial \omega_R}{\partial \eta} = g\ \omega_R\ R_1 - \ell \omega_R \quad ,$$

$$\frac{\partial \Omega}{\partial \eta} = -g\ \Omega\ R_1 \tag{2.28}$$

can be combined to give

$$\frac{\partial}{\partial \eta}(\omega_R \Omega) = -\ell\ \omega_R\ \Omega \quad . \tag{2.29}$$

Equation (2.29) can be integrated at once, with the result

$$(\omega_R \Omega)_\eta = (\omega_R \Omega)_{\eta=0}\ e^{-\ell\eta} \quad . \tag{2.30}$$

Hence, for sufficiently large distances into the amplifier, the detuning parameter vanishes, and the instantaneous carrier frequency of the field becomes $1/2\ \omega_{ba}$. On the other hand, if the incident signal satisfies the condition $\omega = 1/2\ \omega_{ba}$, resonance will be maintained throughout the amplification process.

III. ATOMIC RELAXATION, COHERENT AND INCOHERENT PROPAGATION

Before introducing relaxation terms for the atomic variables, we observe that the Bloch variables R_1, R_2, and R_3 are not fundamental parameters of the problem. The structure of the atomic polarization (2.22) and of the field equations (2.23) suggests the following choice of atomic variables

$$S = \sqrt{1+\gamma^2}\,\frac{R_1}{R_3^e} \quad ,$$

$$C = \sqrt{1+\gamma^2}\left[\frac{R_2}{R_3^e} - \gamma\left(\frac{R_3}{R_3^e} - 1\right)\right] \quad ,$$

$$D = (1+\gamma^2)\,\frac{R_2}{R_3^e} - \gamma^2 \quad , \qquad (3.1)$$

where R_3^e is the equilibrium value of the population difference prior to the arrival of the leading edge of the propagating pulse.

In terms of the new variables the coupled Schrodinger-Maxwell equations take the form

$$\frac{\partial S}{\partial \tau} = (\Gamma\omega_R + \Omega)C + (\omega_R + \Gamma\Omega)D - \Gamma\Omega - \frac{S}{T_2} \quad ,$$

$$\frac{\partial C}{\partial \tau} = -\Omega S - \frac{C}{T_2} \quad ,$$

$$\frac{\partial D}{\partial \tau} = -\omega_R S - \frac{1}{T_1}(D-1) \quad ,$$

$$\frac{\partial \omega_R}{\partial \eta} = G\omega_R S - \ell\omega_R \quad ,$$

$$\frac{\partial \Omega}{\partial \eta} = -G\Omega S - \frac{G}{T_2}\,C \quad , \qquad (3.2)$$

where we have set

$$\Gamma = \frac{\gamma}{\sqrt{1+\gamma^2}} \quad ,$$

$$G = \frac{g\, R_3^e}{\sqrt{1+\gamma^2}} \quad , \tag{3.3}$$

and where we have introduced the phenomenological relaxation terms S/T_2, C/T_2, $(D-1)/T_1$ in the usual fashion. We observe that in resonance ($\Omega = 0$) the set of coupled equations reduces to

$$\frac{\partial S}{\partial \tau} = \omega_R D - \frac{S}{T_2} \quad ,$$

$$\frac{\partial D}{\partial \tau} = -\omega_R S - \frac{1}{T_1}(D-1)$$

$$\frac{\partial \omega_R}{\partial \eta} = G\omega_R S - \ell\omega_R \quad , \tag{3.4}$$

i.e., the in-phase component of the polarization C remains identically equal to zero for all time. In terms of these new variables, the formal analogy between the resonant set of equations for a two-photon amplifier and those derived by Arecchi and Bonifacio is even closer. In fact, if we neglect the relaxation terms, the formal solution of the atomic equations becomes

$$S = \sin\sigma \quad ,$$

$$D = \cos\sigma \quad , \tag{3.5}$$

where

$$\frac{\partial \sigma}{\partial \tau} = \omega_R \quad . \tag{3.6}$$

In the one-photon transition literature, σ (referred to as the pulse area) has played an important role in connection with the description of coherent transient phenomena. In the context of the two-photon transition theory an equivalent central role is played by the area under the pulse intensity. The resonant set of equations (3.4) can be reduced to a single area equation by using Eqs. (3.5) and (3.6). The result of the simple calculation is

$$\frac{\partial^2}{\partial\tau\,\partial\eta}\sigma = G\sin\sigma\frac{\partial\sigma}{\partial\tau} - \ell\frac{\partial\sigma}{\partial\tau} \quad . \tag{3.7}$$

It is convenient to follow the evolution of the total integrated pulse

$$\Sigma(\eta) = \lim_{\tau\to\infty}\sigma(\eta,\tau) = \int_0^\infty \omega_R\, d\tau \quad . \tag{3.8}$$

Upon integrating Eq. (3.7) with respect to τ and letting τ approach infinity, we find

$$\frac{\partial\Sigma}{\partial\eta} = -\ell\Sigma(\eta) + G(1 - \cos\Sigma(\eta)) \quad . \tag{3.9}$$

It is simple to derive precise qualitative and quantitative predictions on the behavior of the total integrated pulse area on the basis of the area equation (3.9).

Consider first the steady state solutions of Eq. (3.9). They must satisfy the transcendental equation

$$1 - \cos\Sigma(\eta) = \frac{\ell}{G}\Sigma(\eta) \quad . \tag{3.10}$$

As shown in Fig. 2, non-trivial solutions to Eq. (3.10) will exist if the ratio ℓ/G is sufficiently small, i.e., if the gain parameter is sufficiently larger than the linear loss coefficient. Multiple roots are also apparent for small values of ℓ/G. Fig. 2 shows the possible intersects of the straight line $\ell/G\ \Sigma$ for different values of the gain to loss ratio G/ℓ. The three cases shown in the figure correspond to $G/\ell = 1$, 2 and 5 respectively. A simple stability argument shows that beginning with the trivial stable solution $\Sigma = 0$, the stable and unstable solutions alternate with one another. The actual value of the asymptotic solution $\Sigma(\infty)$ is determined, by the gain to loss ratio and by the initial value of Σ. Thus, for example, a pulse with an initial area $\Sigma(0) = 3$ propagating in a medium characterized by a gain to loss ratio of 5 will continue to grow until its area reaches the stable asymptotic value 4.76, while a pulse with an input area $\Sigma(0) = 6$ will experience power amplification (as shown by the computer solutions of the full set of equations) with a reduction of the pulse area from the initial value 6 to the asymptotic value 4.76.

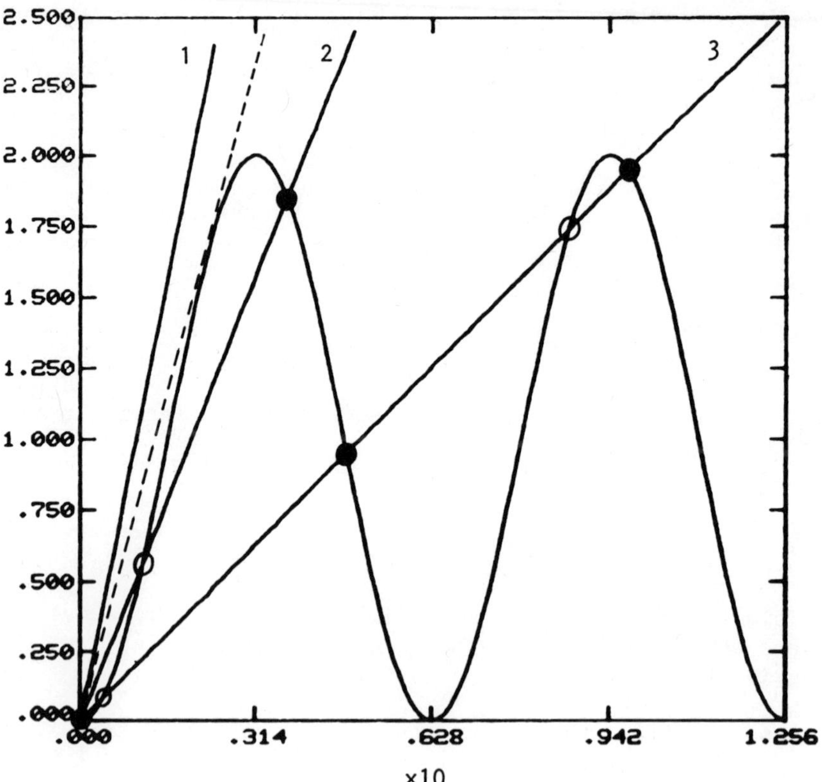

Figure 2. The asymptotic steady state solutions of the area equation (3.9) correspond to the intercepts of the straight lines $\ell/G\ \Sigma(\eta)$ with the curve $1 - \cos \Sigma(\eta)$. The straight lines 1, 2, and 3 have slopes equal to 1, 1/2 and 1/5 respectively. The stable solutions are marked with solid circles. The unstable solutions are marked with open circles. The critical slope (dashed line) is 0.7246.

The general conclusions that can be reached from the analysis of the area equation (3.9) and of its asymptotic solutions are summarized as follows:

i) Two requirements must be met for the propagation of a pulse with an asymptotic area $\Sigma(\infty) \neq 0$. First the gain to loss ratio must be larger than a threshold value (the inverse of the slope of the dashed line in Fig. 2, $G/\ell \geq 1.38$). Secondly, the incident pulse energy ($\Sigma(0)$) must be larger than the first unstable root corresponding to the given choice of G/ℓ. If both conditions are met, the total pulse energy will converge to a stable non-zero value.

ii) The output value of the pulse energy can be larger or smaller than the input value $\Sigma(0)$. In both cases, the solution of the coherent resonant equations shows that power amplification will occur if the conditions (i) are met.

iii) There is no small signal gain.

iv) It is anticipated, and confirmed by the computer simulation, that multiple pulses can be propagated. In fact, we find that if a pulse evolves to an area Σ given by the n-th stable root, the pulse will split into (n-1) distinct pulses.

v) It is also anticipated that no steady state pulse envelope will be found. This is confirmed by the computer simulation, where it is seen that when the threshold conditions are satisfied, the peak power continues to grow, while the pulse duration becomes smaller and smaller as the pulse area approaches its stable asymptotic value.

We consider now the extreme limiting case of very short atomic relaxation times (rate equation limit). From Eq. (3.2) we can see that, if $\partial C/\partial\tau \approx 0$, then

$$\frac{\partial\Omega}{\partial\eta} = 0 \tag{3.11}$$

as well. Hence, in the incoherent propagation limit the detuning parameter becomes independent of η. This is in sharp contrast with the result of Eq. (2.30), which characterizes the coherent propagation limit. In the rate equation limit, the steady state values of the atomic variables are given by

$$S = \frac{T_2\,\omega_R}{1 + T_1T_2\,\omega_R(\omega_R + \Gamma\Omega) + T_2^2\,\Omega(\Gamma\omega_R + \Omega)},$$

$$C = -\frac{T_2^2\,\Omega\,\omega_R}{1 + T_1T_2\,\omega_R(\omega_R + \Gamma\Omega) + T_2^2\,\Omega(\Gamma\omega_R + \Omega)},$$

$$D = \frac{1 + T_2^2\,\Omega(\Gamma\omega_R + \Omega) + \Gamma\Omega\,T_1T_2\,\omega_R}{1 + (\Gamma\omega_R + \Omega)\,T_2^2\,\Omega + (\omega_R + \Gamma\Omega)\,T_1T_2\,\omega_R}. \tag{3.12}$$

The evolution of the field variable ω_R is described by the non-linear transport equation

$$\frac{\partial \omega_R}{\partial \eta} = G \frac{T_2\, \omega_R^2}{1 + T_1 T_2\, \omega_R(\omega_R + \Gamma\Omega) + T_2^2\, \Omega(\Gamma\omega_R + \Omega)} - \ell\omega_R \quad . \tag{3.13}$$

It is easy to see that, even in the rate equation limit, small signal gain is not possible for a two-photon amplifier. Each segment of the propagating pulse can be viewed as propagating independently of the others according to Eq. (3.13). If $\omega_R(\eta = 0,\tau) = \omega_R^o(\tau)$ represents the input field intensity at a particular position from the pulse leading edge, it is easy to see that amplification will occur if

$$\frac{G}{\ell} \frac{T_2\, \omega_R}{1 + T_1 T_2\, \omega_R^o(\omega_R^o + \Gamma\Omega) + T_2^2\, \Omega(\Gamma\omega_R^o + \Omega)} > 1 \quad . \tag{3.14}$$

In resonance ($\Omega = 0$) it is clear that the condition $G/\ell > 1$ is not sufficient for amplification. Indeed, one must require

$$\frac{G}{\ell} \frac{T_2\, \omega_R^o}{1 + T_1 T_2 (\omega_R^o)^2} > 1 \quad , \tag{3.15}$$

which may not be satisfied for ω_R^o sufficiently small. Surprisingly enough, a large signal is also not amplified unless the condition

$$\frac{G}{\ell} \frac{1}{T_1\, \omega_R^o} > 1 \tag{3.16}$$

is satisfied. In the rate equation limit and for resonant propagation, our qualitative analysis points to the following general conclusions:

i) The leading edge of the propagating pulse is always absorbed by the action of the linear loss mechanism.

ii) If the gain to loss ratio is sufficiently large, amplification will occur for those values of the pulse intensity that satisfy Eq. (3.15).

iii) As the pulse intensity $\omega_R(\eta,\tau)$ increases, a steady state condition is reached when

$$\frac{G}{\ell} \frac{T_2 \, \omega_R^{ss}}{1 + T_1 T_2 \, (\omega_R^{ss})^2} = 1 \quad . \tag{3.17}$$

Since the steady state value ω_R^{ss} is only a function of the gain to loss ratio and of the atomic relaxation times, it is predicted that pulse reshaping will take place with the pulse leading and trailing edges becoming sharper and the pulse-top flattening out. On the other hand, pulse splitting will be impossible.

iv) Out of resonance ($\Omega \neq 0$) it is predicted that the maximum pulse growth rate will occur at the intensity dependent value of the detuning

$$\Omega_{max} = - \frac{1}{2} \left(\frac{T_1}{T_2} + 1 \right) \Gamma \omega_R \quad . \tag{3.18}$$

Thus, unlike the case of the single-photon amplifier, the frequency dependence of the instantaneous gain is not symmetric about $\Omega = 0$. This lack of symmetry is also exhibited by the dispersion characteristics of the steady state values of the atomic variables given by Eqs. (3.12).

The details of our computer simulation of the coherent and incoherent regimes are discussed in Sections IV and V.

IV. COMPUTER SIMULATION, COHERENT AND INTERMEDIATE PROPAGATION REGIMES

The set of coupled equations (3.2) have been analyzed with a hybrid computer for a variety of choices of the parameters. We have limited our simulations to the resonant condition, and investigated the following main features of the propagation problem:

i) A verification of the analytic prediction that if the detuning parameter Ω is initially zero, it remains zero for every value of η.

ii) An analysis of the threshold conditions discussed in Section 3 and of the relation between the input pulse area and the propagation of single and multiple pulses in the coherent limit.

iii) A study of the transient pulse modulation occurring even when a single pulse is expected to propagate along the amplifier. In particular, we have focused on the effects of atomic relaxation on the envelope modulation.

iv) Power amplification and approach to equilibrium of the pulse energy. As already pointed out, the pulse peak power will increase above threshold with an increase or decrease of the total energy depending on the gain to loss ratio of the amplifier and on the incident pulse energy.

The physical system is assumed to be excited in a swept excitation mode corresponding to the initial and boundary conditions

$$S(\eta,\tau = 0) = C(\eta,\tau = 0) = 0 \quad ,$$

$$D(\eta,\tau = 0) = 1 \quad ,$$

$$\Omega(\eta = 0,\tau) = 0 \quad ,$$

$$\omega_R(\eta = 0,\tau) = \omega_R(\tau) \quad . \tag{4.1}$$

The initial population is inverted, i.e., the parameter G is positive, and the input pulse envelope is assumed to have the convenient analytic form

$$\omega_R(\tau) = \omega_R^o \sin^2\left(\pi \frac{\tau}{\tau_p}\right) \quad , \tag{4.2}$$

where τ_p denotes the duration of the pulse from the leading to the trailing edge. The longitudinal relaxation time T_1 is taken to be infinite in this simulation, while T_2 has been varied over two orders of magnitude to explore the entire range from coherent to incoherent propagation.

As a result of the choice of boundary conditions for the detuning parameter ($\Omega(\eta = 0,\tau) = 0$) it is expected that both Ω and the in-phase component of the polarization should remain identically equal to zero throughout the entire propagation. This prediction has been verified to excellent accuracy.

The results of our analysis can be grouped into two main classes depending on the choice of the input area $\Sigma(0)$ and of the gain to loss ratio (a summary of the input data is given in Table I). For fixed values of $\Sigma(0)$ and G/ℓ we have varied the ratio T_2/τ_p to simulate the different effects of atomic relaxation on the pulse propagation.

Figure Number	Input Area	G/ℓ	T_2/τ_p	T_1
3	3	2	12.5	∞
4	3	2	1.25	∞
5	3	2	0.38	∞
6	8	2	12.5	∞
7	9.8	5	∞	∞
8	9.8	5	1.25	∞
9	9.8	5	0.13	∞
10	9.8	5	0.038	∞
11	9.8	12	0.038	∞

Table I. Summary of the input data used for the hybrid computer simulation.

In Figs. 3 through 5 we show the evolution of an incident pulse of area $\Sigma(0) = 3$ propagating in a medium characterized by a gain to loss ratio equal to 2. Upon inspection of Fig. 2, we see that for $G/\ell = 2$ and for an input area larger than the threshold value, a single peak is expected to propagate in the coherent limit. This is shown clearly in Fig. 3. By contrast, we see that for the same gain to loss ratio, but with a larger input area ($\Sigma(0) = 8$), considerable envelope modulation is present in the coherent limit (Fig. 6). Since, however, in the coherent limit the asymptotic value of the pulse area $\Sigma(\infty)$ is controlled only by the ratio G/ℓ, the asymptotic values of Σ for the cases displayed in Fig. 3 and 6 have been verified to be the same.

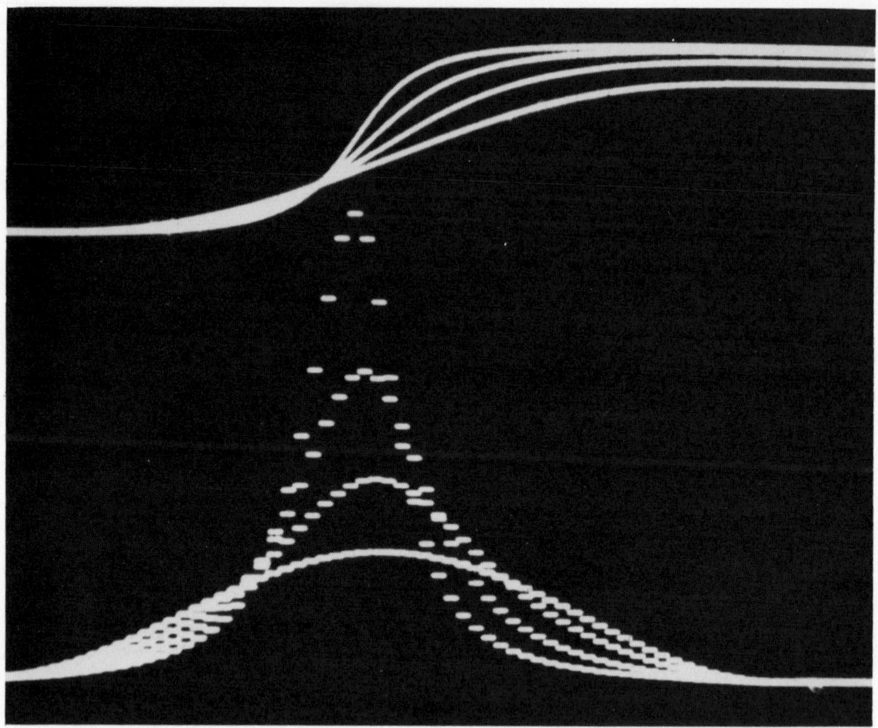

Figure 3. Computer simulation illustrating the evolution of the pulse intensity through the amplifying medium. The different dashed curves represent the intensity envelope in different sections of the amplifier. The solid curves show the behavior of the corresponding integrated areas $\sigma(\eta,\tau)$. The values of σ at the far right give the total integrated area $\Sigma(\eta) = \lim_{\tau\to\infty} \sigma$. The horizontal axis is the local time axis with $\tau = 0$ (leading edge of the pulse) at the far left. The input area is $\Sigma(0) = 3$ and the gain to loss ratio G/ℓ is 2. The value of T_2/τ_p is 12.5.

As the ratio T_2/τ_p is made smaller (Fig. 4) a considerable sharpening of the pulse is observed corresponding to a larger fraction of the incident pulse being below threshold for amplification. For even smaller values of T_2 there is not enough gain in the system to support amplification and the entire pulse is dissipated out by the linear loss mechanism (Fig. 5).

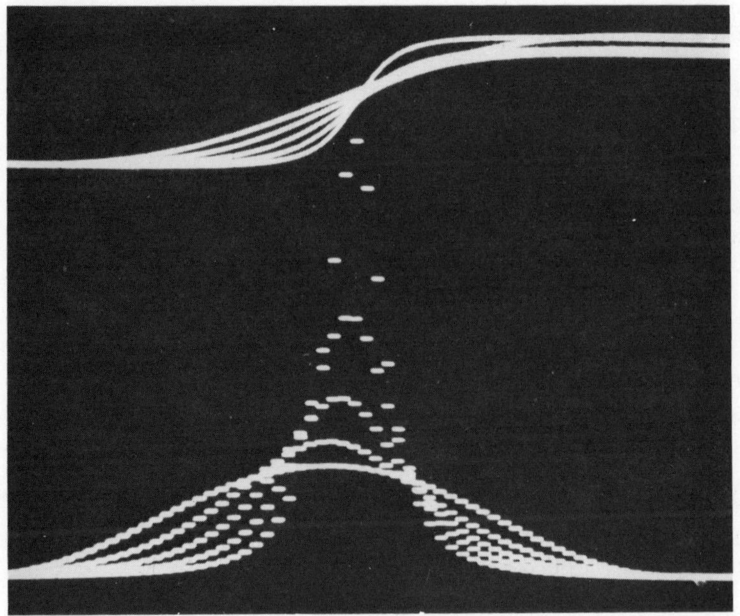

Figure 4. Same as Fig. 3 with $T_2/\tau_p = 1.25$.

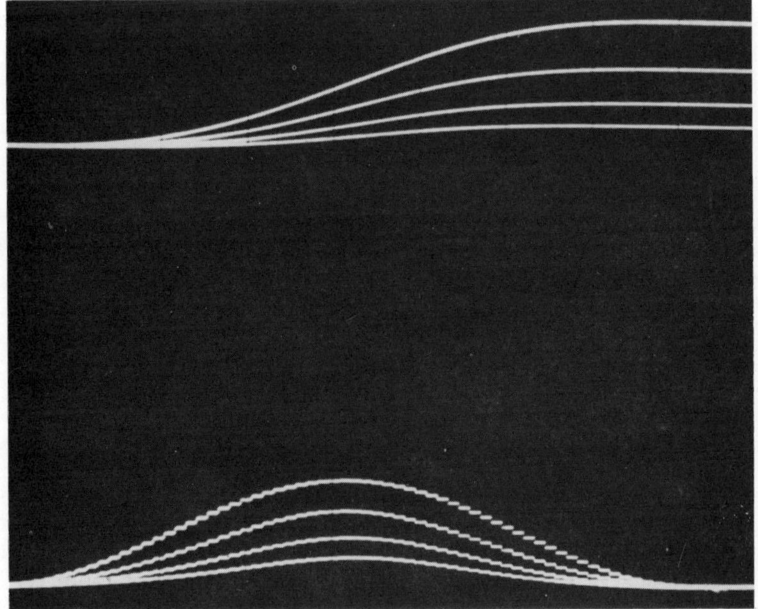

Figure 5. Same as Fig. 3 with $T_2/\tau_p = 0.38$.

For a larger value of the gain to loss ratio, multiple steady state solutions for $\Sigma(\infty)$ become possible. The computer simulation indicates that in the coherent regime the existence of (n+1) stable roots of the transcendental equation (3.10) (including the trivial solution $\Sigma(\infty) = 0$) implies the possibility of propagating pulses which exhibit up to n peaks. The general rule that has emerged out of our analysis is that the pulse whose asymptotic area $\Sigma(\infty)$ corresponds to the n-th stable root will show (n-1) peaks. Thus, for a gain to loss ratio of 5 and for an input area $\Sigma(0) = 9.8$, stable double peak propagation is expected in the coherent limit. This is shown in Fig. 7 where the ratio T_2/τ_p is taken to be infinite.

For decreasing values of T_2 the envelope modulation becomes less and less pronounced until the second peak disappears altogether. This is shown in Figs. 8 and 9 where the values of T_2/τ_p are 1.25 and 0.13, respectively. Upon decreasing T_2 even further, the amplifier can no longer support power amplification and the entire pulse dies off as shown in Fig. 10 ($T_2/\tau_p = 0.038$). The rate equation (3.13) indicates that if the gain G is made sufficiently large to compensate for the decrease in the polarization relaxation time, power amplification becomes possible once again. This is confirmed in Fig. 11 where T_2/τ_p is kept equal to 0.038 (as in Fig. 10), but the gain to loss ratio is increased to 12. In this case the amplified pulse is more symmetric and it evolves in much the same way as in the rate equation limit discussed in Section V.

V. COMPUTER SIMULATION, THE RATE EQUATION LIMIT

In the rate equation limit the evolution of the atomic variables is controlled by the steady state solutions (3.12) and by the transport equation

$$\frac{\partial \omega_R}{\partial \eta} = \frac{G\, T_2\, \omega_R^2}{1 + T_1 T_2\, \omega_R(\omega_R + \Gamma\Omega) + T_2^2\, \Omega(\Gamma\omega_R + \Omega)} - \ell\omega_R \tag{5.1}$$

for the field intensity $\omega_R(\eta,\tau)$. Equation (5.1) indicates that different portions of the pulse envelope evolve independently of one another. This eliminates the possibility of pulse envelope modulation and multiple peak formation which is typical of the coherent propagation regime. In addition, different sections of the input pulse with the same intensity will evolve identically. Thus, a symmetric pulse will evolve symmetrically for all values of η. It is simple to establish the threshold condition for

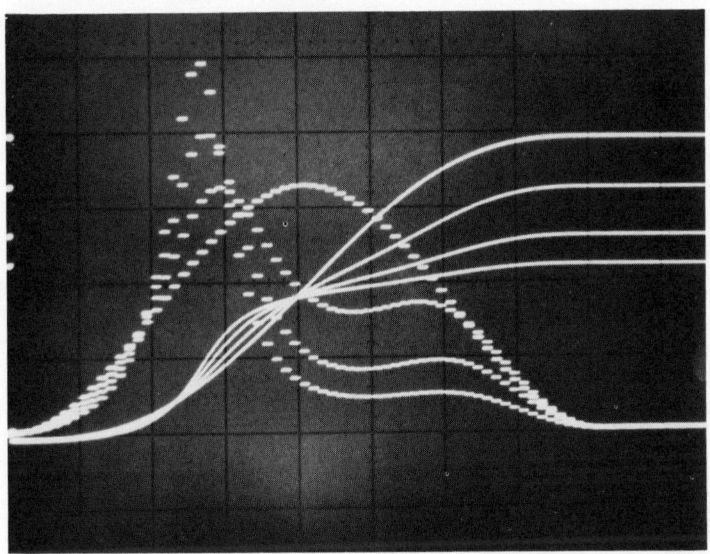

Figure 6. Evolution of the pulse intensity and pulse area corresponding to the initial value $\Sigma(0) = 8$, and to a gain to loss ratio of 5. The ratio T_2/τ_p is 12.5.

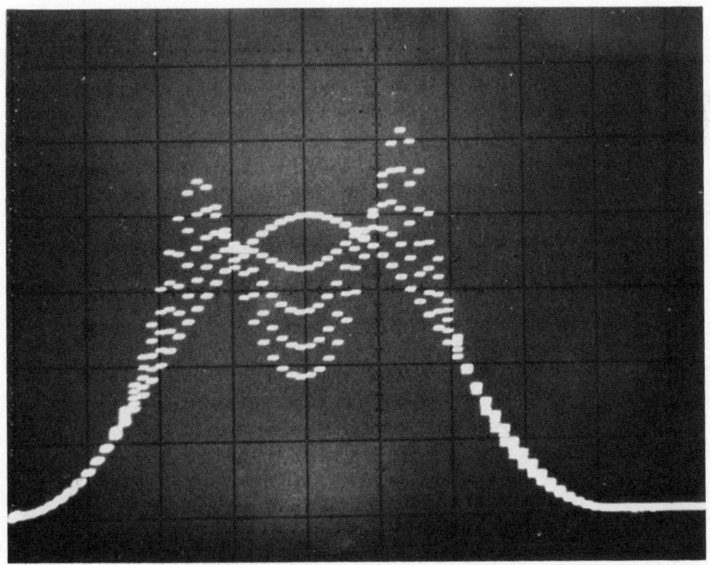

Figure 7. Pulse splitting and stable double-pulse propagation in the coherent limit. The input area is $\Sigma(0) = 9.8$ and the gain to loss ratio is 5. The ratio T_2/τ_p is 12.5.

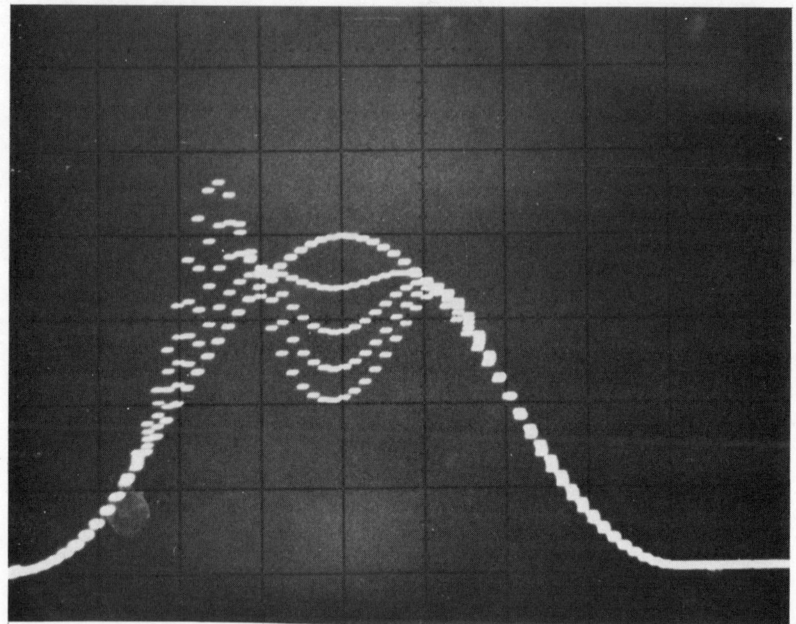

Figure 8. Same as Fig. 7 with $T_2/\tau_p = 1.25$.

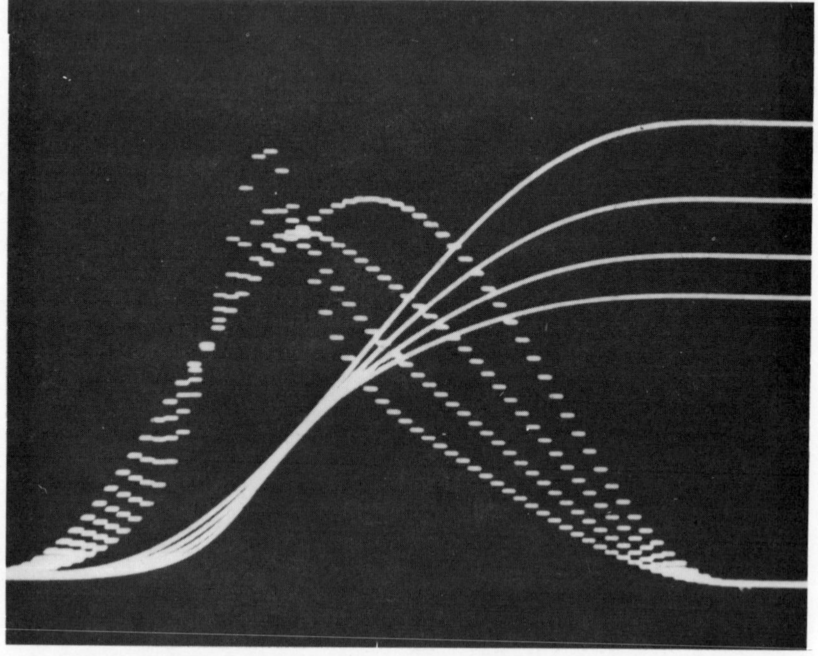

Figure 9. Same as Fig. 7 with $T_2/\tau_p = 0.13$.

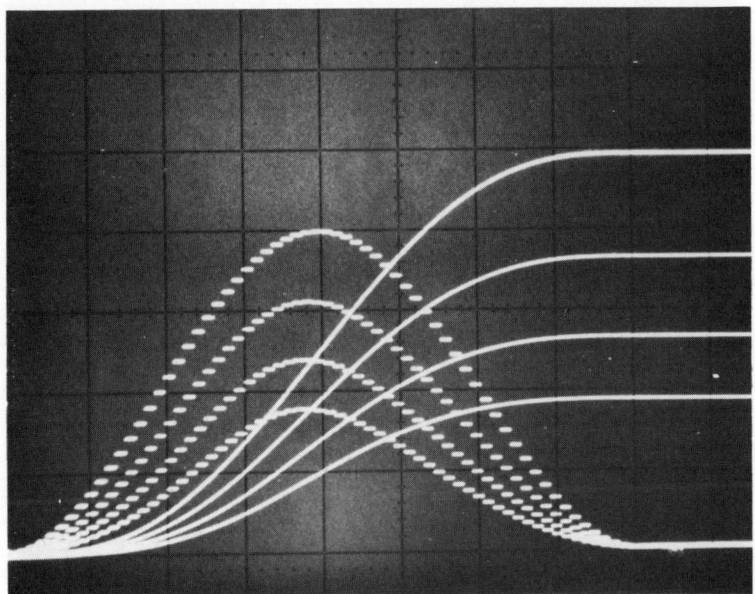

Figure 10. Same as Fig. 7 with $T_2/\tau_p = 0.038$.

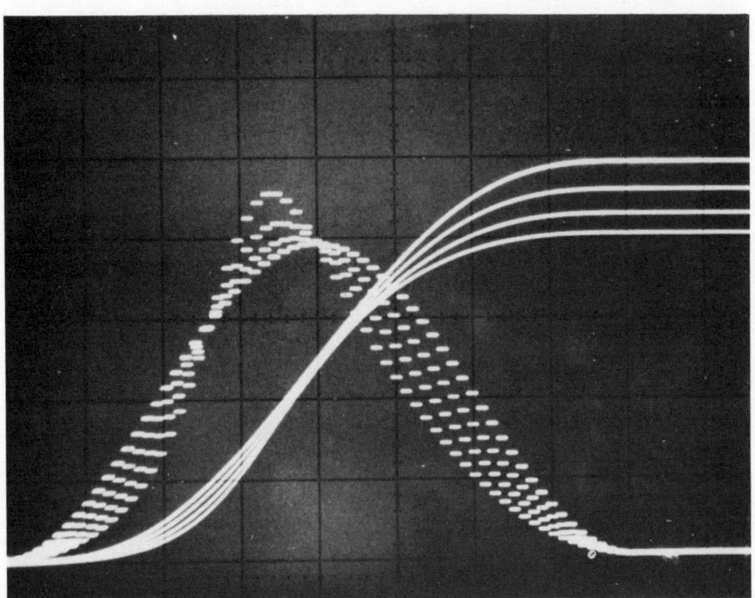

Figure 11. Same as Fig. 10 with a larger gain to loss ratio ($G/\ell = 12$). The amplifier is now above threshold.

amplification from Eq. (3.13). As expected, a sufficiently large value of the gain to loss ratio G/ℓ will not be enough for the pulse to undergo amplification; instead, one must require that the local instantaneous value of ω_R satisfy the inequality

$$\frac{G}{\ell}\,\frac{T_2\,\omega_R}{1 + T_1T_2\,\omega_R(\omega_R + \Gamma\Omega) + T_2^2\,\Omega(\Gamma\omega_R + \Omega)} > 1 \quad . \tag{5.2}$$

Thus, the leading and trailing edges of the pulse are always absorbed. Unlike the case of the coherent propagation limit, the pulse envelope does not sharpen up indefinitely, but rather it approaches a rectangular steady state shape, the pulse height being characterized by the asymptotic value of ω_R which satisfies the condition $\partial\omega_R/\partial\eta = 0$. This qualitative picture is only approximately true, and fails to account for the pulse evolution when the rise time of the leading edge and the decay time of the trailing edge become comparable to T_2.

The threshold intensity ω_R^{th} can be easily computed from Eq. (5.2). The result is

$$\omega_R^{th} = (G/\ell - (T_1/T_2 + 1)\,\Gamma\,T_2\,\Omega)/2T_1 - [(G/\ell - (T_1/T_2 + 1)\,\Gamma\,T_2\Omega)^2 - 4(1 + T_2^2\Omega^2)\,T_1/T_2]^{\frac{1}{2}}/2T_1 \quad . \tag{5.3}$$

The asymptotic steady state value is given by

$$\omega_R^{ss} = (G/\ell - (T_1/T_2 + 1)\,\Gamma\,T_2\,\Omega)/2T_1 + [(G/\ell - (T_1/T_2 + 1)\,\Gamma\,T_2\,\Omega)^2 - 4(1 + T_2^2\Omega^2)\,T_1/T_2]^{\frac{1}{2}}/2T_1 \quad . \tag{5.4}$$

From Eq. (5.3) we conclude that a necessary condition for pulse amplification is

$$\frac{G}{\ell} > \left(\frac{T_1}{T_2} + 1\right)\Gamma\,T_2\,\Omega + 2\sqrt{\frac{T_1}{T_2}\,(1 + T_2^2\,\Omega^2)} \quad . \tag{5.5}$$

A second necessary condition, of course, is the validity of Eq. (5.3) itself.

A detailed computer analysis of the transport equation (5.1) has confirmed the above threshold conditions. A few typical solutions are displayed in Figs. 12 through 17, with the input

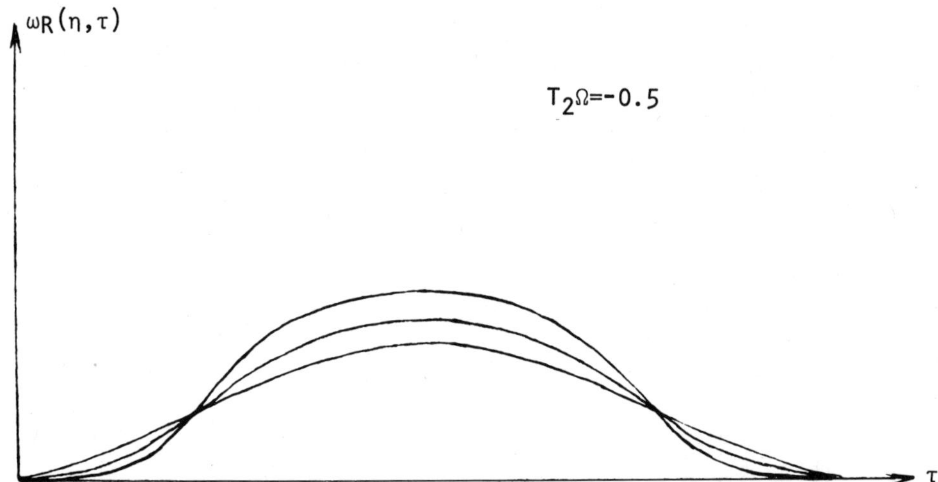

Figure 12. Computer simulation of the field transport equation (5.1) in the rate equation limit. The gain to loss ratio is equal to 5, the ratio T_1/T_2 equals 2, and the parameter Γ is taken to be -0.7 from an estimate based on spectroscopic data of Calcium atoms. The central portion of the input pulse is amplified. The leading and trailing edges are absorbed.

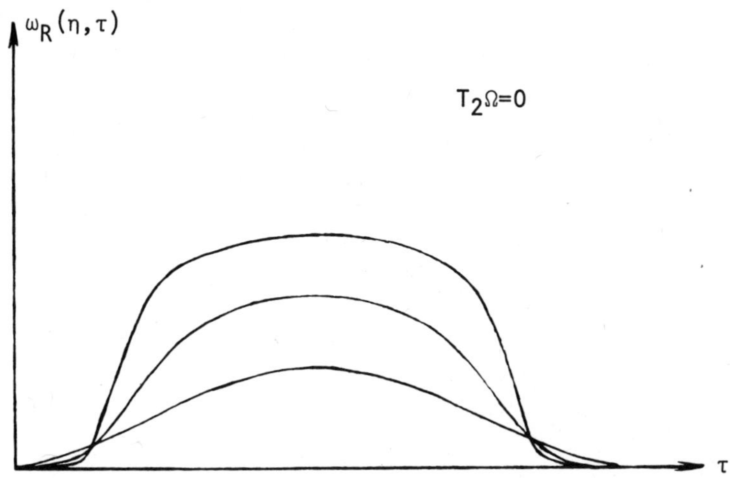

Figure 13. Same as Fig. 12 with the incident pulse in resonance with the active medium.

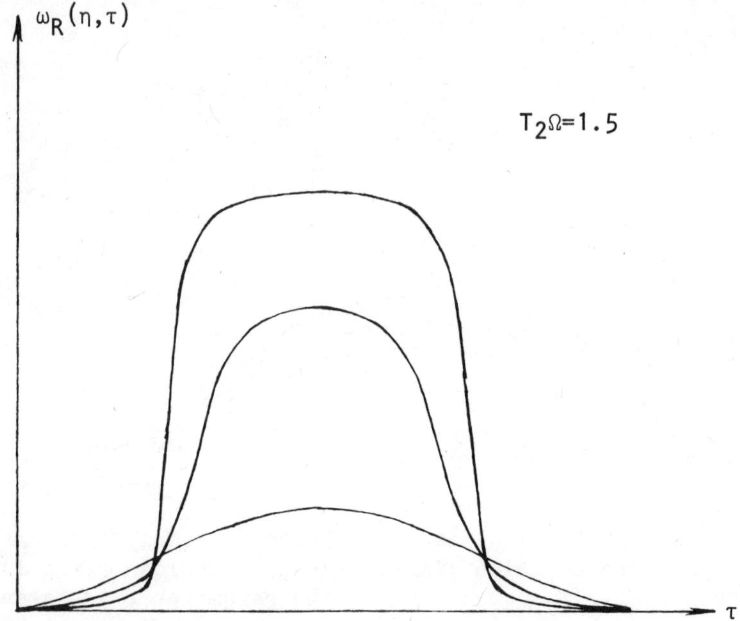

Figure 14. Same as Fig. 13 with $T_2\Omega$ = 1.5.

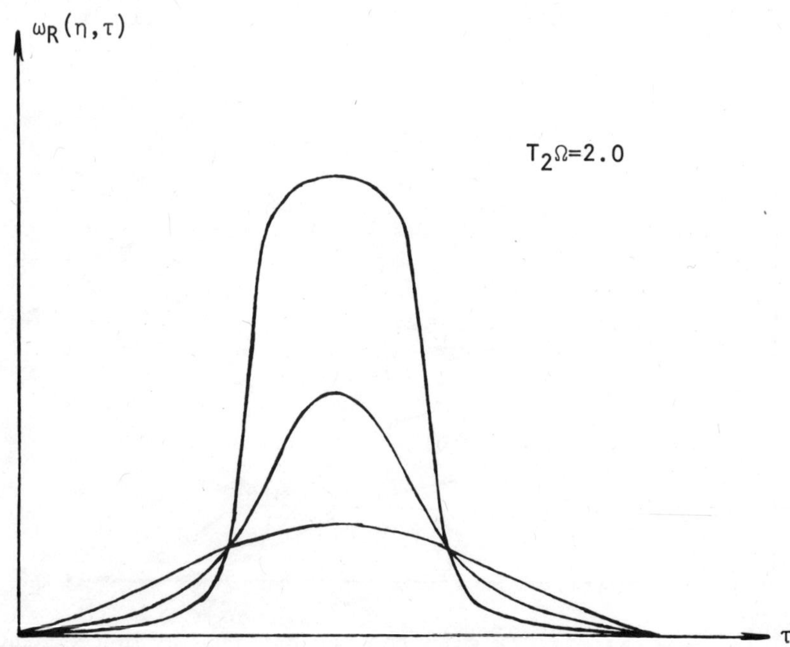

Figure 15. Same as Fig. 13 with $T_2\Omega$ = 2.0.

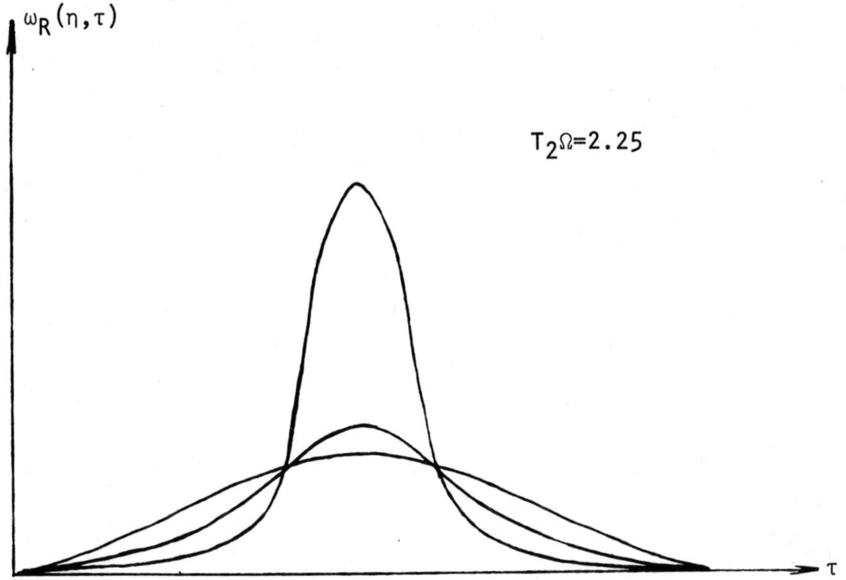

Figure 16. Same as Fig. 13 with $T_2\Omega$ = 2.25.

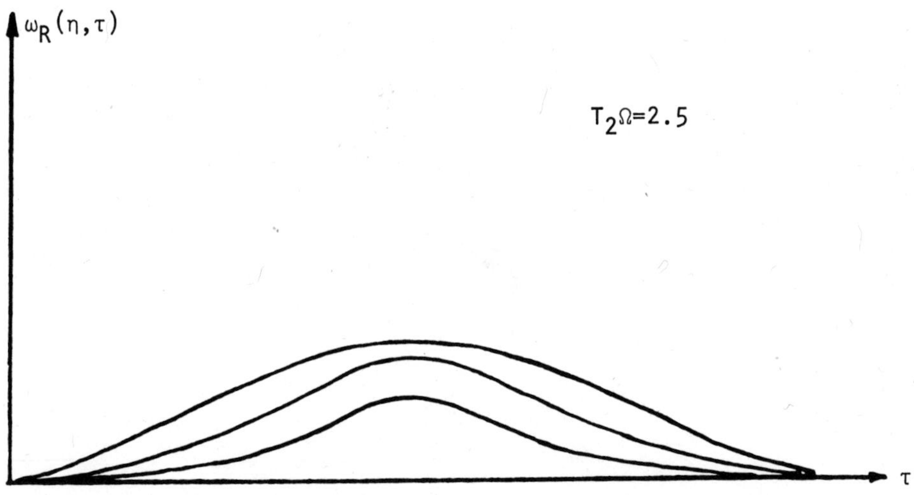

Figure 17. Same as Fig. 13 with T_2 = 2.5. The entire pulse is below threshold and no amplification occurs.

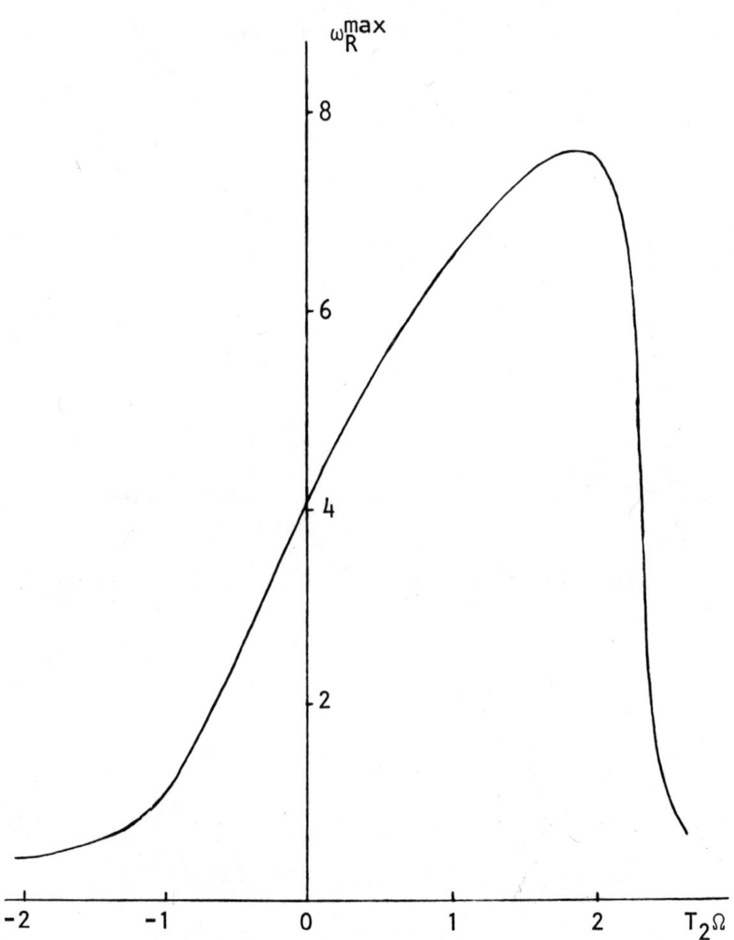

Figure 18. The maximum pulse intensity at the output of a fixed length of amplifier is plotted as a function of the detuning parameter $T_2\Omega$. The gain to loss ratio is 5. The ratio T_1/T_2 is 2 and the parameter Γ equals -0.7.

pulse plotted together with two amplified (or absorbed) pulses further down along the amplifier (a summary of the input data is given in Table II). The gain to loss ratio, the ratio T_1/T_2, and the parameter Γ have been kept fixed, while the detuning Ω has been varied on either side of the resonant value $\Omega = 0$.

Figure Number	T_2	Γ	T_1/T_2	G/ℓ
12	-0.5	-0.7	2	5
13	0			
14	1.5			
15	2.0			
16	2.25			
17	2.5			

Table II. Summary of the input data used for the computer simulation of the transport equation (5.1).

The most obvious feature of these solutions is the lack of symmetry in the amplification process for incident pulses that are removed from resonance by equal amounts in opposite directions. This is made explicit in Fig. 18 where the maximum pulse intensity from a fixed length of amplifying medium is plotted as a function of detuning. It is clear that the maximum amplification occurs for values of the detuning that are different from zero. In the present computer simulation, the parameter Γ is negative. If we had chosen a positive value for Γ, the maximum amplification would have occurred for negative values of the detuning Ω.

References

(1) A. M. Prokhorov, Science, 149, 828 (1965).

(2) P. P. Sorokin, N. Braslau, IBM J. Res. Dev. 8, 177 (1964).

(3) S. R. Hartmann, IEEE J. Quant. Electron. QE-4, 802 (1968); E. M. Belenov, I. A. Poluektov, Sov. Phys. JETP 29, 754 (1969); M. Takatsuji, Phys. Rev. B2, 340 (1970), Ibid A4, 808 (1971); M. Takatsuji, Physica 51, 265 (1971); D. Grischkowski, Phys. Rev. Lett. 24, 866 (1970); N. Tan-no, K. Yokoto, H. Inaba, Phys. Rev. Lett. 29, 1211 (1972).

(4) See for example D. Grischkowski, in Laser Applications to Optics and Spectroscopy, Vol. II of Physics of Quantum Electronics Series, Eds. S. F. Jacobs, M. Sargent III, J. F. Scott, M. O. Scully (Addison-Wesley 1975).

(5) L. E. Estes, L. M. Narducci, B. Shammas, Lettere al Nuovo Cimento, Serie 2, 1, 175 (1971).

(6) R. L. Carman, Phys. Rev. A12, 1048 (1975).

(7) F. T. Arecchi, R. Bonifacio, IEEE J. Quant. Electron. QE-1, 169 (1965).

(8) L. Allen and J. M. Eberly, Optical Resonance and Two-Level Atoms, Wiley, New York (1975), Chapter 2.

THEORETICAL DEVELOPMENT OF THE FREE-ELECTRON LASER*

F. A. Hopf† and P. Meystre

Department of Physics and Optical Sciences Center

University of Arizona, Tucson, Arizona 85721

I. INTRODUCTION

The stimulated emission of radiation of a relativistic beam of electrons in a transverse magnetic field has been analyzed theoretically by several authors[1-6], and has very recently been demonstrated experimentally at Stanford[7,8] by J. Madey and coworkers.

In a recent series of papers[5,6], we have developed a classical theory of the free-electron laser which has enabled us to investigate the small signal regime and saturation properties of this device in a particularly simple way. In this article we present a self-contained and more complete description of this work, and extend it to discuss in detail the electron dynamics in a FEL.

This paper is composed of five sections. Section II is a formal development in which we derive our working equations. In Section III, we compute the small signal gain, which is obtained by expanding the Boltzmann distribution function in powers of the product of the incident and scattered field, and keeping contributions up to first order only.

The strong-signal regime is best analyzed in terms of a set of "generalized Bloch equations", obtained as the result of an harmonic expansion discussed in Section IV. This formalism

*Work supported by the U. S. Energy Research and Development Administration.

†Supported by the National Science Foundation, Math GT # 43020.

allows one to show that the effects that lead to gain (i.e., electron recoil) play a relatively minor role compared to other non-linear effects, e.g., a considerable spread of the electron distribution function. The implications of this spread on the performance of a FEL are discussed in Section V.

II. FORMAL DEVELOPMENT

In an FEL, a beam of relativistic electrons of energy $E = \gamma mc^2$ is passed through a helictical magnetic field and produces stimulated emission of radiation[7,8]. In the highly relativistic limit of the FEL, the Weizsäcker-Williams approximation[9] is used, in which the static magnetic field of period λ_q is simulated by a fictitious incident EM field of wavelength $\lambda_i = (1 + v/c)\lambda_q$ propagating in the opposite direction to the electron beam. In the electron rest-frame, the problem can be understood as that of stimulated Thompson scattering. The relevant part of the emitted radiation is in the direction of the electron-beam (backscattering), since the Doppler up-shift is maximum under this configuration. In the laboratory frame the wavelength λ_s of the backscattered radiation is given for $\gamma \gg 1$ by

$$\lambda_s \simeq \lambda_i/4\gamma^2 \quad , \tag{1}$$

and can be tuned continuously by changing the energy of the incident electron beam (i.e., by changing γ). We find it convenient to study this problem directly in the laboratory frame, and to stay in a space-time representation. The FEL is then described classically by the coupling of the collisionless relativistic Boltzmann equation[10]

$$\frac{df}{dt} = \frac{\partial f}{\partial t} + \dot{x}_i \frac{\partial f}{\partial x_i} + \dot{P}_i \frac{\partial f}{\partial P_i} = 0 \quad , \tag{2}$$

to Maxwell's equations.

Here $\vec{P}$ is the canonical momentum, $\vec{x}$ the position, and a dot expresses the total derivative with respect to time. The total number of electrons N(t) is

$$N(t) = \int d^3x \int d^3P f(\vec{x},\vec{P},t) \quad . \tag{3}$$

The Boltzmann equation is coupled via the transverse current $\vec{J}_T$ to the wave equation

$$\nabla^2\vec{A} - \frac{1}{c^2}\frac{\partial^2\vec{A}}{\partial t^2} = -\mu_o \vec{J}_T \quad , \tag{4}$$

where $\vec{A}$ is the vector potential, and

$$\vec{J}_T(\vec{x},t) \equiv e\int d^3P\vec{v}_T f(\vec{x},\vec{P},t) \quad . \tag{5}$$

Here, e is the electron charge, and $\vec{v}_T$ the transverse component of the electron velocity.

In order to simplify the set of Equations (2), (4), we consider a model in which the electromagnetic field is transverse and depends on z and t only, i.e.,

$$\vec{A} = \vec{A}_T(z,t) \quad . \tag{6}$$

Therefore, the single electron Hamiltonian reads

$$\mathcal{H} = \gamma mc^2 = \left[c^2(\vec{P}_T - e\vec{A}_T)^2 + c^2p_z^2 + m^2c^4\right]^{1/2} \quad , \tag{7}$$

where p_z is now the mechanical momentum of the electron along the z-direction ($p_z = \gamma mv_z$) and x and y are cyclic variables, so that the transverse part of the canonical momentum is a constant of motion. Moreover, since the electrons enter the cavity with a velocity which is relativistic along the z-direction, the transverse velocity spread of the electron distribution is completely negligible, and we can set

$$\vec{P}_T = 0 \quad , \tag{8a}$$

or

$$\vec{p}_T = \vec{P}_T - e\vec{A}_T = -e\vec{A}_T \quad . \tag{8b}$$

By Eq. (7) we also have

$$\gamma \simeq \left[1 + (p_z/mc)^2\right]^{1/2} \quad , \tag{9}$$

where we have neglected the electron mass-shift $(e\vec{A}/mc)^2 \ll (p_z/mc)^2$. Equation (8) allows us to reduce the original three-dimensional problem to a much simpler one-dimensional problem.

To show that, we factorize the electron distribution function as

$$f(\vec{x},\vec{P},t) = \left[u(r) - u(r-a)\right]\delta^{(2)}(\vec{P}_T)h(z,p_z,t) \quad , \tag{10}$$

where u(r) is the Heaviside function (circular beam) and $\delta^{(2)}$ the two-dimensional delta-function.

Using Eq. (8), we can reexpress the Boltzmann equation as

$$\frac{\partial f}{\partial t} - \frac{e\vec{A}_T}{m\gamma}\nabla_T f + \frac{p_z}{m\gamma}\frac{\partial f}{\partial z} - \frac{e^2}{m\gamma}\frac{\partial}{\partial z}\left(\frac{\vec{A}_T^2}{z}\right)\frac{\partial f}{\partial p_z} = 0 \quad , \tag{11}$$

where we have also used the Hamilton equation of motion

$$\frac{dp_z}{dt} = \frac{dP_z}{dt} = \frac{\partial \mathcal{H}}{\partial z} = -\frac{e^2}{m\gamma}\frac{\partial}{\partial z}\left(\frac{A_T^2}{2}\right) \quad . \tag{12}$$

We now integrate over the transverse dimensions:

$$\int d^2\vec{r}_T \frac{A_T(z,t)}{m\gamma}\nabla_T f = \int_0^{2\pi} d\theta \int_0^a drr\left[A_x\frac{\partial f}{\partial x} + A_y\frac{\partial f}{\partial y}\right] \quad ,$$

where a is the radius of the electron beam. Assuming a cylindrical symmetry $(\partial f/\partial\theta = 0)$, we have

$$\frac{\partial f}{\partial x} = \frac{\partial f}{\partial r}\cos\theta \quad , \quad \frac{\partial f}{\partial y} = \frac{\partial f}{\partial r}\sin\theta \quad , \tag{13}$$

and therefore

$$\int d^2\vec{r}_T \frac{\vec{A}_T(\vec{r},t)}{m\gamma}\nabla_T f = 0 \quad . \tag{14}$$

The Boltzmann equation reduces then exactly to

$$\frac{\partial h}{\partial t} + \frac{p_z}{m\gamma}\frac{\partial h}{\partial z} = \frac{c^2}{m\gamma}\frac{\partial}{\partial z}\left(\frac{\vec{A}_T^2}{2}\right)\frac{\partial h}{\partial p_z} \quad , \tag{15}$$

and the Maxwell equation becomes, by integrating over the transverse dimension of the cavity,

$$\left(\frac{\partial^2}{\partial z^2} - \frac{1}{c^2}\frac{\partial^2}{\partial t^2}\right)\vec{A}_T = \frac{e^2 F}{mc\varepsilon_o}\vec{A}_T\int_{-\infty}^{\infty} dp_z \frac{h(z,p_z,t)}{\gamma} \quad , \tag{16}$$

where $F = a^2/b^2$ is a filling factor, and b is the radius of the cavity.

We note that the potential $\vec{A}_T^2$ which occurs on the right-hand side of the reduced Boltzmann equation is the same as the interaction term in the Klein-Gordon Hamiltonian, and hence, the nature of the interaction is the same in both quantum-mechanical and classical theories, as it should be. Furthermore, the right-hand side of the Maxwell equation is proportional to a density times an electric field. This is exactly the same as in usual scattering problems, where the d'Alembertian of the electric field is proportional to the second derivative of the polarization, which in turn is proportional to a density times the electric field. Hence, this problem is nothing more than the usual classical scattering problem, complicated only by the fact that we deal here with relativistic particles.

III. SMALL-SIGNAL REGIME[5]

We know of no manner in which the set of Equations (15) and (16) can be solved exactly. If one wishes to find the small-signal gain, it is sufficient to expand $h(z,p_z,t)$ in powers of $|A_i A_s|$ and solve (11) and (12) self-consistently.

Let

$$h = h^{(0)} + h^{(1)} + h^{(2)} + \ldots \quad ,$$

where $h^{(0)}$ is given by

$$\frac{\partial h^{(0)}}{\partial t} + \frac{p_z}{m\gamma}\frac{\partial h^{(0)}}{\partial z} = 0 \quad , \tag{17}$$

and $h^{(n)}$ by

$$\frac{\partial h^{(n)}}{\partial t} + \frac{p_z}{m\gamma}\frac{\partial h^{(n)}}{\partial z} = \frac{e^2}{mc\varepsilon_o}\frac{\partial}{\partial z}\left(\frac{\vec{A}_T^2}{2}\right)\frac{\partial h^{(n-1)}}{\partial p_z} \quad . \tag{18}$$

Keeping only the terms up to $h^{(1)}$ gives the small-signal theory.

It is sufficient here to keep only two modes of the field, namely the incident mode (which, in the Weizsäcker-Williams approximation, simulates the static periodic magnetic field), and the scattered mode.

We take the field to be circularly polarized, and hence, $\vec{A}$ reads

$$\vec{A}_T = \hat{e}_-\left[A_i e^{-i(\omega_i t + k_i z)} + A_s e^{-i(\omega_s t - k_s z)}\right] + c.c. \quad , \tag{19}$$

where $\hat{e}_\pm = (\hat{x} \pm i\hat{y})/\sqrt{2}$ and $\omega_i = ck_i$ $(\omega_s = ck_s)$ is the frequency of the incident (scattered) field. We also make the slowly varying amplitude and phase approximation, and neglect the depletion of the incident field (i.e., we take the static field to be constant).

Taking the electrons to be injected at a constant rate inside the cavity, we find

$$h^{(0)} = n_e F(p_z) \quad , \tag{20}$$

where n_e is the electron density, and $F(p_z)$ is the initial electron momentum distribution. The first order correction to h is

$$h^{(1)}(z,p_z,t) = \frac{-e^2 K n_e A_i^* A_s}{p_z}\frac{dF}{dp_z}\left(\frac{e^{-i\mu z} - 1}{\mu}\right)e^{-i\Delta\omega(t - z/v_z)} + c.c. \quad , \tag{21}$$

where

$$\Delta\omega = \omega_s - \omega_i \quad ,$$

$$K = k_s + k_i \quad ,$$

$$\mu = \Delta\omega/v_z - K \quad . \tag{22}$$

Introducing h into the Maxwell equation, and integrating over the length L of the cavity, we obtain the small-signal gain g, i.e.,

$$dI_s/dt = gI_s \quad , \tag{23}$$

where

$$g(\sec^{-1}) = \frac{-4\pi^2 r_o^2 F n_e m K L}{k_i^2 k_s} \int dp_z F(p_z) \frac{d}{dp_z} \left[\frac{1}{\gamma p_z} \frac{\sin \eta}{\eta} \right]^2 \quad , \tag{24}$$

and

$$\eta = \mu L/2 \quad , \tag{25}$$

$I_i(I_s)$ being the incident (scattered) flux

$$I_{i(s)} = 2c^3 \varepsilon_o k_{i(s)}^2 \, |A_{i(s)}|^2 \quad , \tag{26}$$

and r_o the classical radius of the electron. The $\sin \eta/\eta$ term appearing in the gain formula is a consequence of the finite length of the cavity. As pointed out by Sukhatme and Wolff[4], this finiteness implies that the momentum conservation law is relaxed, so that electrons with a momentum within this cavity bandwidth can scatter light from ω_i to ω_s. It is therefore natural to consider two cases, namely the large cavity and small cavity limits. In the large cavity limit, the electron momentum distribution is very broad compared to the cavity bandwidth, which can therefore be approximated by a δ-function. The small cavity limit corresponds to the opposite case, where the initial electron momentum distribution function $F(p_z)$ can be taken as a δ-function

$$F(p_z) = \delta(p_z - p_o) \quad . \tag{27}$$

This corresponds to the limit of an homogeneously broadened medium. We shall consider only this later case, since it corresponds to the condition of the Stanford experiment[7,8].

Using Eq. (27), the gain formula becomes

$$g(\sec^{-1}) = \frac{2\pi^2 r_o^2 F n_e c(k_s^2 - k_i^2)L^2}{mk_s k_i^2 \gamma_o^5 v_o^3} I_i \frac{d}{d\eta}(\sin\eta/\eta)^2$$

$$+ \frac{2v_o}{cL(k_s - k_i)}(\sin\eta/\eta)^2 \gamma_o^2 \left[1 + (v_o/c)^2\right] , \quad (28)$$

where

$$p_o = m\gamma_o v_o \quad .$$

The second term inside the bracket is of the order

$$(\sin\eta/\eta)^2 \lambda_i/2\pi L \ll 1 \quad . \quad (29)$$

Therefore we can neglect this term and obtain (using Eq. (1) and noting that $k_s \gg k_i$ for $\gamma \gg 1$):

$$g(\sec^{-1}) \simeq 64\pi^2 r_o^2 F \frac{n_e}{mc^2} \frac{k_i^{1/2}}{k_s^{3/2}} L^2 I_i \frac{d}{d\eta}(\sin\eta/\eta)^2 \quad . \quad (30)$$

The point $\eta = 0$ corresponds to the condition for exact conservation of momentum and, at this point, there is no net gain. For $\eta < 0$ which corresponds to electrons with a velocity $v > v_o$, the net result is a gain. This is the exact equivalent of the Stokes line in Raman scattering. For $\eta > 0$ the net result is absorption (anti-stokes line).

The maximum gain results for $\eta \simeq -1/2\ \pi$, and its value is

$$g_{max}(\sec^{-1}) \simeq \frac{4\times(256)}{\pi} r_o^2 F \frac{n_e}{mc^2} \frac{k_i^{1/2}}{k_s^{3/2}} L^2 I_i \quad . \quad (31)$$

The gain line given by Eq. (30) can be understood intuitively from the following argument.

We observe from the coupled Maxwell-Boltzmann equations that the electrons interact with the field through a potential V(z,t) of the form

$$V(z,t) \propto A_i^* A_s \exp i(Kz - \Delta\omega t) + c.c. \quad . \tag{32}$$

This "walking" potential is moving in the same direction as the electrons, at the velocity

$$\beta_s \equiv \frac{v_s}{c} = \frac{\Delta\omega}{cK} = \frac{k_s - k_i}{k_s + k_i} \quad . \tag{33}$$

We can express η in terms of this velocity β_s as

$$\eta = \frac{cKL}{2v_z} (\beta_s - \beta) \quad . \tag{34}$$

Hence η is a measure of the velocity detuning between the electrons and the walking potential[11]. We then have the following picture. If $\beta > \beta_s$, the electrons are "trapped" by the potential, and, <u>on the average</u>, are decelerated and transfer energy to the field, leading to gain. If $\beta < \beta_s$, the electrons, <u>on the average</u>, are accelerated, and the net result is absorption. We emphasize that this statement applies to the average electron velocity, but not to a single electron, which can be either accelerated, or decelerated, in both cases, depending upon the phase with which it enters the walking potential. This point is of importance in order to understand the spread of the electron distribution, which we will discuss later on.

We note that this mechanism of amplification is closely related to the Landau damping phenomena[12].

The classical gain expression is in agreement with that obtained by a fully quantum-mechanical treatment[3,4] under the following circumstances: In the quantum-mechanical approach, one has to evaluate the finite difference $[F(p_o) - F(p_z)]/(p_o - p_f)$, where $p_o(p_f)$ is the initial (final) electron momentum, and[4]

$$\Delta p_z \equiv p_o - p_f = \hbar(k_i + k_s) \simeq \hbar k_s \quad .$$

We argue that if Δp_z is much smaller than the difference in momentum δp_z between, say, the point of zero gain and the point of maximum gain, this finite difference can be replaced by a derivative. In this case, the quantum-mechanical and classical gain formulas are identical.

We can obtain an estimate of δp_z by first considering the difference $\delta\beta = \beta_s - \beta_o$ between the point of maximum gain and the point of zero gain. It is easily shown to be

$$\delta\beta \simeq \frac{\pi}{L(k_i + k_s)} \simeq \frac{\lambda_s}{2L} \quad .$$

Since $p_z = \gamma\beta mc$, we have

$$\frac{\delta p_z}{p_z} = \frac{1}{\beta(1 - \beta^2)} \delta\beta \simeq \gamma^2 \delta\beta \quad ,$$

or

$$\delta p_z \simeq \gamma^3 mc(\lambda_s/2L) \quad .$$

The quantum-mechanical gain reaches the classical limit if $\hbar k_s \ll \gamma^3 mc(\lambda_s/2L)$, or, with eq. (1)

$$\lambda_s \gg (32\pi)^2 (\lambda_c L)^2/\lambda_i^3 \quad ,$$

where $\lambda_c = \hbar/mc$ is the Compton wavelength. In the Stanford experiment, for $L \simeq 5$ m and $\lambda_i \simeq 1$ cm, this implies that quantum-mechanical corrections should be completely negligible for $\lambda_s \gg 10^{-14}$ m, i.e., the classical treatment should give the correct result up to the hard X-ray regime.

IV. STRONG SIGNAL REGIME[6]

It would be tempting to expand $h(z,p_z,t)$ to higher orders to compute the saturation, keeping, for example, terms up to third order in the field strength as in the usual procedures of laser theory. However, this perturbative expansion diverges. To approach the strong-signal theory, it is necessary to use some other expansion for $h(z,p_z,t)$.

To this end, we express $h(z,p_z,t)$ as an expansion in harmonics of the "walking" potential:

$$h(z,p_z,t) = n(z,p_z,t) + \sum_{m=1}^{\infty} \left(ig_m \, e^{im(\Delta\omega t - Kz)} + c.c. \right) \quad . \tag{35}$$

This harmonic expansion presents several advantages over the perturbative one. First, each term contains the saturation to all orders in the field, and we do not encounter the divergence problems associated with the power expansion. Second, a computer analysis shows that the expansion (35) can be truncated at m = 1 without introducing noticeable corrections in the small signal regime of the FEL. Finally, with this truncation the Boltzmann equation (15) can be re-expressed at steady state ($\partial_t n = \partial_t g_1 = 0$) in terms of the following set of equations:

$$\frac{\partial R_1}{\partial \zeta} + pR_2 = \frac{-\partial R_3}{\partial p} \quad , \tag{36a}$$

$$\frac{\partial R_2}{\partial \zeta} - pR_1 = 0 \quad , \tag{36b}$$

$$\frac{\partial R_3}{\partial \zeta} = - \frac{\partial R_1}{\partial p} \quad , \tag{36c}$$

subject to the boundary conditions

$$R_1(0,p) = R_2(0,p) = 0 \quad ,$$

$R_3(0,p)$ prescribed by the initial electron momentum distribution. (37)

These equations contain, in addition, a perturbation expansion in p which is based on the observation that since the small signal gain is finite only in the region

$$-2\pi \lesssim \mu L \lesssim 2\pi \quad , \tag{38}$$

where

$$\mu = 2\eta/L = \Delta\omega/v_z - K \quad , \tag{39}$$

we can expand μ about $p_s = mc\beta_s\gamma_s$.

We can find that p is approximately given by

$$p \simeq -\mu\ell = -2\eta(\ell/L) \quad . \tag{40}$$

Also, the $R_i(\zeta,p)$ are dimensionless functions that are related to n and g_1 through

$$R_1 = (mc\sqrt{2}/4n_e\sigma\beta_s\gamma_s)(g_1 + g_1^*) \quad , \tag{41a}$$

$$R_2 = -i(mc\sqrt{2}/4n_e\sigma\beta_s\gamma_s)(g_1 - g_1^*) \quad , \tag{41b}$$

$$R_3 = (mc/2n_e\sigma\beta_s\gamma_s)n \quad , \tag{41c}$$

where

$$\sigma^2 = mc/4\sqrt{2}\gamma_s^4\beta_s^2e^2A_iA_s \quad . \tag{42}$$

The scaling coefficients serve to eliminate all explicit field dependence from Eq. (36).

The dimensionless length ζ is expressed in terms of the scale length ℓ as

$$\zeta = z/\ell \quad .$$

We will discuss the meaning of the length $\ell = mc\gamma_s^2\beta_s/2^{1/4}eK(A_iA_s)^{1/2}$ later. Before doing this, let us note that the set of equations (36) presents a striking resemblance to the optical Bloch equations (13), where R_3 would be the population inversion, and R_1 the polarization. However, it differs from them in two respects.

First, the signs on the right-hand side of Eqs. (36a) and (36c) are opposite in the optical Bloch equations; and second, the right-hand side of the Bloch equations contains R_3 and R_1, rather than their derivatives. This difference in structure lies in the fact that in a free-electron laser, the gain is not proportional to the electron distribution function. It is its derivative with respect to p_z (rather than the distribution itself) which plays the role of an inversion.

To complete this development, we still have to express Maxwell's Equation (16) in terms of the new variables R_i. Noting that it is consistent with the perturbation expansion in p to let

$$\gamma(dp/dp_z) \simeq \gamma_s(dp/dp_z)/p_z = p_s = 2\sigma\gamma_s^2\beta_s$$

on the RHS of Maxwell's equation, we obtain

$$\frac{d}{dt}\,|A_s|^2 = \frac{-\mathcal{F}n_e e^4 K^2 L^2 |A_i|^2 |A_s|^2}{k_s \varepsilon_o \gamma_s^5 \beta_s^2 (mc)^3} \left(\frac{\ell}{L}\right)^3 \int_0^{L/\ell} d\zeta \int_{-\infty}^{\infty} dp R_1(\zeta,p) \quad . \tag{43}$$

The double integral on the RHS of Equation (43) can be re-expressed in a way that shows that the gain is explicitly connected to the recoil of the electrons. Integrating by parts and using the fact that $R_1(\zeta,\pm\infty) = 0$, we find, using Equations (36), that

$$\delta p \equiv \int_0^{L/\ell} d\zeta \int_{\infty}^{\infty} d\, R_1(\zeta,p) = \int_{\infty}^{\infty} dp p[R_3(L/\ell,p) - R_3(0,p)] \quad . \tag{44}$$

Since R_1 and R_2 contribute in a rapidly varying time-dependent fashion to the electron distribution function, the RHS of Eq. (44) can be understood as the time average of the difference between the average final energy and the average initial energy of the electrons.

In the small-signal regime, where standard perturbation theory can be used, we find that the recoil δp is then given by

$$\delta p = \frac{1}{4}\left(\frac{L}{\ell}\right)^3 \frac{d}{d\eta}\left(\frac{\sin\eta}{\eta}\right)^2 \quad . \tag{45}$$

This gives the small signal gain derived in Section III.

In the small-cavity limit, the gain is then given in terms of the maximum small-signal gain g_{max} as

$$g(A_i,A_s,p_0,L) = g_{max}(A_i,p_0,L) \times \mathcal{S}(A_i,A_s,p_0,L) \quad , \tag{46}$$

where the saturation function $\mathcal{S}$ is the double integral

$$\mathcal{S}(A_i,A_s,p_0,L) = \frac{\pi^3}{4}\left(\frac{\ell}{L}\right)^3 \int_0^{L/\ell} d\zeta \int_{-\infty}^{\infty} dp R_1(\zeta,p) \quad ,$$

and p_o is the initial electron momentum.

In Fig. 1, we present a numerical computation of the saturation function $\mathcal{S}$ as a function of η and $\sqrt{\mathcal{R}} \equiv (\sqrt{2}/\pi^2)(L/\ell)^2$. This figure was computed in the small-cavity limit. The vertical axis corresponds to the small-signal regime ($\sqrt{\mathcal{R}} = 0$). The working point of the free-electron laser should be chosen to be about $\eta = -\pi/2$, and for this value $\mathcal{S}$ reaches the value 0.5 for $\sqrt{\mathcal{R}} \simeq 9$.

The numerical results demonstrate that our scaling coefficient correctly describes the saturation by the requirement $\mathcal{R} \sim 1$ or $\ell \sim L$. We would like now to derive ℓ (to within numerical factors) by a heuristic argument that will serve to make connection with the physics of the saturation, using the scaling relations of ordinary lasers.

Let us suppose that the mechanism of saturation is a deceleration of the electrons through the gain line to the point of zero gain. For example, we assume that the electrons are subject to recoil only, and that the spread of their momentum distribution is negligible. The maximum amount of energy that a single electron can transfer to the field is

$$\Delta E = \left(\frac{\lambda_s}{L}\right) \gamma^3 mc^2 \quad . \tag{48}$$

In the limit that the scattered flux S_s becomes large, the gain is limited by the maximum energy flux S_{ex} available from the electron beam per unit time:

$$S_{ex} = \frac{\mathcal{F} n_e c^2}{L} \Delta E \quad . \tag{49}$$

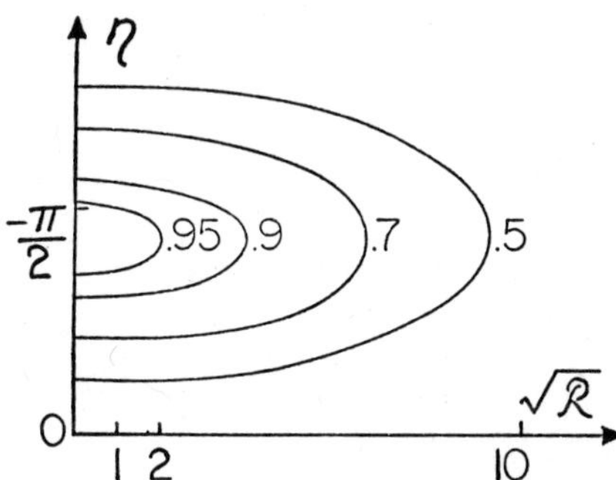

Figure 1. Plot of the equisaturation curves S = constant as a function of $\sqrt{R}$ and η. The curves are labeled with their value of $S = 0.95$, 0.9, 0.7, and 0.5. These results were obtained in the small cavity limit, and the ratio of the electron momentum distribution width to the cavity bandwidth was taken to be 1/20.

Defining (as usual) the saturation flux S_{sat} as the ratio of S_{ex} to the maximum small signal gain g_{max},

$$S_{sat} = S_{ex}/g_{max} \quad , \tag{50}$$

we find that the scale length ℓ is related to the saturation flux through the relation

$$(\sqrt{2}/\pi^2)(L/\ell)^2 = (S_s/S_{sat})^{1/2} \quad . \tag{51}$$

The left-hand side of Eq. (51) is precisely the parameter $\sqrt{R}$ used in the numerical computation, and is equal to 1 for $S_s = S_{sat}$. We note that the right-hand side is $(S_s/S_{sat})^{1/2}$, rather than S_s/S_{sat}. This is due to the fact that the field is excited coherently by the electron beam.

It follows that the maximum field extractable from a free-electron laser (i.e., the output field when the laser is in the saturation regime) is

$$A_{s,max} \cong \left(\frac{\lambda_s}{L}\right)^2 \frac{(mc)^2\gamma^4}{e^2 A_i} ,$$

which gives a saturation flux

$$S_{sat} \cong (\pi^2/2^{12}) \frac{m^2c^6}{L^4 r_o^2 S_i} (\lambda_i/\lambda_s)^2 . \qquad (52)$$

With use of the numerical values of the Stanford experiment[7], Eq. (52) gives a maximum extractable field on the order of $E_{s,max} \simeq 10^7$ V/m (i.e., $S_{sat} \sim 10^7$ W/cm^2) for a static magnetic field of 2.4 kG.

Such a high saturation flux prohibits continuously sustained laser operation in the saturation regime. Typically, the breakdown threshold of laser material is on the order of 1 J/cm^2, limiting the width of the pulse to about 100 nsec.

Furthermore, in contrast to what happens in usual lasers, the electrons that contribute to the build-up of the pulse are not the same as those which are involved in saturation. Since the transit time ($\tau \simeq 17$ nsec) of the electrons through the cavity is much shorter than the build-up time of the pulse, most electrons interact with the electromagnetic field in the small-signal regime.

Since the efficiency of the laser goes roughly as the ratio of the power to the saturation power, the per-shot efficiency of a free electron laser will be very small. In order to achieve appreciable efficiency, it has been suggested that one should recycle the electron beam from one shot to the next. In a description of the problem taking into account the electron recoil only, this would appear to be straightforward, since one could simply replace the energy lost through recoil in each cycle of the accelerator. However, as we show in the next section, such a description is incomplete at best, and the real picture is quite different. Indeed, the effects that lead to small-signal gain in a FEL are almost negligibly small compared to other effects which occur together with them!

V. ELECTRON DYNAMICS

In the previous section, we developed the dynamics of the FEL in terms of a set of generalized Bloch equations. In this section we develop an approximate analytic solution for the functions R_i $(i = 1,2,3)$. We take p_o to be the center of the initial distribution function (i.e., the center of R_3 $(\zeta = 0,p)$). In the small cavity limit, we have then

$$R_3(0,p) = \delta(p - p_o) \quad . \tag{53}$$

The amount by which this initial distribution spreads out depends upon the magnitude of the field. In the small-signal regime, the spread due to the field is small compared to p_o, so that the domain of interest in p remains extremely small. This suggests that it might be interesting to replace p by p_o in the terms in Eq. (36) that do not contain derivatives. When that is done, Eq. (36) leads directly to the Klein-Gordon equation,

$$\frac{\partial^2 R_1}{\partial \zeta^2} - \frac{\partial^2 R_1}{\partial p^2} + p_o^2 R_1 = 0 \quad , \tag{54}$$

whose solution is well known[15]. With the boundary conditions (37) we then obtain

$$R_1(\zeta,p) = -\frac{1}{2} H(\zeta,p) \frac{\partial}{\partial p} J_o\left[p_o\sqrt{\zeta^2 - (p - p_o)^2}\right]$$
$$- \frac{1}{2}\left\{\delta\left[\zeta + (p - p_o)\right] - \delta\left[\zeta - (p - p_o)\right]\right\} \quad ,$$

$$R_2(\zeta,p) = -\frac{1}{2} H(\zeta,p) \int_{|p_o(p-p_o)|}^{p_o\zeta} dx \frac{\partial}{\partial p} J_o\left[\sqrt{x^2 - p_o^2(p - p_o)^2}\right]$$

$$+ \frac{p_o}{2} \tilde{H}(\zeta,p) \quad ,$$

$$R_3(\zeta,p) = \frac{p_o}{2} H(\zeta,p) \int_{|p_o(p-p_o)|}^{p_o\zeta} dx\, J_o\left[\sqrt{x^2 - p_o^2(p - p_o)^2}\right]$$

$$+ \frac{1}{2} H(\zeta,p) \frac{\partial}{\partial \zeta} J_o\left[p_o\sqrt{\zeta^2 - (p - p_o)^2}\right]$$

$$+ \frac{1}{2}\left\{\delta\left[\zeta + (p - p_o)\right] + \delta\left[\zeta - (p - p_o)\right]\right\} \quad , \tag{55}$$

where

$$H(\zeta,p) = \begin{cases} 1 & |p - p_o| < \zeta \\ 0 & |p - p_o| > \zeta \end{cases} \tag{56a}$$

and

$$\tilde{H}(\zeta,p) = \begin{cases} H(\zeta,p) & p > p_o \\ -H(\zeta,p) & p < p_o \end{cases} \quad , \tag{56b}$$

J_o being the zeroth order Bessel function.

One can get a reasonable qualitative picture of the distribution function by ignoring all contributions from the background terms containing Bessel functions. To quantify this remark we define the energy spread Δp as the root of the second moment of the electron distribution function:

$$|\Delta p| = \left[\int dp (p - p_o)^2 R_3\right]^{1/2} \quad , \tag{57}$$

where, as in the case of the recoil, we neglect the contribution from the rapidly varying terms R_1 and R_2. The delta functions alone give $\Delta p = \zeta$, and the complete solution is

$$\Delta p = \frac{2}{p_o} \sin \frac{p_o \zeta}{2} \quad . \tag{58}$$

This spread reduces to the delta function's contribution for small ζ. The contribution from the background can be readily determined from the difference between the exact answer and the delta function's contribution. The point of maximum gain corresponds to the maximum of the sine function (i.e., $p_o L/2\ell = -\eta = \pi/2$), in which case the exact spread at the output is given by

$$\Delta p = (2/\pi)(L/\ell) \quad . \tag{59}$$

This is about 35% smaller than the spread obtained from the delta function alone. Thus, the main part of $R_3(\zeta,p)$ is still the two delta-functions, and the Bessel function contribution can be omitted if one is interested in the qualitative behavior of the free electron laser only.

The splitting of $R_3(\zeta,p)$ is independent of the initial distribution, which is in complete agreement with the numerical results.

To simplify the discussion let us assume that the spread in energy is cumulative from one shot of the laser to the next. This would be the case if the emittance ($\Delta E \Gamma t \propto \Delta p_z \Delta z$) of the electron beam were conserved in the storage ring, since the duration in time $\Delta t = c\Delta z$ of the pulse of electrons is fixed by the requirements of maximizing peak power while avoiding material damage. Note, however, that emittance is not exactly conserved due to synchrotron radiation in the storage rings, so that our arguments serve only to establish the importance of the spread effect. In Table I we have summarized the results of the previous discussion by giving the spread $\Delta p(\Delta E)$ in scaling (p) and real (E) units. The real units are written in terms of S/S_{sat}, which is found from the ratio L/ℓ in Eq. (51). All of the quantities are evaluated at the output ($z = L$), and for $p_o(E_o)$ evaluated at the point of maximum gain. The last row shows the ratio of recoil to spread, and one sees that for $S < S_{sat}$ it is the spread which is dominant. The numerical calculations show that the effects are comparable for $S \sim S_{sat}$, so that the ratio is valid for large signals. Since fields of 10^7 watts/cm^2 can be reached only briefly in the FEL, the dominant effect on the distribution function will be the spread.

Table I.

Process	Physical Units	Scaled Units
Center of initial electron distribution, relative to the zero gain condition	$E_o = \gamma^3 mc^2 (\lambda_s/2L)$	$p_o = \pi(\ell/L)$
Electron recoil	$\delta E = (\lambda_s/L)\gamma^3 mc^2 (S/S_{sat})$	$\delta p = (4/\pi^3)(L/\ell)^3$
Electron spread	$\Delta E = (1/\sqrt{2})(\lambda_s/L)\gamma^3 mc^2 (S/S_{sat})^{1/2}$	$\Delta p = (2/\pi)(L/\ell)$
Recoil to spread ratio	$\frac{\delta E}{\Delta E} = \sqrt{2}\ (S/S_{sat})^{1/2}$	$\frac{\delta p}{\Delta p} = (2/\pi^2)(L/\ell)^2$

In light of this, it is important to note that a reduction in gain can occur either from saturation (i.e., recoil) when $\delta E \sim E_o$, or from reaching the "large cavity" limit when the total spread $\Delta E_{tot} \sim E_o$. If the spread is cumulative from shot to shot, the number of cycles before the large cavity limit is reached is given by

$$N \simeq E_o/\Delta E \quad . \tag{60}$$

The efficiency in any one cycle is given by $\delta E/\gamma mc^2$, so that the net efficiency of N such cycles is

$$\eta = N\delta E/\gamma mc^2 \simeq (E_o/\gamma mc^2)(S/S_{sat})^{1/2} \quad . \tag{61}$$

The ratio $E_o/\gamma mc^2$ is the efficiency that one would get on a CW basis if the FEL were operated in the saturation regime (i.e., assuming $\delta E \simeq E_o$). This quantity can be easily evaluated from Table I, and is $\sim 2 \times 10^{-3}$ for $\lambda_s = 10.6\ \mu$ and $\gamma = 40$. The quantity $(S/S_{sat})^{1/2}$ is the reduction in efficiency due to the spread effect. Since the vast majority of electrons will interact with the light in the small signal regime, it is clear that the spread effect should be of major concern in the operation of the FEL.

ACKNOWLEDGMENTS

We are indebted to Dr. M. O. Scully, Dr. W. H. Louisell, and H. Al-Abawi for their important contributions to this work. We also acknowledge numerous fruitful discussions with Dr. W. E. Lamb, Jr., and Dr. G. L. Lamb, Jr.

REFERENCES

1. H. Dreicer, Phys. Fluids 7, 735 (1964).

2. R. H. Pantell, G. Soncini, and H. E. Puthoff, IEEE J. Quantum Electronics 4, 905 (1965).

3. J. M. J. Madey, J. Appl. Phys. 42, 1906 (1971).

4. V. P. Sukhatme and P. W. Wolff, J. Appl. Phys. 44, 2331 (1973).

5. F. A. Hopf, P. Meystre, M. O. Scully, and W. H. Louisell, Optics Comm. 18, 413 (1976).

6. F. A. Hopf, P. Meystre, M. O. Scully, and W. H. Louisell, Phys. Rev. Letters 37, 1342 (1976).

7. L. R. Elias, W. H. Fairbank, J. M. J. Madey, H. A. Schwettmann, and T. J. Smith, Phys. Rev. Letters 36, 717 (1976).

8. D. A. G. Deacon, L. R. Elias, J. M. J. Madey, H. A. Schwettmann, and T. J. Smith, VII Winter Symposium on high power lasers, Park City (Utah, 1977).

9. W. Heitler, The Quantum Theory of Radiation, (Oxford University Press, 1954).

10. I. Lerche, in Relativistic Plasma Physics, Ed. by O. Buneman and W. B. Pardo, (Benjamin, N.Y., 1968), p. 159.

11. In reality, this walking potential is moving quite fast, since $k_s \gg k_i$ and therefore $\beta_s \simeq 1$. W. E. Lamb, Jr., suggested that it might be more appropriate to call it a "jogging" potential.

12. See for instance J. D. Jackson, Classical Electrodynamics (Wiley, N.Y., 1962).

13. See for instance L. Allen and J. H. Eberly, Optical Resonance and Two-Level Atoms (Wiley-Interscience, N.Y., 1975).

14. H. Al-Abawi, F. A. Hopf, and P. Meystre, unpublished.

15. P. R. Garabedian, Partial Differential Equations, (Wiley, N.Y., 1964).

FREE ELECTRON LASERS*

John M. J. Madey and David A. G. Deacon†

High Energy Physics Laboratory

Stanford University, Stanford, California 94305

Abstract: Lasers based on the stimulated emission of radiation by free electrons in a spatially periodic magnetic field offer a unique potential for tunable operation at high power and high efficiency. Substantial advances have been made recently in the theoretical analysis of this class of device and in the operation in our laboratory of a 10 μ free electron laser amplifier. These results and the implications for the design of a practical device will be reviewed.

I. INTRODUCTION

In the free electron laser, a beam of relativistic electrons is made to pass through a periodic transverse magnetic field (Figure 1). Radiation travelling through the field with the electron beam can either be amplified or absorbed depending on the relationship of the wavelength to the electron energy, the field period, and the magnetic field strength. A device of this kind has recently been operated as an amplifier at a wavelength of 10 microns.[1] The device will operate as an oscillator if the interaction region is made a part of an optical resonant cavity.[2]

* Work supported in part by the Air Force Office of Scientific Research, Contract F49620-76-C-0018.

† National Research Council of Canada graduate fellow.

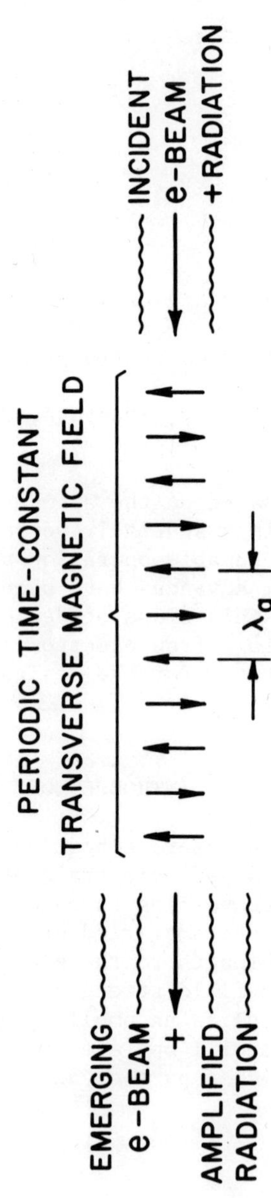

Figure 1. In the Free Electron Laser, the interaction between a beam of free electrons and a spatially periodic transverse magnetic field is used to amplify optical radiation.

Figure 2 illustrates the relationship between gain and absorption and the electron energy. Figure 2a indicates the dependence on electron energy of the 10.6 micron spontaneous power radiated by the electron beam in the periodic field. It is convenient to define the energy which maximizes the spontaneous power at the wavelength λ as the resonance energy. Theoretically, the resonance energy for radiation in the forward direction in a circularly polarized magnetic field is given by:

$$E = \gamma_o mc^2 \approx \left\{ \frac{\lambda_q}{2\gamma^2} \left[1 + \left(\frac{1}{2\pi}\right)^2 \frac{\lambda_q^2 r_o B^2}{mc^2} \right] \right\}^{1/2} , \quad (1)$$

where λ_q is the magnet period, r_o the classical electron radius, mc^2 the electron mass-energy, and B the amplitude of the magnetic field in the interaction region. In the Stanford experiment, the magnet period was 3.2 cm and the magnetic field strength was 2.4 kilogauss. The spontaneous power at 10 microns peaked at 24 MeV. Figure 2b shows the measured gain at 10.6 microns. As can be seen, there is net gain at electron energies slightly higher than the resonant energy and net absorption at lower energies. A fraction of the electrons' energy of the order of 0.2% was extracted as radiation in this experiment without evidence of saturation.[1]

The rationale for interest in this class of device lies in the range of wavelengths at which laser action should be possible and the potential for high power output and efficiency. The most promising configuration involves the use of an electron storage ring (Figure 3). Sufficient current has been attained in existing storage rings to sustain laser operation at wavelengths as short as 1000 Å. Given the saturation characteristics of the amplification mechanism, there is a well defined potential for an average power output of the order of 100 Kwatts in the visible.[3]

Historically, the first reference to devices of this sort was made by Motz[4] in 1951. A quantum analysis of the gain was published by Madey[5] in 1971 and the experimental effort at Stanford was begun in 1972. While Stanford's experimental effort is unique, several groups have contributed to the development of the theory. A listing of the relevant contributions is provided in the references.[6]

Our purpose in this discussion is to review the models which can be used to analyze the operation of the free electron laser, to comment on the circumstances in which these models can be used, and to discuss some of the factors affecting stability in storage ring lasers of the kind illustrated in Figure 3. The development and selection of appropriate models for the interactions is important for both practical and academic reasons. Detailed information on the gain, saturation characteristics and electron energy fluctuations

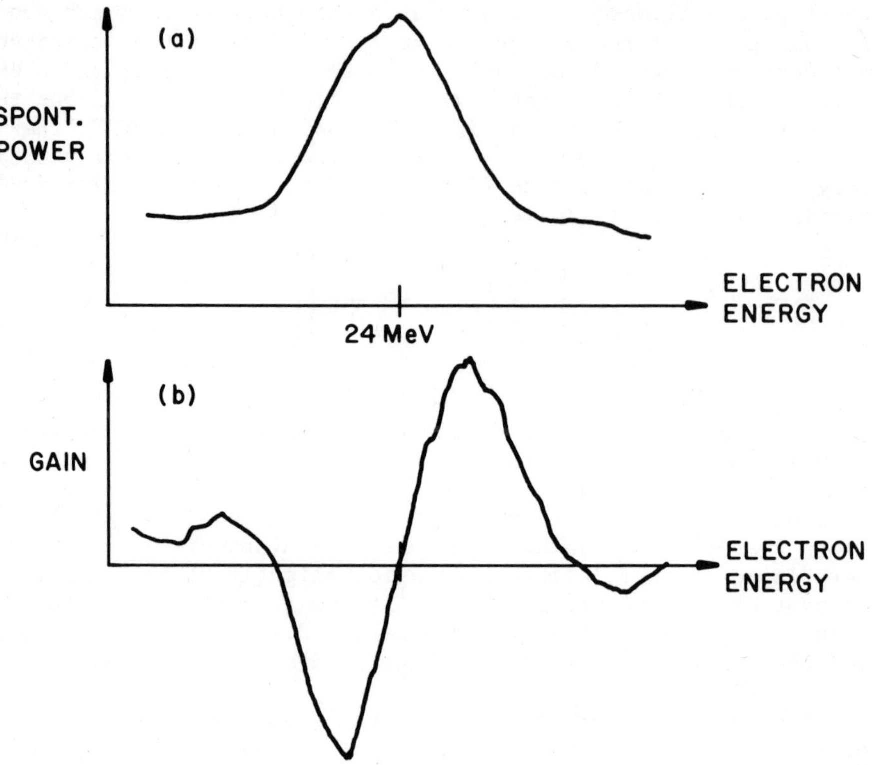

Figure 2. Figure 2(a) shows the experimental relationship between the spontaneous power emerging from the interaction region at 10.6 microns and the electron energy. Figure 2(b) shows the dependence of the experimental gain per pass at 10.6 microns on the electron energy. The helix field amplitude was 2.4 kilogauss and the field period was 3.2 cm. The gain reached 7% at an electron current of 70 mA.

is required for the assessment of the potential of the device and to support the development of the electron accelerators and storage rings required for laser operation. On the academic side, many lasers and many electron devices have been proposed and it is of evident interest to determine the relationship of the free electron laser to these pre-existing ideas.

With regard to the development of a model for the interaction, quantum electrodynamics provides, at least in principle, the framework for a comprehensive theory for the device. The problem is that QED is complicated and there are relatively few cases in which

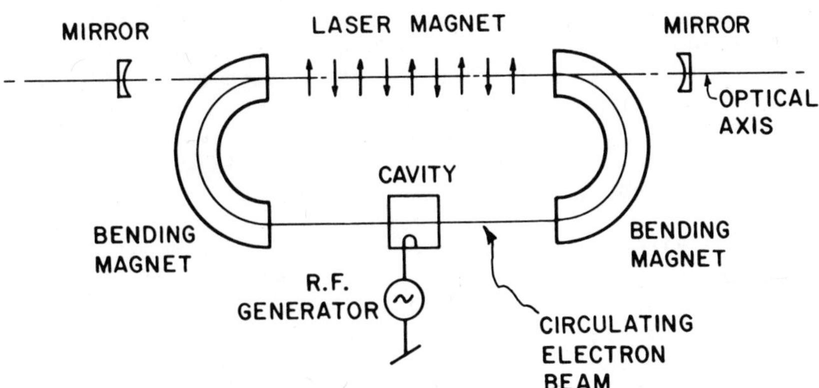

Figure 3. A storage ring Free Electron Laser Oscillator (schematic drawing).

exact solutions are possible. So approximations must be devised. Since several approximations are possible, a critical appraisal must be made to determine comparative reliability and the circumstances under which each can be applied.

II. THE CLASSICAL APPROXIMATION

The classical approximation has traditionally been applied to the analysis of free electron devices, and the language of this approximation permeates the subject. In the classical approximation, the commutators of the operators for the electromagnetic field and the electrons are assumed to be small and the analysis proceeds by integration of the classical equations of motion. The classical approximation is particularly useful at long wavelengths and at high power, but breaks down at short wavelengths and low power.

The evolution of the dynamical variables in the classical approximation is completely determined by the initial conditions. In the free electron laser, the initial conditions would include the specifications of the time of arrival and the momentum of the electrons and the amplitude and first derivatives of the field at the entrance to the interaction region. For fixed electron momentum and field amplitude the classical trajectory of the electrons is determined by the phase of the field at the time of entry.

For the purpose of this discussion considerations can be restricted to the case in which the interaction between electrons in the beam can be neglected. This neglects space-charge effects, and, more fundamentally, neglects the effects of changes in the radiation field due to radiation and absorption by the other electrons in the beam. The restriction limits the discussion to the case in which the gain per pass is small. The restriction is not particularly serious since this is the regime in which FEL oscillators are most likely to operate.

It is an elementary problem to integrate the equations of motion under these conditions. Figure 4 shows the result of a numerical integration for the conditions in the Stanford 10 micron experiment. Depending on the phase, an electron can either gain or lose energy as radiation. It is found that if the electron momentum is properly chosen the electrons will, on the average, transfer energy to the field. The mean radiated energy is found to be proportional to the square of the field amplitude and the corresponding gain is comparable to that estimated in the other approximations. The periodic variation in radiated energy with phase leads to velocity modulation and bunching, effects familiar from the theory of microwave electronic devices.

The data in Figure 4 is for a power density of 10^4 W/cm^2. Numerically the mean energy of the electron was reduced by 0.005% during the interaction. The rms spread in energy was 0.03% at this power level. The spread in energy due to velocity modulation is linear in the amplitude of the field.

III. APPLICABILITY OF THE CLASSICAL APPROXIMATION

Classical analyses of this sort are based on the presumption that the phase and amplitude of the field in the interaction region can be fixed by specification of the initial conditions. But this is only a simplifying assumption. In actuality, specifications of the initial condition is not merely insufficient but also inconsistent with specification of the field in the interaction region. According to the quantum theory of fields, our ability to measure the field or, alternatively, to predict the field on the basis of past measurements, is limited. The uncertainty in the field will set a limit to the applicability of the classical approximation.

The operators for the various components of the field do not, in general, commute when applied at points in space-time which can be connected by a light signal.[7] The commutation relations can be used to develop a set of uncertainty products. Analogous to the Heisenberg uncertainty product $\Delta p \cdot \Delta x \approx \hbar$, these products set limits to the precision with which the components of the field can

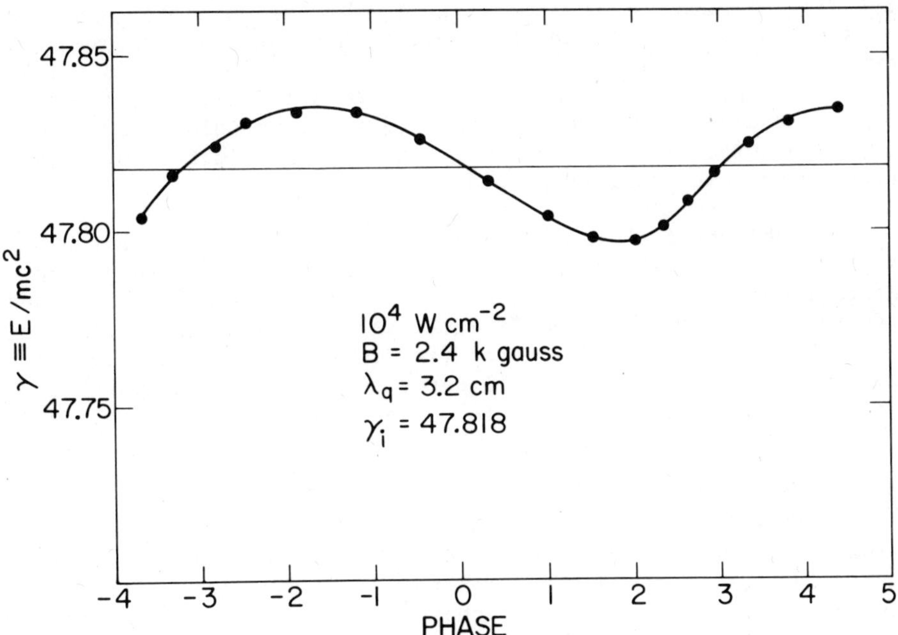

Figure 4. This figure illustrates the relationship between the electron energy at the output of the interaction region and the phase of the $\underset{\sim}{\varepsilon}$ field relative to the electron's transverse velocity in the last period of the magnet. The data points indicate the results of a numerical integration of the equations of motion in which the phase of the $\underset{\sim}{\varepsilon}$ field at the entrance to the interaction region was varied in steps of (π/6) radians. The straight line indicates the initial electron energy. The physical parameters were chosen to match the condition in the Stanford 10 micron experiment.[1]

be specified at neighboring points in space and time. The existence of these uncertainty products leads to a remarkable conclusion: if, in the spirit of the classical approximation, an attempt was made to exactly specify the initial conditions, the result would be to reduce, rather than to improve, our state of knowledge concerning the field within the interaction region.

A similar problem is associated with the boundary conditions. In the Free Electron Laser, the interaction region is enclosed by conducting boundaries. The imposition of metallic boundary conditions fixes the value of the tangential electric field and the normal magnetic field at the surface. From the perspective of the preceding paragraphs, the specification of the boundary condition leads to an uncertainty in the field within the boundaries. A more conventional appraisal of the problem can be formulated in terms of the zero point fluctuations. The boundaries define the normal modes of the electromagnetic field and the zero point fluctuations of the quantized normal modes will introduce a fluctuating term in any measurements of the field.

The object of the classical analysis of the Free Electron Laser is the determination of the electron's trajectories through the interaction region. This objective evidently can not be attained when the uncertainty in the field acting on the electrons is comparable to the mean value of the field. An estimate of the power level at which this occurs will serve to determine the point at which the classical approximation breaks down. In what follows, we estimate the uncertainty in the field due to the zero point fluctuations.

The scatter in a measurement of the field will depend on the volume and time over which the field is averaged. The field acting on an electron in the interaction region is the local field, hence, the averaging volume should be chosen comparable to the microscopic dimensions of the electron. The necessity to average the field over time arises from the high frequency fluctuations: we are interested only in those fluctuations whose frequency is comparable to the stimulating radiation field. The uncertainty is determined solely by the averaging time if the averaging volume is small in comparison to the rest frame wavelength.

For simplicity, we compute the fluctuations in the electron rest frame. The operator for the electric field is:

$$\underset{\sim}{\varepsilon}(\underset{\sim}{x},t) = \frac{1}{\sqrt{V}} \sum_{K,\alpha} \sqrt{\frac{\hbar\omega}{2}} \Big[a_{K,\alpha}\, \underset{\sim}{\varepsilon}^{\alpha}\, e^{i\,\underset{\sim}{k}\cdot\underset{\sim}{x} - \omega t} - a^{+}_{K,\alpha}\, \underset{\sim}{\varepsilon}^{\alpha}\, e^{-i\,\underset{\sim}{K}\cdot\underset{\sim}{x} + \omega t} \Big] \quad , \tag{2}$$

where V is the volume defined by the boundary conditions and $\omega = c|\underset{\sim}{K}|$. We define the average field by:

$$\overline{\varepsilon}(\underset{\sim}{x},t) = \frac{1}{2a\pi^2 b^3 \tau} \int dt' \int d\underset{\sim}{x}'\, e^{-i\omega_o t}\, e^{-(t-t')^2/2\tau^2}\, e^{-|\underset{\sim}{x}-\underset{\sim}{x}'|^2/2b^2} \times \underset{\sim}{\varepsilon}(\underset{\sim}{x}',t') \quad , \tag{3}$$

where ω'_o is the rest-frame frequency of the stimulating radiation field, τ the averaging time, and b the radius of the averaging volume. The mean-squared uncertainty in the field is given by the vacuum expectation value of $\bar{\varepsilon}^* \cdot \bar{\varepsilon}$.

For the free electron laser, the time scale is bounded by the time for the electrons to pass through the interaction region. In considerations involving the evolution of the electron beam within the interaction region, the pertinent length is the magnet period, and the averaging time can appropriately be taken as the time to traverse this length. But in the electron rest frame, the Lorentz-contracted magnet period is equal, approximately, to the wavelength of the radiation, hence, $\tau \approx 2\pi/\omega'_o$. The radius of the averaging volume could appropriately be taken either as the classical electron radius or the Compton wavelength. In either case, the averaging volume would be small compared to the rest frame wavelength and the mean-squared uncertainty in the field is given by:

$$<0|\frac{1}{2}(\bar{\varepsilon}^* \cdot \bar{\varepsilon} + \bar{\varepsilon} \cdot \bar{\varepsilon}^*)|0> \approx \frac{1}{\sqrt{\pi}} \frac{1}{c^3} \frac{\hbar\omega'^3_o}{\tau} \approx \frac{1}{2\pi^{3/2}c^3} \hbar\omega'^4_o \qquad (4)$$

To complete the analysis, this result needs to be compared to the mean squared amplitude of the stimulating radiation field. For a circularly polarized plane wave in free space the power density is given by the magnitude of Poynting's vector:

$$|\underset{\sim}{S}| = \frac{c\varepsilon^2}{4\pi} \quad . \qquad (5)$$

For the classical approximation, the stimulating field should be large in comparison to the zero point fluctuation. Equivalently:

$$|\underset{\sim}{S}| >> \frac{c}{4\pi} <0|\frac{1}{2}\bar{\varepsilon}^* \cdot \bar{\varepsilon} + \bar{\varepsilon} \cdot \bar{\varepsilon}^*)|0> = \frac{1}{8\pi^{5/2}c^2} \hbar\omega'^4_o \qquad (6)$$

To transform this result to the laboratory and the electron frame, we note that in a free electron laser the k-vector of the radiation and the electron velocity are co-parallel so that

$$S_{lab} = (1 + \beta^*)\gamma^{*2} \, S_{ERF} \quad ,$$

where $\beta^* c$ is the longitudinal velocity of the electrons in the lab frame and $\gamma^{*2} = (1 - \beta^{*2})^{-1}$. We then have the condition for the classical approximation:

$$S_{lab} >> \frac{1}{4\pi^{3/2}} \frac{\hbar\omega_o^3}{(1+\beta^*)\lambda_q} , \qquad (7)$$

where ω_o is the lab frame frequency and λ_q is the magnet period.

The form of Equation (7) is Lorentz covariant. In fact, its form is unique: it is the only covariant product with the units of power density which can be formed from h, c, ω_o, and λ_q. Less certainty can be ascribed to the numerical coefficients in the equation. It is apparent that the effect of the fluctuations has been over-estimated because the sum in Equation (2) includes all the components of polarization whereas only the transverse components of ε will interact with the electrons. Also, the choice of averaging time, though reasonable, is obviously not unique. But the basic order of magnitude of the zero-point fluctuations is not in doubt.

Numerically, the power density required for the classical approximation is low at long wavelengths but quite substantial at short wavelengths:

$$\begin{aligned} S_{lab} >> \; & 1.6 \times 10^{-4} \text{ watts/cm}^2 \text{ at } 10\ \mu \\ & 1.6 \times 10^{2} \qquad\qquad \text{at } 1000\ \text{Å} \\ & 1.6 \times 10^{8} \qquad\qquad \text{at } 10\ \text{Å} \end{aligned} \qquad (8)$$

These figures are for the 3.2 cm magnet period in the Stanford experiment. When the stimulating field is small in comparison to the zero point fluctuations, the motion of an electron in the interaction region will be dominated by the fluctuations. It will no longer be possible to predict from the initial condition whether a given electron will emit or absorb radiation as it moves through the interaction region.

The magnitude of the zero point fluctuations clearly defines the necessity for a quantum analysis at short wavelengths. Even at 10 microns, there may be insufficient power density at turn-on (when the power is of the order of the spontaneous power) to justify the classical approximation. In the Stanford 10 micron experiment, the power density at turn on was $\sim 10^{-6}$ watts/cm^2, four orders of magnitude below the level required for the classical approximation.

In the classical approximation, velocity modulation and bunching occur as a consequence of the periodic reversal of the phase of the field. Electrons which enter the interaction region at half cycle intervals will alternately be accelerated and retarded. It is not clear that velocity modulation will persist when the amplitude of the field becomes uncertain. It might be hypothesized that phase reversal would occur independent of the uncertainty and that bunching would proceed unhindered. That this does not occur can be seen from the expectation value of the operator $\bar{\varepsilon}^*(\underset{\sim}{x}, t + \Delta t) \cdot \bar{\varepsilon}(\underset{\sim}{x},t)$ where $\bar{\varepsilon}$ has been defined according to Equation (3). We have:

$$< 0|\bar{\varepsilon}^*(\underset{\sim}{x}, t + \Delta t) \cdot \bar{\varepsilon}(\underset{\sim}{x},t)|0 > \sim e^{-\Delta t^2/4\tau^2} \quad . \tag{9}$$

Phase coherence does not extend beyond the averaging time.

IV. A USEFUL QUANTUM APPROXIMATION

Several small signal quantum analyses of the gain have been published.[6] These analyses proceed by calculating the difference between the transition rates for stimulated emission and absorption and, with the exception of Colson's work, are limited to the case in which fewer than one photon per electron is radiated during the interaction. To extend these analyses to higher power and find the statistics of the electron energy loss, an estimate must be developed for the probability amplitude for multiple photon emission and absorption.

Given that the phase and magnitude of the field are uncertain, the possibility must be allowed that an electron will both emit and absorb radiation during the interaction. The amplitude for a particular sequence of events in which N + K photons are emitted and K photons are absorbed can be expressed as the product of the amplitude for the electron to propagate to the first event times the amplitude for that event times the amplitude to propagate to the second event and so on. The ordered integral of this product over the length of the interaction region yields the net amplitude for the emission of (N + K) and the absorption of (K) photons. The sum over K of all such amplitudes yields the net amplitude for the emission of N photons. The square of this last amplitude yields the probability distribution function.

Evaluation of the exact amplitude would yield a general solution, but the exact problem is formidable. An approximate result can be obtained by means of some judicious approximations. The most important of these is the "chaotic field" approximation according to which emission and absorption occur as statistically independent events. In this approximation, the amplitude for emission of radiation in an interval along the trajectory of an electron is unaffected by the occurrence of emission and absorption in the adjoining intervals. The approximation follows from the discussion of the zero-point fluctuations in Section III.

The probability distribution in the chaotic field approximation has been evaluated. The magnet period and length were chosen to match the conditions in the 10 micron gain experiment. As shown in Fig. 5 (top), the electrons radiate a portion of their energy if E is above the resonance energy E_o, and absorb a similar amount if E is below E_o. This result reproduces what we obtain from the small-signal analysis which gives a curve similar to Fig. 2b. Both the size and the shape of this curve change as the interaction strength is increased by, for example, increasing the magnetic field strength. As saturation is approached, the energy loss becomes equal to the energy offset $E - E_o$.

For a given energy, we find that the probability distribution about the mean number of emitted photons has a finite statistical width. Figure 5 (bottom) shows this distribution for a number of interaction strengths. Saturation is again evident, and to a good approximation the statistical fluctuation in the number of radiated photons per electron has the value

$$\langle N^2 \rangle - \langle N \rangle^2 \approx 1600 \langle N \rangle^{1/2} \tag{10}$$

for the conditions described. The general character of the results is shown in Fig. 6, which is a contour map of the log of the probability. The vertical axis is energy offset and the horizontal axis is energy loss in the form of radiation. Figure 5 (top) is then a trace of the ridgeline of this contour graph, presenting the mean energy loss; and Fig. 5 (bottom) is formed from a horizontal cut through the ridge at a given offset energy. A series of these plots has been generated for the range of values of the interaction strength, and form a complete specification of the operation of the laser.

In this approximation, gain is an interference effect. The amplitudes for emission and absorption interfere destructively off resonance. Above the resonance energy, interference is minimized when the electrons transfer energy to the field and this leads to an enhancement in the probability for radiation. The situation is similar below resonance except that interference is minimized when the electrons are accelerated by the field. Hence, absorption predominates for electron energies below the resonant energy. There is nothing analogous in this approximation to velocity modulation and bunching.

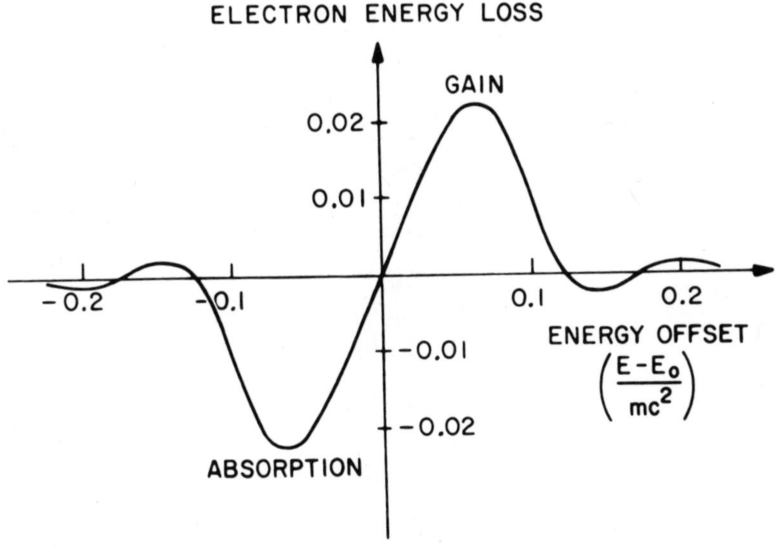

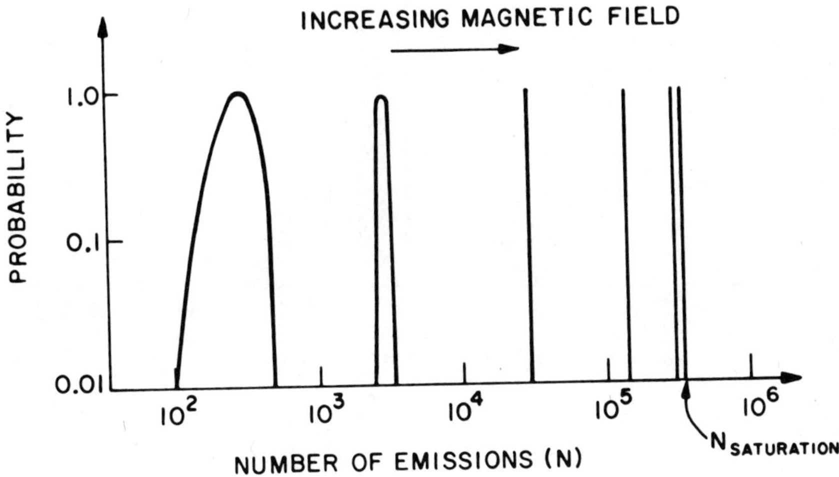

Figure 5. Top: Theoretical predictions of the Chaotic Field Approximation. Electron energy loss is plotted against energy offset, both in units of mc^2. Compare with the experimental results, Fig. 2b. Bottom: The probability distribution of emitted photons is shown for several values of the interaction strength. The energy offset is 0.05 mc^2.

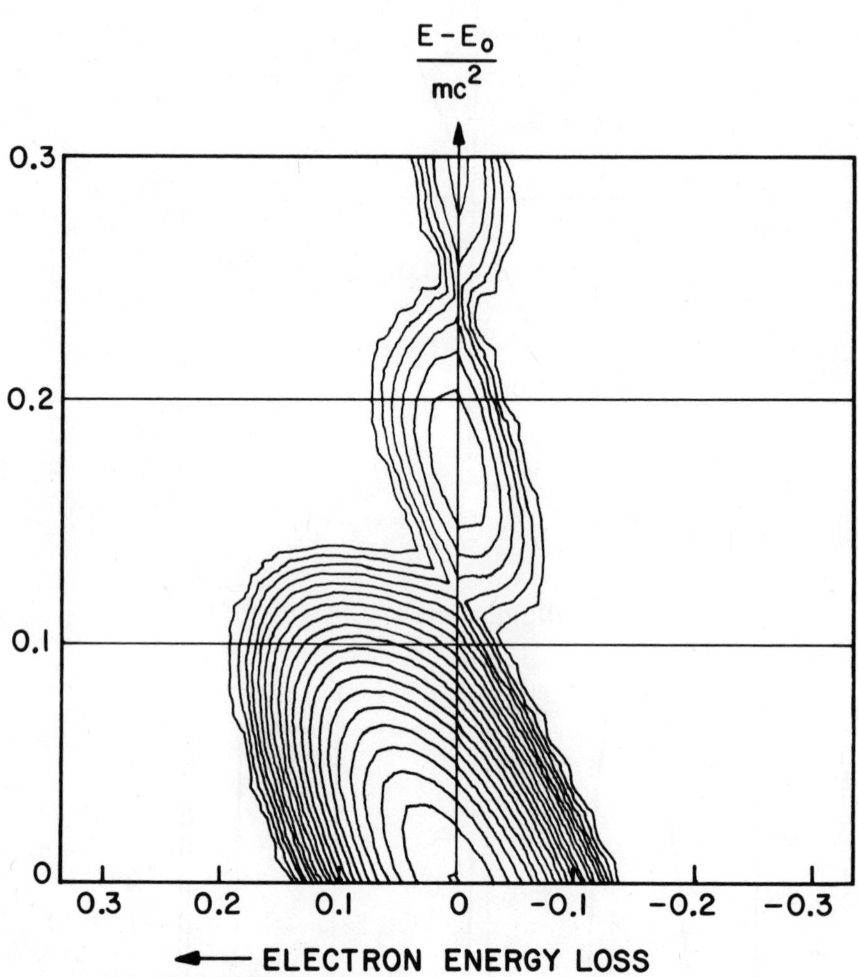

Figure 6. Contour map of the log of the probability as a function of energy offset vertically, and energy loss horizontally, both in units of mc^2. This map is antisymmetric with respect to energy offset, so only the high energy half is shown. The results in Fig. 5 are taken from a series of these graphs which completely specify the operation of the laser. See the discussion in Section IV.

The chaotic field approximation is clearly complimentary to the classical approximation. The classical approximation fails at short wavelengths and low power when the amplitude and phase of the field become uncertain while the chaotic field approximation fails at long wavelengths and high power when the phase determines whether a given electron will preferentially emit or absorb radiation.

V. ELECTRON BEAM STABILITY IN A STORAGE RING

A fundamental question of interest in the development of this class of laser is the source of the relativistic electron beam required for operation. Use of an electron storage ring would be particularly attractive. The idea was illustrated in Figure 3. The ring would provide, at once, an economical and compact means to generate the high current required for higher power operation at optical wavelengths and also the means to achieve high overall efficiency.[3]

A peak circulating current in excess of 10 amps was attained in the Stanford-Princeton ring at 500 MeV: this current would be sufficient to sustain laser operation at wavelengths as short as 1000 Å. The prospect for operation at high power arises from the high circulating power of the stored electron beam. A fraction equal to 0.25% of the electrons' energy was converted to radiation in a single pass through the interaction region in the Stanford 10 micron experiment. For a storage ring with a 100 MeV, 0.5 amp average current beam, an extraction ratio of 0.25% would yield an average power output of 125 kilowatts at 6200 Å.

The prospect for operation at high efficiency arises from the recirculation of the electron beam. The RF field accelerating the beam within the ring would need to supply only the energy extracted from the beam as radiation. Thus, the efficiency would be limited only by the efficiency of the high level system and the power consumed by the ring magnets, controls, and instrumentation. Estimates of the attainable overall efficiency exceed 20% for 100 kilowatts optical power output.

The central question in the analysis of this concept concerns the stability of the circulating electron beam at high power. In a conventional storage ring the statistical fluctuations in the electron energy are damped by radiation losses leading to beams with an energy spread of the order of parts in 10^4.[8] Installation of a periodic magnet and operation of the laser will certainly alter the electron distribution and the question is whether the electron beam quality will remain within acceptance limits.

In general, both the mean energy and the energy spread of an electron beam passing through the interaction region will be altered. The change in the mean energy of the beam can be compensated by the RF accelerating field in the ring. The problem of beam spread is more subtle. A general analysis would involve consideration of the characteristics of the guide and focussing magnets for the ring and also the time dependence of the accelerating field. These considerations are critical to the determination of the bunch length and the emittance of the circulating electron beam, but are neglected in the discussion which follows. Our purpose here is to establish the physical mechanism responsible for stability in the laser and to obtain an order of magnitude estimate of the equilibrium energy spread.

It is clear from both the classical and chaotic field approximation that an initially monoenergetic electron beam will emerge from the interaction region with a finite spread in energy. But the radiated energy is also energy dependent and electrons on the high energy side of the electron distribution can lose substantially more energy than the electrons on the low energy side. This effect would tend to narrow the energy distribution. The equilibrium energy spread will be determined by the balance between the two effects.

The equilibrium energy spread can be estimated using the statistical Green's function (Fig. 7). If $G(\gamma,\gamma')$ is the output distribution function for a monoenergetic input electron beam of energy $\gamma' mc^2$, and $F(\gamma')$ is the distribution function for the "equilibrium" distribution at the input to the interaction region, the distribution at the output end will be $F'(\gamma)$, where:

$$F'(\gamma) = \int G(\gamma,\gamma')\, F(\gamma')\, d\gamma' \quad . \tag{11}$$

If it is assumed that the remainder of the ring simply accelerates the beam by $\Delta\gamma mc^2$, then the condition for equilibrium yields an integral equation for $F(\gamma)$:

$$F(\gamma) = F'(\gamma - \Delta\gamma) = \int G(\gamma - \Delta\gamma,\ \gamma')\, F(\gamma')\, d\gamma' \quad . \tag{12}$$

The form of the Green's function will be determined by the operating conditions and the magnet geometry. The dependence on magnet geometry is a subject of particular interest. The width of $G(\gamma,\gamma')$ will reflect both the statistical fluctuations in the number of radiated photons and the energy modulation apparent in Figure 4.

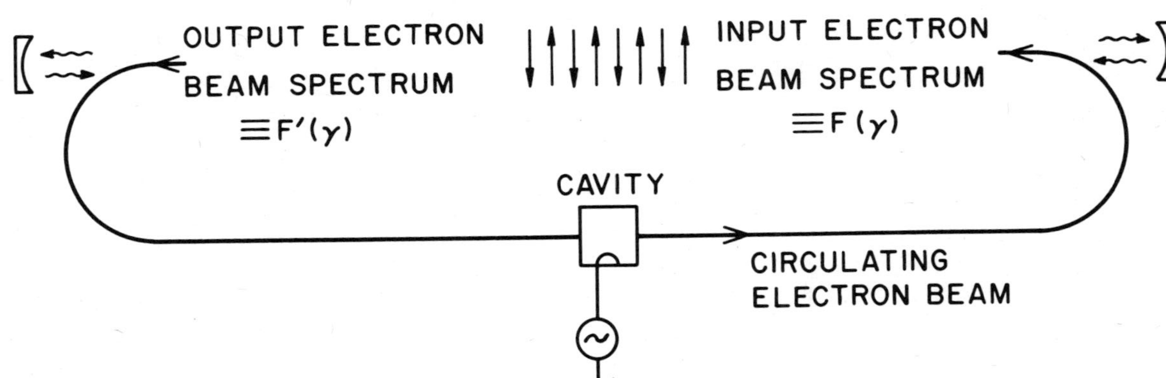

Figure 7. In a storage ring, the beam emerging from the periodic magnet would be re-accelerated and returned to the interaction region. The statistical Green's function can be defined as the output distribution for a nomoenergetic input electron beam, e.g., if $F(\gamma) = \delta(\gamma - \gamma')$, then $F'(\gamma) = G(\gamma,\gamma')$. The functions F, F', and G will depend on the power density and spectrum of the field in the interaction region at equilibrium.

The statistical fluctuations degrade the phase space of the electron beam but this is not true of the classical energy modulation. On a microscopic scale, momentum remains correlated with position in the classical approximation and it is possible, at least in principle, to design a magnet which exploits the correlation to suppress the modulation. The effect of a two period helix spaced at the end of the main magnet at a distance chosen to minimize the energy spread is illustrated in Figure 8.

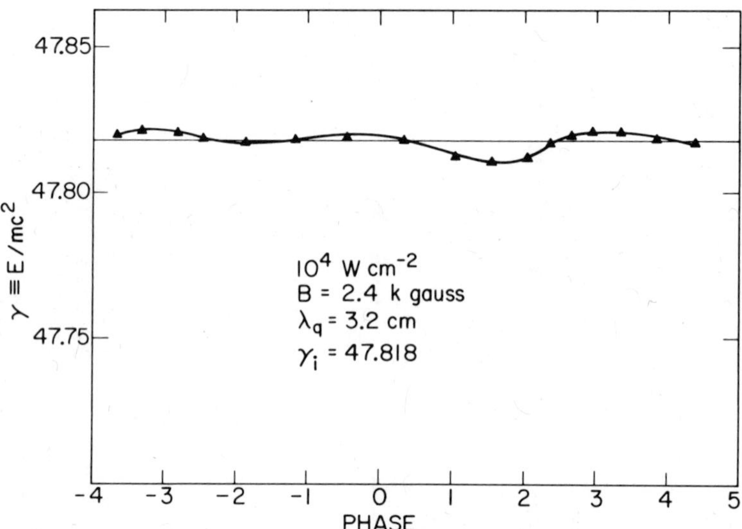

Figure 8. The conditions are the same as in Figure 4 except that the electron beam and the radiation emerging from the interaction region have been run through a period helical magnet whose spacing from the main magnet has been adjusted to minimize the peak-to-peak spread in energy. The horizontal axis now indicates the phase of the ε field relative to the electron transverse velocity in the laser period of the short magnet.

The form of F can be estimated from the first and second moments of G. The first moment of G reflects the dependence of the mean radiated energy on the initial energy. Theory and experiment indicate that the difference

$$\gamma' - \left[\int \gamma \, G(\gamma,\gamma') \, d\gamma \right] \tag{13}$$

will have the form shown in Figure 2b. For the purpose of this discussion, the difference can be taken to be proportional to the difference between the initial electron energy, $\gamma_i \; mc^2$, and the resonance energy $E_o = \gamma_o \; mc^2$:

$$\gamma' - \left[\int \gamma \; G(\gamma,\gamma') \; d\gamma\right] \approx \alpha(\gamma' - \gamma_o) \quad . \tag{14}$$

This is reasonable as long as the electron energy remains in the region between the gain and absorption peaks. The width of G can be defined in the standard way as the difference between the second moment and the square of the first moment:

$$\int \gamma^2 G(\gamma,\gamma') \; d\gamma - \left[\int \gamma G(\gamma,\gamma') \; d\gamma\right]^2 \equiv \sigma^2(\gamma') \quad . \tag{15}$$

For the purpose of this approximation, σ is taken to be independent of γ'. From the theoretical results plotted in Fig. 6, it is apparent that this approximation is very good in the region mentioned above.

From the steady state requirement that the width of the output distribution $F'(\gamma)$ equal the width of the input distribution $F(\gamma)$, it is straightforward to calculate the equilibrium width which we define as Σ:

$$\Sigma^2 \equiv \int \gamma^2 F(\gamma) \; d\gamma - \left[\int \gamma F(\gamma) \; d\gamma\right]^2 = \frac{\sigma^2}{2\alpha + \alpha^2} \quad . \tag{16}$$

The width of F depends only on the first and second moments of G. As previously noted, σ^2, the width of G, will depend on the magnet geometry. Approximate upper and lower bounds on the width of $F(\gamma)$ can be obtained from the widths calculated for the constant period helix in the classical and chaotic field approximations. The chaotic field approximation yields only the statistical fluctuations while the classical approximation provides a worst case estimate of the spread due to energy modulation:

$$7 \times 10^{-5} \lesssim \frac{\Sigma}{\gamma} \lesssim 1.1 \times 10^{-3} \quad . \tag{17}$$

The upper and lower limits are calculated for the conditions in Figure 4, e.g., for a power density of 10^4 watts/cm^2 at 10 μ. The upper limit is insensitive to the power density up to saturation, while the lower limit decreases slowly with power density.

For laser operation, the contribution of the electron energy spread to the gain linewidth needs to be kept small in comparison to the broadening, due to the finite length of the interaction region. The same condition sets limits to the radius and the angular divergence of the electron beam in the interaction region. The fractional energy spread should not exceed (1/2N), where N is the number of magnet periods. Numerically, for the 160 period magnet assumed in the derivation of the limits in Equation (17), Σ/γ should be maintained below 0.3%. The range of values estimated in Equation (17) is consistent with this limit. At the same time, the spread in energy should not be too small: the spread must be sufficiently large to maintain the bunch length at the desired value. The minimum acceptable spread has yet to be determined.

VI. SUMMARY

The recent experiments by the Stanford Free Electron Laser group have established that the device is capable of providing useful gain at optical wavelengths. The devices are interesting because of the tuning range, efficiency, and the potential for high power output.

The small signal gain for the interaction in the "low gain" limit was first estimated using the quantum theory and the Weizsacker-Williams approximation. The existence of gain in this approximation is due to the difference in the frequencies for emission and absorption. The theory yields a result for the gain close to the experimental value.

Free electron devices are analyzed conventionally using classical electrodynamics, and it has been established that the classical theory yields approximately the same gain equation as the small signal quantum theory. In the classical theory, gain is due to the re-organization of charge within the electron beam which increases the net work done by the electron on the field. To what extent is the classical approximation justified in the analysis of the device at optical wavelengths?

The classical approximation is applicable when the commutators of the operators for the dynamical variables of the electron and the electromagnetic field can be neglected. In classical electrodynamics, the amplitude and phase of the field and the position and momentum of the charged particles can be specified simultaneously with arbitrary precision. But the non-commuting variables can <u>not</u> be simultaneously defined. The consequences of this are well known for particle dynamics. The Heisenberg uncertainty principle $\Delta p \cdot \Delta x \sim \hbar$ sets a limit to the applicability of classical dynamics at small distances and small momenta. The corresponding uncertainty relation for the electric field limits the precision with which the field amplitude can be defined when the field interacts with a localized charged particle. As in the case of classical particle dynamics, the classical approximation for the field is inapplicable unless the classical field amplitude is large in comparison to the uncertainty. The power density required to justify the classical approximation ranges from 10^{-2} watts/cm^2 at 10 μ to 10^{10} watts/cm^2 at 10 Å.

The classical notion of an interaction in which energy is transferred continuously from the electron beam to the field, or the reverse, is therefore inappropriate at short wavelengths and low power. At short wavelengths, specific notice needs to be taken of the possibility that an individual electron can both emit and absorb radiation from the field. The uncertainty in the field is a quantum effect, and a quantum analysis of the gain is clearly required when the uncertainty becomes significant.

The early quantum analyses were limited to the case of single photon emission and absorption. A more general analysis is required for multiple photon emission. The "chaotic field" approximation follows from the assumption that emission and absorption occur as statistically independent events. The approximation leads to an estimate of the mean number of radiated (or absorbed) photons, the statistical fluctuations, and the saturation characteristics. Gain is seen to occur as an interference effect. While the electrons, on the average, will transfer energy to the field if the initial energy is above the resonance energy, radiated energy and position are uncorrelated in this approximation, and there is therefore no velocity modulation or bunching.

The uncertainty in the electric field at short wavelengths clearly limits the applicability of the classical concepts in the analysis of the interaction. There are also engineering implications. The spread in energy of the circulating electrons in a storage ring laser will depend on the fluctuations in the energy radiated by the electrons in the periodic magnet. In the classical approximation, the radiated energy is correlated with the phase and this correlation can be exploited to control the spread in energy. But the quantum fluctuations are statistical in origin and set a definite lower bound to the attainable energy spread.

The possibility of stability in a recirculating system arises from the strong energy dependence of the radiated energy. Electrons moving through the field with too much energy radiate more energy than those with an energy deficiency. The damping due to this effect is very strong in comparison to the damping due to incoherent synchrotron radiation in an ordinary storage ring. The estimated equilibrium energy spread in a recirculating system is consistent in order of magnitude with the spread required for laser operation.

References

(1) L. R. Elias, et al., Phys. Rev. Letts. 36, 717 (1976).

(2) D. A. G. Deacon, et al., to be published in Phys. Rev. Letts.

(3) L. R. Elias, et al., "A Discussion of the Potential of the Free Electron Laser as a High Power Tuneable Source of Infrared, Visible and Ultraviolet Radiation," published in the Proceedings of the Synchrotron Radiation Facilities Quebec Summer Workshop, (Université Laval, Quebec, Canada, 15-18 June 1976).

(4) H. Motz, J. Appl. Phys. 22, 527 (1951); H. Motz and M. Nakamura, Ann. Phys. (N.Y.) 7, 84 (1959); H. Motz and M. Nakamura, Proceedings of the Symposium on Millimeter Waves, Microwave Research Institute Symposia Series, Vol. IX, (New York: Interscience, 1960), p. 155; H. Motz, W. Thon, and R. N. Whitehurst, J. Appl. Phys. 24, 826 (1953).

(5) J. M. J. Madey, J. Appl. Phys. 42, 1906 (1971).

(6) H. Dreicer, Phys. Fluids, 7, 735 (1964); R. H. Pantell, G. Soncini and H. E. Puthoff, IEEE J. Quantum Electron. 4, 905 (1968); V. P. Sukhatme and P. W. Wolff, J. Appl. Phys. 44, 2331 (1973); P. Sprangle, V. L. Granatstein and L. Baker, Phys. Rev. A, 12, 1697 (1975); F. A. Hopf, et al., Optics Commun., 18, 413 (1976); F. A. Hopf, et al., Phys. Rev. Letts. 37, 1342 (1976); W. B. Colson, Phys. Letts. 59A, 187 (1976); A. Hasegewa et al., Appl. Phys. Letts. 29, 542 (1976); T. Kwan, J. M. Dawson and T. Lin, to be published in Phys. Fluids; A. Gover to be published in Appl. Phys. (Springer-Verlag).

(7) W. Heitler, The Quantum Theory of Radiation (The Clarendon Press, Oxford, 3rd Edition) p. 76.

(8) M. Sands, in Physics with Intersecting Storage Rings, B. Touschck, ed. (Academic Press, N.Y.: 1971) p. 257.

CLASSICAL AND SEMICLASSICAL TREATMENT OF THE PHASE TRANSITION IN DICKE MODELS*

R. Gilmore† and C. M. Bowden

Quantum Physics, Physical Sciences Directorate

US Army Missile Research and Development Command

Redstone Arsenal, AL 35809

Abstract: A physically intuitive and mathematically transparent method is presented for determining the range of values of the coupling constants for which a phase transition is to be expected in Dicke models. This "classical" method provides information about the free energy and order parameters at $T = 0$. A second method, as physically transparent and involving no mathematics more difficult than the diagonalization of a 2×2 matrix, is then presented. This "semiclassical" method is sufficient to determine the free energy and the system order parameters as a function of temperature. It also determines the critical temperature of the phase transition, if any occurs. The "semiclassical" method provides values for intensive parameters which are rigorously correct in the thermodynamic limit.

I. INTRODUCTION

The recent proof that systems described by a Dicke Hamiltonian[1] can undergo a phase transition at some well-defined critical temperature[2] has stimulated a great deal of subsequent interest. However, even with the later simplifications,[3-7] the mathematics involved in the description of this process is not straightforward. It is the purpose of the present work to present an analysis of the phase transition which is both mathematically and physically transparent.

*Work partially supported by the US Army Research Office, Durham, North Carolina, Grant DAHC04-72-0001.

†Permanent Address: Physics Department, University of South Florida, Tampa, Florida 33620.

In Section II we introduce the "classical ansatz." In this ansatz, both the field and atomic subsystems are considered as classical. It is then a simple matter to estimate the ground state energy, and from that estimate, to determine the conditions on the coupling constants under which a phase transition is to be expected.

The "classical ansatz" leads to an accurate description of the thermodynamic properties of these models at $T = 0$. It leaves unanswered three important questions: (1) what is the free energy as a function of temperature; (2) what are the values of the order parameters as a function of temperature; (3) what is the critical temperature for phase transitions? To treat these questions, the "semiclassical ansatz" is introduced in Section III. With this ansatz, these three questions are answered completely and accurately for the simple Dicke model. The calculation involves nothing more difficult than the diagonalization of a 2×2 matrix. In Section III we also discuss why the "semiclassical ansatz" leads to a free energy per particle equal to the exact quantum mechanical free energy to order $(\ell n\ N)/N$ which vanishes in the thermodynamic limit $N \to \infty$, $V \to \infty$, N/V = constant.

In Section IV, we study the multimode Dicke model. The "classical" method indicates the conditions under which a phase transition is to be expected. The "semiclassical" method provides information on the behavior of the free energy and the order parameters as a function of temperature. It also determines the critical temperature for a phase transition. The same methods are followed in Section V, where we study the Dicke model Hamiltonian including also the A^2 term and the counter-rotating terms.

Throughout this paper we use $\beta = 1/kT$, where k is Boltzmann's constant.

II. CLASSICAL TREATMENT OF THE DICKE HAMILTONIAN

The Dicke model[1] Hamiltonian describing the interaction of a single mode of the radiation field with N identical 2-level atoms is, in the long wavelength approximation and with $\hbar = 1$

$$H_D(\lambda) = \omega\, a^\dagger a + \varepsilon \sum_{k=1}^{N} 1/2\ \sigma_k^z + \frac{\lambda}{\sqrt{N}} \sum_{k=1}^{N} \left(a^\dagger \sigma_k^- + a\sigma_k^\dagger\right) \quad . \tag{2.1}$$

The operators $a^\dagger$, a are photon creation and annihilation operators for a single mode of the electromagnetic field. They obey the commutation relations

$$[a^\dagger a \;,\; a^\dagger] = +a^\dagger$$
$$[a^\dagger a \;,\; a\,] = -a \qquad [\cdot, I] = 0 \quad . \tag{2.2}$$
$$[a \;,\; a^\dagger] = I$$

The operators $\sigma_k^{\pm;z}$ are Pauli spin operators which describe the internal properties of the kth atom. These operators obey the commutation relations

$$\left[\frac{1}{2}\sigma_j^z \;,\; \sigma_k^\pm\right] = \pm\sigma_k^\pm \,\delta_{jk}$$
$$[\sigma_j^+ \;,\; \sigma_k^-] = \sigma_k^z \,\delta_{jk} \quad . \tag{2.3}$$

Each photon has energy ω; each atom has internal energy level spacing ε in the absence of interactions. The coupling constant describing the strength of the interaction is $\lambda/\sqrt{N}$. The factor $1/\sqrt{N}$ comes from the normal mode expansion[1] of the electromagnetic field

$$A(\underline{x},t) = \sum_{\underline{k},\sigma} \sqrt{\frac{2\pi hc^2}{V\omega_k}} \left(\underline{\varepsilon}_{\underline{k}\sigma}\, a^\dagger_{\underline{k}\sigma}\, e^{-i\underline{k}\cdot\underline{x}} + \underline{\varepsilon}^*_{\underline{k}\sigma}\, a_{\underline{k}\sigma}\, e^{+i\underline{k}\cdot\underline{x}} \right) \tag{2.4}$$

and the imposition of the constant density assumption $N/V = \rho$ = constant. Following Dicke, we initially make the rotating wave approximation (no terms of the form $a^\dagger\sigma^+$, $a\sigma^-$) and neglect the $\underline{A}\cdot\underline{A}$ term.

It is not at all apparent that the system described by (2.1) undergoes a phase transition for sufficiently large values of the coupling constant λ. To see that this might be so, we introduce the "classical ansatz." In the thermodynamic limit $N \to \infty$, $V \to \infty$, $N/V = \rho$ = constant, we might expect the ground state of (2.1) to behave classically in some sense. To the extent that this is so, we can replace the field operators $a, a^\dagger$ and the atomic shift operators by their ground state expectation values:

$$a \to \langle g;\lambda | a | g;\lambda\rangle = \mu\sqrt{N}$$

$$a^\dagger \to \langle g;\lambda | a^\dagger | g;\lambda\rangle = \mu^*\sqrt{N} \tag{2.5a}$$

$$\sigma_k^- \to \langle g;\lambda|\sigma_k^-|\, g;\lambda\rangle = \nu$$

$$\sigma_k^+ \to \langle g;\lambda|\sigma_k^+|\, g;\lambda\rangle = \nu^* \quad . \tag{2.5b}$$

Here $|g;\lambda\rangle$ represents the ground state, which depends on the coupling constant λ. The complex quantities μ, ν have a natural interpretation as order parameters for the field and atomic subsystem, respectively.

We have explicitly included the factor $\sqrt{N}$ for subsequent convenience. These complex quantities are order parameters for the field and atomic subsystems. In addition, each set of atomic operators obeys the equation

$$\left(\frac{1}{2}\sigma_k^z\right)^2 + \frac{1}{2}\left(\sigma_k^+\sigma_k^- + \sigma_k^-\sigma_k^+\right) = J_{op}^2 \begin{array}{l}\nearrow j(j+1) \text{ quantum limit} \\ \searrow j^2 \text{ classical limit}\end{array} \tag{2.6}$$

In anticipation of the classical behavior of the atomic systems, we shall choose the classical limit in (2.6) with $2j + 1 = 2$ and set

$$\frac{1}{2}\sigma_k^z = -\sqrt{\left(\frac{1}{2}\right)^2 - \nu^*\nu} \quad . \tag{2.5b'}$$

The minus sign has been chosen in (2.5b') because, in the absence of interaction ($\lambda = 0$), the ground state energy of each atomic system is $-\varepsilon/2$.

With the "classical ansatz," the Hamiltonian (2.1) assumes the form

$$H_D(\lambda;\mu,\nu)/N = \omega\mu^*\mu - \varepsilon\sqrt{\frac{1}{2}^2 - \nu^*\nu} + \lambda(\mu^*\nu + \mu\nu^*) \quad . \tag{2.7}$$

It is clear from (2.7) why there may be a phase transition.[8] In the absence of interaction, the ground state has no photons in the field mode and all atoms in their ground state: $(\mu,\nu) = (0,0)$. The ground state energy per atom is $-\varepsilon/2$. For small values of λ, the ground state energy is unchanged. However, if λ is sufficiently large, it may be energetically favorable for a large number of photons to be present in the field mode ($\mu\sqrt{N} \gg 1$) and for each atom to be somewhat excited ($\nu \neq 0$), because the energy return from the interaction term $\lambda(\mu^*\nu + \mu\nu^*)$ may be more negative than the energy cost $\omega\mu^*\mu - \varepsilon\sqrt{(1/2)^2 - \nu^*\nu}$ from the field and atomic contributions.

When this can occur the system ground state is an ordered state $\mu \neq 0$, $\nu \neq 0$ rather than the disordered state $(\mu,\nu) = (0,0)$.

In order to estimate the value of the coupling constant λ for which there may be a changeover from disordered to ordered ground state, we compute the free energy at $T = 0$. The free energy per particle of (2.1) is its minimum eigenvalue divided by N, or approximately the minimum energy of (2.7). The values of the parameters (μ,ν) for which (2.7) has stationary values are easily determined

$$\partial(2.7)/\partial\mu^* = \omega\,\mu + \lambda\,\nu = 0$$

$$\partial(2.7)/\partial\nu^* = \frac{\varepsilon\,\nu}{2\sqrt{\left(\frac{1}{2}\right)^2 - \nu^*\nu}} + \lambda\,\mu = 0 \quad . \tag{2.8}$$

These coupled nonlinear equations may be treated by eliminating either μ or ν

$$\left[\frac{\varepsilon}{2\sqrt{\left(\frac{1}{2}\right)^2 - \nu^*\nu}} - \frac{\lambda^2}{\omega}\right]\nu = 0 \quad . \tag{2.9}$$

Equation (2.9) always possesses the trivial solution $\nu = 0 \Rightarrow \mu = 0$, called the disordered state. However, a nontrivial solution (ordered state) is possible if the factor within the brackets vanishes. This can only occur when $\lambda^2/\omega \geq \varepsilon$. The vanishing of this factor determines the magnitude but not the phase of ν. The values of (2.7) computed on the solution branches of (2.8) or (2.9) which make (2.7) stationary, are

$$H_D(\lambda;\mu,\nu)/N = -\frac{1}{2}\,\varepsilon \qquad (\mu = 0,\ \nu = 0) \tag{2.10a}$$

$$= -\frac{1}{2}\,\varepsilon \times \frac{1}{2}\left[\frac{\lambda^2}{\varepsilon\omega} + \frac{\varepsilon\omega}{\lambda^2}\right] \text{ for } \frac{\lambda^2}{\varepsilon\omega} > 1 \text{ and}$$
$$\left(\mu = -\frac{\lambda}{\omega}\,\nu,\ |\nu| = \frac{1}{2}\left[1 - (\varepsilon\omega/\lambda^2)^2\right]^{1/2}\right) \quad . \tag{2.10b}$$

It is clear that whenever the ordered state can exist, (2.7) assumes its minimum value when evaluated at the nontrivial solution of (2.9).

The order parameters μ,ν are not independent, but are related by (2.8). In order to express (2.7) in terms of an independent set of variables, one of the equations of (2.8) may be used to eliminate either μ,μ^* or ν,ν^*:

$$H_D(\lambda;\nu) = -\frac{\lambda^2}{\omega}\nu^*\nu - \varepsilon\sqrt{\left(\frac{1}{2}\right)^2 - \nu^*\nu} \quad . \tag{2.11}$$

To investigate the stability properties of (2.11) on solutions of (2.9) it is necessary to compute the eigenvalues of the matrix

$$\frac{\partial^2 H_D(\lambda,\nu = y_1 + iy_2)}{\partial y_i \partial y_j} \quad .$$

This matrix of mixed second partial derivatives is called the stability matrix of $H_D(\lambda;\nu)$.

The stability matrix, evaluated on the solution branch $\nu = 0$, has two positive eigenvalues when $\lambda^2/\omega < \varepsilon$, two zero eigenvalues when $\lambda^2/\omega = \varepsilon$, and two negative eigenvalues when $\lambda^2/\omega > \varepsilon$. For a nontrivial solution of (2.9) to exist, $\lambda^2/\omega > \varepsilon$. When evaluated on the nontrivial solution, the stability matrix has one positive and one zero eigenvalue. Thus, when (2.9) has a nontrivial solution, (2.7) and (2.11) assume their minimum values on that nontrivial solution.

These results can easily be understood by visualizing the shape of (2.11). For $\lambda^2/\varepsilon\omega > 1$, it has the shape of a circularly symmetric sombrero, with a peak in the center at $\nu = 0$ and a circle of minimum values at $|\nu| \neq 0$ determined by (2.9). As λ decreases, the peak shrinks and the circle of minimum values moves towards the center. As $\lambda^2/\varepsilon\omega \to 1^+$ the central local maximum and the ring of minimum values coalesce as $|\nu| \to 0$. For $\lambda^2/\varepsilon\omega < 1$, there is no longer a central peak but a central dip, and (2.11) is roughly paraboloidal in shape (Fig. 1).

When $\lambda^2/\varepsilon\omega > 1$ and the stability matrix is evaluated on the ring of minimum values, the radial eigenvalue is positive while the azimuthal eigenvalue is zero. The zero eigenvalue shows clearly the rotational symmetry of (2.11) and (2.7). This rotational invariance is usually called "gauge" invariance, and is clearly shown in Fig. 1.

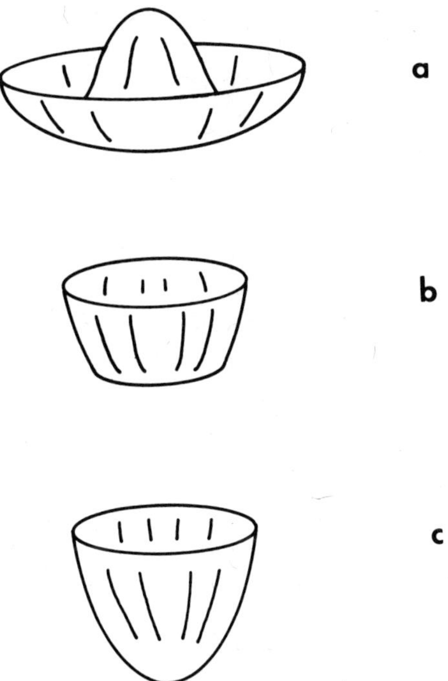

Figure 1. Shape of the energy function (2.11) for three values of the coupling constant λ. a) For $\lambda^2 > \varepsilon\omega$, the function (2.11) looks like a sombrero, with a central peak and a ring of minimum values whose radius $|\nu|$ is determined by the non-zero solution of (2.9). b) for the critical value $\lambda_c^2 = \varepsilon\omega$ the central peak and minimum ring have just coalesced. The function behaves like $|\nu|^4$ near $\nu = 0$. c) For $\lambda^2 < \varepsilon\omega$, the function is paraboloidal in shape.

III. SEMICLASSICAL TREATMENT OF THE DICKE HAMILTONIAN

The "classical ansatz" (2.5) provides an extremely useful mechanism for estimating the values of the coupling constants for which a phase transition can be expected. In fact, the "classical ansatz" applied to the Dicke Hamiltonian (2.1) is equivalent to a study of its equilibrium statistical mechanical properties at T = 0. This is because the free energy at T = 0 is equal to the minimum eigenvalue of the Hamiltonian. Thus, the "classical ansatz" leaves three very important questions unanswered:

1. What is the free energy F/N as a function of T?

2. What are the order parameters as a function of T?

3. What is the critical temperature as a function of λ?

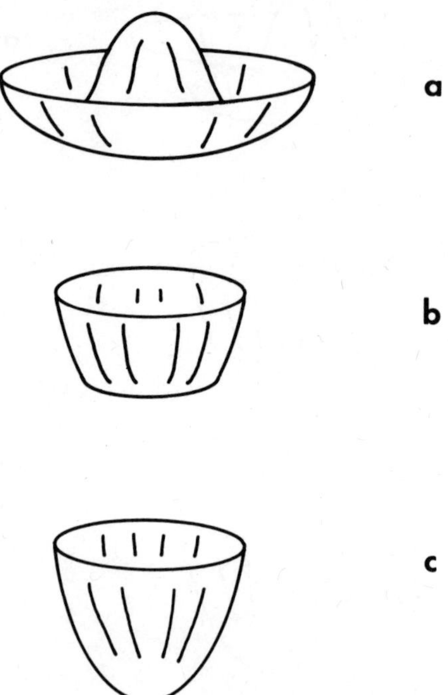

Figure 2. Shape of the free energy potential (3.4) as a function of temperature T when $\lambda^2 > \varepsilon\omega$. a) For $\beta > \beta_c$, the potential (3.4) looks like a sombrero, with a central peak and a ring of minimum values whose radius $|\mu|$ is determined by the non-zero solution of (3.5). b) For the critical value $\beta = \beta_c$, with β_c determined by (3.7), the central peak and minimum ring have just coalesced. The potential behaves like $|\mu|^4$ near $\mu = 0$. c) For $\beta < \beta_c$, the potential is paraboloidal in shape.

These questions are conveniently and correctly answered by making the "semiclassical ansatz." This involves replacing only the field operators by their expectation values: $a \rightarrow \mu\sqrt{N}$. In short, the "semiclassical ansatz" is (2.5a) of the "classical ansatz," (2.5). With this ansatz, (2.1) becomes

$$H_D(\lambda;\mu)/N = \omega\mu^*\mu + \sum_{k=1}^{N} h(k) \tag{3.1}$$

$$h(k) = \frac{1}{2}\varepsilon\sigma_k^z + \lambda(\mu^*\sigma_k^- + \mu\sigma_k^+) = \begin{bmatrix} \frac{\varepsilon}{2} & \lambda\mu \\ \lambda\mu^* & -\frac{\varepsilon}{2} \end{bmatrix}_k \quad . \tag{3.2}$$

The free energy is determined from

$$e^{-\beta F} = \mathrm{Tr}\ e^{-\beta\ H_D(\lambda;\mu)} = e^{-\beta N\omega\mu^*\mu}\left(\mathrm{Tr}\ e^{-\beta h(k)}\right)^N \quad . \tag{3.3}$$

The trace over the 2 × 2 matrices in (3.3) can easily be taken in terms of the eigenvalues $\pm\theta$, $\theta = [(\varepsilon/2)^2 + |\lambda\mu|^2]^{1/2}$ of the 2 × 2 matrix in (3.2). Then

$$F/N = \omega\mu^*\mu - \frac{1}{\beta}\ \ell n\ 2\cosh\beta\theta \quad . \tag{3.4}$$

As with (2.7) and (2.11), the order parameter μ must be chosen to minimize the value of this potential. The necessary conditions are

$$\partial(3.4)/\partial\mu^* = \omega\mu - \frac{\lambda^2\mu}{2\theta}\tanh\beta\theta = 0 \quad . \tag{3.5}$$

This always has the trivial solution $\mu = 0$. It may also possess a nontrivial solution $\mu \neq 0$ if

$$\omega - \frac{\lambda^2}{2\theta}\tanh\beta\theta = 0 \quad . \tag{3.6}$$

Equation (3.6) can be satisfied only if $\lambda^2/\varepsilon \geq \omega$ and, in that case, only for temperatures less than a critical temperature determined from

$$\omega - \frac{\lambda^2}{\varepsilon}\tanh\beta_c\varepsilon/2 = 0 \quad . \tag{3.7}$$

When these conditions are met, the nonzero magnitude of μ is uniquely determined by (3.6), but the phase of μ is not determined. The values of the <u>function</u> (3.4) evaluated at its stationary points, are

$$F/N = -\frac{1}{\beta}\ \ell n\ 2\cosh\beta\varepsilon/2 \qquad \mu = 0 \tag{3.8a}$$

$$\begin{aligned} F/N &= \omega\mu^*\mu - \frac{1}{\beta}\ \ell n\ 2\cosh\beta\left[(\varepsilon/2)^2 + \lambda\mu^*\mu\right]^{1/2} \qquad \mu \neq 0,\ \beta \geq \beta_c \\ &= \frac{\omega}{\lambda^2}\left[\theta^2 - (\varepsilon/2)^2\right] - \frac{1}{\beta}\ \ell n\ 2\cosh\beta\theta \quad , \end{aligned} \tag{3.8b}$$

where θ satisfies (3.6). Since the thermal equilibrium state of a physical system is the state which minimizes the free energy, we see that the free energy for the Dicke model (2.1) is (3.8a) for $T \geq T_c$ and (3.8b) for $T < T_c$. The zero temperature limit of (3.8b) is (2.10b).

When the stability matrix for the function (3.4) is computed, it is found that the eigenvalues on the disordered branch $\mu = 0$ are (+,+) for $T > T_c$. On the ordered branch $\mu \neq 0$, the eigenvalues are (+,0) for $T < T_c$.

Once again, these results can easily be visualized by the sombrero analogy (Fig. 2). For fixed $\lambda^2/\varepsilon > \omega$, at $T = 0$ the potential (3.4) looks like a sombrero with a central peak and a ring of minimum values. As the temperature is increased, the height of the central peak diminishes and the ring shrinks. As $T \to T_c^-$, the central peak and the ring of minimum values coalesce. For $T > T_c$ there is only one stationary point at $\mu = 0$, and the shape of (3.4) is roughly paraboloidal. Since the physical free energy is the minimum of (3.4), it is obtained by evaluating (3.4) on the ring of minimum values, when the ring exists, or at the central dip when (3.6) has no solutions.

The "semiclassical ansatz" provides a mechanism for computing the order parameter $<a> = \mu\sqrt{N}$ directly from (3.5). The order parameter $<\sigma^-> = \nu$ may be computed simply:

$$\nu = <\sigma^-> = \mathrm{tr}\ \sigma^-\ e^{-\beta h(k)}/\mathrm{tr}\ e^{-\beta h(k)}$$

$$= -\frac{\lambda\mu(\sinh\ \beta\theta)/\theta}{2\ \cosh\ \beta\theta} = -\frac{\lambda\mu}{2\theta}\tanh\ \beta\theta \quad . \tag{3.9}$$

The "semiclassical ansatz" provides an estimate to the free energy F/N determined from the full quantum mechanical Hamiltonian (2.1), which is accurate to order $\ell n\ N/N$. This difference vanishes in the thermodynamic limit. We may ask why the "semiclassical ansatz" provides such good results.

The free energy must be computed directly from (2.1) by performing the trace over both the atomic and the field states. The trace over the field states can be simply[3] and accurately[4] estimated by introducing the quantum analogs of the classical field states, the so-called coherent states.[9] The trace is then

$$e^{-\beta F} = \sum_{\substack{2^N \text{ atomic} \\ \text{states}}} \frac{N}{\pi} \iint d^2\mu \; e^{-\beta N \omega \mu^* \mu} \; e^{-\beta \sum_{k=1}^{N} h(k)} \quad . \tag{3.10}$$

The summation in (3.10) is exactly the sum performed in (3.3). Moreover, the remaining integral has the form $\int\int e^{-\beta N \phi(\mu)} d^2\mu$. The principle contributions to this integral come from regions where $\phi(\mu)$ has its minimum value. In fact, for ϕ bounded below, the value of this integral as $N \to \infty$ is asymptotically $e^{-\beta N \phi_o}$, where ϕ_o is the minimum value which ϕ assumes. This result is also true if the integral is carried out over an n (finite) dimensional space, provided the set of points on which ϕ assumes its minimum value is bounded. As a result

$$F/N = \min_{\mu \text{ complex}} \left[\omega \mu^* \mu - \frac{1}{\beta} \ln 2 \cosh \beta\theta\right] + O\left(\frac{\ln N}{N}\right) \quad . \tag{3.11}$$

It is for this reason that the "semiclassical ansatz" gives the correct free energy to order $\ln N/N$ in the thermodynamic limit.

If the system of N identical atoms has — not two levels of importance but rather is a spin multiplet with 2j+1 equally spaced levels — only minor modifications are required in the preceeding analyses. The 2 × 2 Pauli spin matrices $1/2\ \sigma^z$, $\sigma^{\pm}$ must be replaced by the (2j+1) × (2j+1) angular momentum matrices J_z, $J_{\pm}$ in (2.1) and (3.2). In the "classical ansatz," $1/2 \to j$ in (2.5b') and the ground state is an ordered state if $(2j)\lambda^2/\varepsilon\omega > 1$. The trace in (3.3) becomes $\sinh(2j+1)\beta\theta/\sinh\beta\theta$, equations (3.4) and (3.8b) are changed by the substitution $2\cosh\beta\theta \to \sinh(2j+1)\beta\theta/\sinh\beta\theta$, and (3.5)-(3.7) are changed by the substitution $\tanh\beta\theta \to 2jB_j(\beta\theta)$, where $B_j(x)$ is the jth Brillouin function.

The three questions posed at the beginning of this section are completely answered by using the "semiclassical ansatz" as follows: question 1 by (3.8); question 2 by (3.5) or (3.6) and (3.9); question 3 by (3.7).

IV. MULTIMODE SYSTEMS

The Dicke Hamiltonian describing the interaction of n modes of the electromagnetic field with N identical 2-level atoms in the long wavelength approximation is

$$H_D'(\lambda) = \sum_{j=1}^{n} \omega_j a_j^\dagger a_j + \varepsilon \sum_{k=1}^{N} \frac{1}{2} \sigma_k^z$$

$$+ \frac{1}{\sqrt{N}} \sum_{k=1}^{N} \left(\sum_{j=1}^{n} \lambda_j^* a_j^\dagger \sigma_k^- + \sum_{j=1}^{n} \lambda_j a_j \sigma_k^+ \right) . \qquad (4.1)$$

The operators a_j, $a_j^\dagger$ describe the jth independent field mode and $[a_i, a_j^\dagger] = \delta_{ij}$.

We will treat $H_D'(\lambda)$ first using the "classical ansatz" to determine the conditions on the coupling constants λ_i for which a phase transition is to be expected. Then we will apply the "semiclassical ansatz" to determine the critical temperature, free energy, and order parameters.

The "classical ansatz" for (4.1) is obtained from (2.5) with the additional assumption $a_j \to \mu_j \sqrt{N}$, $j = 1,2,\ldots,n$:

$$H_D'(\lambda;\mu_j,\nu)/N = \sum_{j=1}^{n} \omega_j \mu_j^* \mu_j - \varepsilon \sqrt{\left(\frac{1}{2}\right)^2 - \nu^*\nu}$$

$$+ \sum_{j=1}^{n} \left(\lambda_j^* \mu_j^* \nu + \lambda_j \mu_j \nu^* \right) . \qquad (4.2)$$

The coupled equations are

$$\partial(4.2)/\partial\mu_j^* = \omega_j\mu_j + \lambda_j^* \nu = 0 \qquad j = 1,2,\ldots,n$$

$$\partial(4.2)/\partial\nu^* = \frac{\varepsilon\nu}{2\sqrt{\left(\frac{1}{2}\right)^2 - \nu^*\nu}} + \sum_{j=1}^{n} \lambda_j \mu_j = 0 . \qquad (4.3)$$

The nonlinear equation for ν is

$$\left[\frac{\varepsilon}{2\sqrt{\frac{1}{2}^{2} - \nu^{*}\nu}} - \sum_{j=1}^{n} \frac{\lambda_j^{*}\lambda_j}{\omega_j}\right]\nu = 0 \quad . \tag{4.4}$$

The potential (4.2), as a function of ν, is

$$H(\lambda;\nu)/N = -\left(\sum_{j=1}^{n} \frac{\lambda_j^{*}\lambda_j}{\omega_j}\right)\nu^{*}\nu - \varepsilon\sqrt{\left(\frac{1}{2}\right)^{2} - \nu^{*}\nu} \quad . \tag{4.5}$$

Equations (4.4) and (4.5) are identical with (2.9) and (2.11) under the indentification

$$\frac{\lambda^{2}}{\omega} \rightleftarrows \sum_{j=1}^{n} \frac{\lambda_j^{*}\lambda_j}{\omega_j} \quad . \tag{4.6}$$

As a result, (4.1) is expected to exhibit a phase transition for

$$\sum_{j=1}^{n} \frac{|\lambda_j|^{2}}{\omega_j} > \varepsilon \quad .$$

With the "semiclassical ansatz" applied to (4.1), the free energy is

$$F/N = \sum_{i=1}^{n} \omega_i\mu_i^{*}\mu_i - \frac{1}{\beta}\ln 2\cosh\beta\theta \tag{4.7}$$

$$\theta^{2} = (\varepsilon/2)^{2} + \left|\sum_{i=1}^{n} \lambda_i\mu_i\right|^{2} \quad . \tag{4.8}$$

The minimum value of (4.7) is determined from the equations

$$\partial(4.7)/\partial\mu_i^{*} = \omega_i\mu_i - \lambda_i^{*}\left(\sum_{j=1}^{n} \lambda_j\mu_j\right)\frac{1}{2\theta}\tanh\beta\theta = 0 \quad . \tag{4.9}$$

Solving for μ_i and constructing the sum $\sum_{i=1}^{n} \lambda_i\mu_i$ leads to the nonlinear equation

$$\left\{1 - \left(\sum_{i=1}^{n} \frac{\lambda_i^* \lambda_i}{\omega_i}\right) \tanh \beta\theta\right\} \sum_{j=1}^{n} \lambda_j \mu_j = 0 \quad . \tag{4.10}$$

This is identical with (3.5) under the identification (4.6). Thus, the thermodynamic properties of the single mode and multimode Dicke Hamiltonians (2.1) and (4.1) are identical under the identification (4.6).

V. A^2 AND COUNTER-ROTATING TERMS

We consider next a more complicated Hamiltonian. This involves the original single mode Dicke Hamiltonian without the rotating wave approximation and including the $\underline{A}\cdot\underline{A}$ term. The Hamiltonian is

$$H_{DD}(\lambda) = \omega a^\dagger a + k(a^\dagger + a)^2 + \varepsilon \sum_{k=1}^{N} \frac{1}{2} \sigma_k^z$$

$$+ \frac{\lambda}{\sqrt{N}} \sum_{k=1}^{N} \left(a^\dagger \sigma_k^- + a\sigma_k^\dagger + r^* a^\dagger \sigma_k^+ + ra\sigma_k^-\right) \quad . \tag{5.1}$$

We treat the resonant and counterrotating terms asymmetrically because in general they will couple to the external bath with different relaxation rates and mechanisms. It is convenient to assume r is real. The classical form of (5.1) is ($\mu = x_1 + ix_2$; $\nu = y_1 + iy_2$)

$$H_{DD}(\lambda;\mu,\nu)/N = (\omega + 4k)\, x_1^2 + \omega x_2^2 - \varepsilon \sqrt{\left(\tfrac{1}{2}\right)^2 - y_1^2 - y_2^2}$$

$$+ 2\lambda\,(1 + r)\, x_1 y_1 + 2\lambda\,(1 - r)\, x_2 y_2 \quad . \tag{5.2}$$

The coupled nonlinear equations obtained by setting all first derivatives of (5.2) equal to zero are

$$\begin{bmatrix} \omega + 4k & \lambda\,(1 + r) \\ \lambda\,(1 + r) & \dfrac{\varepsilon}{2\sqrt{\left(\frac{1}{2}\right)^2 - \nu^*\nu}} \end{bmatrix} \begin{bmatrix} x_1 \\ y_1 \end{bmatrix} = 0 \quad . \tag{5.3a}$$

$$\begin{bmatrix} \omega & \lambda\,(1-r) \\ \lambda\,(1-r) & \dfrac{\varepsilon}{2\sqrt{\left(\frac{1}{2}\right)^2 - \nu^*\nu}} \end{bmatrix} \begin{bmatrix} x_2 \\ y_2 \end{bmatrix} = 0 \quad . \tag{5.3b}$$

Once again, it is somewhat simpler to treat the system (5.3) by eliminating $\mu = x_1 + ix_2$ or $\nu = y_1 + iy_2$:

$$\left\{ \frac{\varepsilon}{2\sqrt{\left(\frac{1}{2}\right)^2 - \nu^*\nu}} - \frac{\lambda^2(1+r)^2}{\omega + 4k} \right\} y_1 = 0 \tag{5.4a}$$

$$\left\{ \frac{\varepsilon}{2\sqrt{\left(\frac{1}{2}\right)^2 - \nu^*\nu}} - \frac{\lambda^2(1-r)^2}{\omega} \right\} y_2 = 0 \quad . \tag{5.4b}$$

The Hamiltonian (5.2) is then

$$H_{DD}(\lambda;\nu)/N = -\frac{\lambda^2(1+r)^2}{\omega + 4k}\, y_1^2 - \frac{\lambda^2(1-r)^2}{\omega}\, y_2^2 - \varepsilon\sqrt{\left(\frac{1}{2}\right)^2 - y_1^2 - y_2^2} \quad . \tag{5.5}$$

Equations (5.3) have exactly the same form as (2.8) and (5.4) as (2.9). They can therefore be treated identically. Equations (5.4) always possess the trivial solution $y_1 = 0$, $y_2 = 0$. If $\lambda^2(1 + r)^2/(\omega + 4k) > \varepsilon$, the nontrivial solution ($y_1 \neq 0$, $y_2 = 0$) is possible. The magnitude, but not the sign, of $y_1 \neq 0$ is determined uniquely from (5.4a). This solution is called the Real branch. Similarly, if $\lambda^2(1 - r)^2/\omega > \varepsilon$, the nontrivial solution ($y_1 = 0$, $y_2 \neq 0$) is possible. Such a solution is called the Imaginary branch. The value of $y_2 \neq 0$ is determined up to sign by (5.4b).

It is possible to have a nontrivial solution ($y_1 \neq 0$, $y_2 \neq 0$) only if

$$\left(\frac{1+r}{1-r}\right)^2 = 1 + 4(k/\omega) \quad . \tag{5.6}$$

When this occurs, the Hamiltonian (5.1) is equivalent to (2.1) with renormalized parameters under a canonical transformation.[6] The discussion then reduces to that of Section II.

The value of $H_{DD}(\lambda;\mu,\nu)/N$, evaluated on each of the branches, is

1. Disordered branch ($y_1 = 0$, $y_2 = 0$):

$$H_{DD}(\lambda;\mu,\nu)/N = -\frac{\varepsilon}{2} \times \frac{1}{2}\,[1 + 1]$$

2. Real branch ($y_1 \neq 0$, $y_2 = 0$) if $\lambda^2(1 + r)^2 > \varepsilon\,(\omega + 4k)$:

$$H_{DD}(\lambda;\mu,\nu)/N = -\frac{\varepsilon}{2} \times \frac{1}{2}\left[\frac{\lambda^2(1 + r)^2}{\varepsilon\,(\omega + 4k)} + \frac{\varepsilon\,(\omega + 4k)}{\lambda^2(1 + r)^2}\right]$$

3. Imaginary branch ($y_1 = 0$, $y_2 \neq 0$) if $\lambda^2\,(1 - r)^2 > \varepsilon\omega$:

$$H_{DD}(\lambda;\mu,\nu)/N = -\frac{\varepsilon}{2} \times \frac{1}{2}\left[\frac{\lambda^2(1 - r)^2}{\varepsilon\omega} + \frac{\varepsilon\omega}{\lambda^2(1 - r)^2}\right] .$$

The value is minimum on the disordered, real, or imaginary branch depending on which of 1, $\lambda^2(1 + r)^2/(\omega + 4k)\varepsilon$, or $\lambda^2(1 - r)^2/\omega\varepsilon$ is maximum. In particular, if an ordered branch exists, the energy is lower on the ordered branch than on the disordered branch.

This analysis can be carried out in terms of the eigenvalues of the stability matrix for (5.5). If only the trivial solution is possible, the eigenvalues of this matrix are (+,+) on the disordered branch. If only one ordered branch exists, the eigenvalues are (+,+) on that branch, and (+,-) on the disordered branch. If both ordered branches exist, the eigenvalues are (+,+) on the branch with the lowest energy, (+,-) on the other ordered branch, and (-,-) on the disordered branch.

Once again, these results are simple to visualize with a sombrero analogy. This time the potential (5.5) looks like a soggy sombrero. For λ sufficiently large there is a peak in the center ($\nu = 0$), with the front and back of the rim dipping lower than the right and left side if the real branch has lower free energy, and vice versa if $\lambda^2(1 - r)^2/\varepsilon\omega > \lambda^2(1 + r)^2/\varepsilon(\omega + 4k)$. For the sake of concreteness, we take the real axis through the nose and the real branch with lower energy than the imaginary. As the coupling constant is decreased, the height of the central peak diminishes. As $\lambda^2(1 - r)^2/\varepsilon\omega \to 1^+$, the central peak and the saddle points over the ears coalesce. For smaller values of λ the imaginary branch ceases to exist, and as $\lambda^2(1 + r)^2/\varepsilon(\omega + 4k) \to 1^+$, the central "peak" (now really a saddle) and the minima in front and back coalesce. For smaller values of λ, the potential assumes a roughly paraboloidal shape, with only one stationary point, a global minimum at $\nu = 0$, corresponding to the disordered state as the system thermodynamic ground state.

Next, we apply the "semiclassical ansatz" to the Hamiltonian (5.1). With $\mu = x_1 + ix_2$ and r real, the free energy is easily calculated:

$$F/N = (\omega + 4k)\, x_1^2 + \omega x_2^2 - \frac{1}{\beta} \ln 2 \cosh \beta\theta \tag{5.7}$$

$$\theta^2 = (\varepsilon/2)^2 + \lambda^2(1 + r)^2\, x_1^2 + \lambda^2(1 - r)^2\, x_2^2 \quad . \tag{5.8}$$

The minimum of F/N is determined from

$$\partial(5.7)/\partial x_1 = \left[(\omega + 4k) - \frac{\lambda^2(1 + r)^2}{2\theta} \tanh \beta\theta \right] x_1 = 0 \tag{5.9r}$$

$$\partial(5.7)/\partial x_2 = \left[\omega - \frac{\lambda^2(1 - r)^2}{2\theta} \tanh \beta\theta \right] x_2 = 0 \quad . \tag{5.9i}$$

These equations are identical in form to (3.5); their analysis proceeds identically.

If $\lambda^2(1 + r)^2/\varepsilon(\omega + 4k) > 1$, a real branch can exist below a critical temperature T_1 determined by

$$1 - \frac{\lambda^2(1 + r)^2}{\varepsilon\ (\omega + 4k)} \tanh \frac{1}{2} \beta_1 \varepsilon = 0 \quad . \tag{5.10r}$$

If $\lambda^2(1 - r)^2/\varepsilon\omega > 1$, an imaginary branch can exist below a critical temperature T_2 determined by

$$1 - \frac{\lambda^2(1 - r)^2}{\varepsilon\omega} \tanh \frac{1}{2} \beta_2 \varepsilon = 0 \quad . \tag{5.10i}$$

In the first case the order parameters $(x_1 \neq 0,\ x_2 = 0)$ are determined by the nontrivial solution of (5.9r), in the second, $(x_1 = 0,$

$x_2 \neq 0$) is determined by the nontrivial solution to (5.9i). In order for both ($x_1 \neq 0$, $x_2 \neq 0$) simultaneously, (5.6) must be satisfied, in which case (5.1) is equivalent to (2.1) as previously discussed,[6] so the discussion of this section reduces to that of Section III.

If ordered branches can exist, the free energy (5.7) is minimum on whichever ordered branch can exist at the higher temperature $T < \text{Max}\ (T_1, T_2)$. For $T \geq \text{Max}\ (T_1, T_2)$, only the disordered branch can exist, and the free energy is minimum on that branch. For all temperatures, the free energy is given by (3.8), where $\beta_c \to \min\ (\beta_1, \beta_2)$ and $\lambda^2/\varepsilon\omega \to \text{Max}\ \lambda^2(1 + r)^2/\varepsilon(\omega + 4k),\ \lambda^2(1 - r)/\varepsilon\omega$.

The shape of the free energy surface (5.7) is as described below Eq. (5.6). Decreasing the coupling constant has the same effect as increasing the temperature, as discussed at the end of Section III.

It has recently been shown[10] that the Thomas-Kuhn-Reiche sum rule requires $\lambda^2(1 + r)^2/\varepsilon(\omega + 4k) < 1$, so a real ordered solution ($x_1 \neq 0$) to (5.9r) can never exist. However, it is possible[6,7] for an imaginary branch ($x_2 \neq 0$) to exist, and $r \to 0$; the critical temperature and ordered state behavior is exactly the same as given for (2.1). In the absence of counterrotating terms ($r = 0$), the thermodynamic properties of (5.1) are independent of whether or not the A^2 terms are included.[11]

VI. DISCUSSION AND CONCLUSIONS

The "classical ansatz" (2.5) provides a very useful mechanism for determining the values of the coupling constants for which a phase transition can be expected in various Dicke models (2.1), (4.1), (5.1). However, this ansatz only provides information about the free energy and the order parameters in the $T \to 0$ limit.

To determine the temperature dependence of the free energy and the order parameters, as well as to locate the critical temperature, the "semiclassical ansatz" was introduced in Section III. With this ansatz, the free energy and the order parameters for the Hamiltonians (2.1), (4.1) and (5.1) were simply computed in Sections III, IV, and V. In each case, the critical temperatures were determined by "gap equations." We also discussed why the "semiclassical ansatz" provides an estimate of F/N which is accurate to order $\ell n\ N/N$, when a finite number of field modes is present.

The "classical" and "semiclassical" ansätze are deeply related. Of course, the former is the $T \to 0$ limit of the latter. But the relation goes yet deeper. The "classical ansatz" allows us to construct an energy surface (free energy at $T = 0$), and to study this potential function for possible phase transitions as a function of decreasing coupling constant λ. The "semiclassical ansatz" allows us to construct a free energy for $T > 0$ and to study this potential function for possible phase transitions as a function of decreasing inverse temperature β for fixed value of the coupling constant λ. The behavior (change in shape) of the energy surface as a function of decreasing λ is identical with the behavior of the free energy as a function of decreasing β, as described in Sections II and III, Figs. 1 and 2, and in Section V.

The relationship between the "classical" prediction of λ-dependent phase transitions and the "semiclassical" prediction of β-dependent phase transitions is embodied in the "Crossover

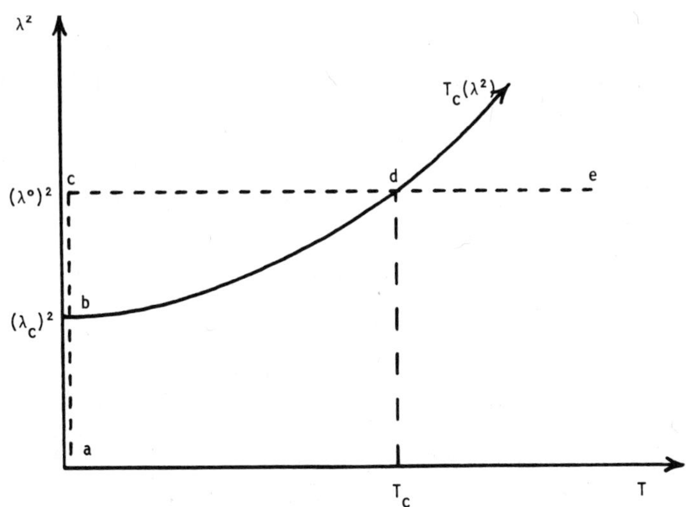

Figure 3. If there is a crossover of ground state energies at λ_c^2 between $\lambda^2 = 0$ and $\lambda^2 = (\lambda°)^2$ (path abc), there will be a phase transition at T_c between $T = 0$ and $T = \infty$ (path cde). The "classical ansatz" determines whether a "phase transition" occurs for increasing λ^2 at $T = 0$. The "semiclassical ansatz" determines the $\lambda^2 - T$ duality and locates the critical temperature as a function of λ^2.

Theorem." Let $H(N,\underline{\lambda})$ be the Hamiltonian for an N-particle system, with coupling constants $\lambda_1,\lambda_2,\ldots$. Assume $H(N,\underline{\lambda})$ has a discrete spectrum and that the energy eigenvalues $E_1(\underline{\lambda})$, $E_2(\underline{\lambda})$, ..., listed in order of increasing energy for $\underline{\lambda} = \underline{0}$, are analytic functions of $\underline{\lambda}$. If $E_1(\underline{\lambda} = \underline{0})$ is nondegenerate and if there is some value $\underline{\lambda}^\circ$ of $\underline{\lambda}$ for which an energy level crossover has occurred, in the sense that $E_k(\underline{\lambda}^\circ) < E_1(\underline{\lambda}^\circ)$ for some $k > 1$, the system described by $H(N;\underline{\lambda})$ will exhibit at least one phase transition when $\underline{\lambda} = \underline{\lambda}^\circ$. The function of the "classical ansatz" is to determine if there is a crossover; the function of the "semiclassical ansatz" is to exhibit analytically the $\lambda - \beta$ duality and to determine the line of critical points $\beta_c(\lambda)$ in the $\lambda - \beta$ plane. The application of this theorem to the Dicke Hamiltonian is shown in Fig. 3.

ACKNOWLEDGEMENTS

One of the authors (R.G.) thanks Prof. D. Speiser for the hospitality extended at the Université Catholique de Louvain, where much of this work was done. He also thanks Prof. R. Brout for useful discussions, and for pointing out that the "semiclassical ansatz" was widely applied to the Dicke model at the time the model was proposed.

References

(1) R. H. Dicke, Phys. Rev. 93, 99 (1954).

(2) K. Hepp and E. H. Lieb, Ann. Phys. (N.Y.) 76, 360 (1973).

(3) Y. K. Wang and F. T. Hioe, Phys. Rev. A7, 831 (1973).

(4) K. Hepp and E. H. Lieb, Phys. Rev. A8, 2517 (1973).

(5) R. Gilmore and C. M. Bowden, Phys. Rev. A13, 1898 (1976).

(6) R. Gilmore and C. M. Bowden, J. Math. Phys. 17, 1617 (1976).

(7) R. Gilmore, J. Math. Phys. 18, 17 (1977).

(8) K. Hepp and E. H. Lieb, Helv. Phys. Acta 46, 573 (1973).

(9) R. J. Glauber, Phys. Rev. 130, 2529 (1963); 131, 2766 (1963).

(10) K. Rzażewski, K. Wódkiewicz, and W. Żakowicz, Phys. Rev. Letters 35, 432 (1975).

(11) The claim to the contrary was made in Ref. 10.

DISCUSSIONS AT THE COOPERATIVE EFFECTS MEETING, REDSTONE ARSENAL, ALABAMA, 1-2 DECEMBER 1976

Marek J. Konopnicki* and Albert T. Rosenberger**

*Department of Physics and Astronomy, University of Rochester

**Department of Physics, University of Illinois

COOPERATIVE EFFECTS MEETING

AGENDA

Wednesday Afternoon, 1 December 1976, Bldg 5250, Room A115, Redstone Arsenal, Alabama

1:00	Introduction and Special Announcements	C. M. Bowden Redstone Arsenal, AL
	SESSION I: Experimental Superradiance in FIR CHAIRMAN: C. M. Bowden, Redstone Arsenal, AL	
1:10	FIR Superradiance: An Experimental Point of View	M. S. Feld MIT
1:45	Experiments in FIR Superradiance	T. A. DeTemple U. of Illinois
2:20	Coffee Break	
	SESSION II: Experiments in Superradiance in Metal Vapors CHAIRMAN: R. Gilmore, U. of South Florida	
2:30	Cooperative Effects in Metal Vapors	S. R. Hartmann Columbia U.

3:05 Quantum Beat Superfluorescence in Cs — H. M. Gibbs, Bell Labs

3:40 Coffee Break

3:50 Single Pulse Superfluorescence in Cs — Q. H. F. Vrehen, Philips Research Labs, Netherlands

SESSION III: Cooperative Effects in Plasmas and Self-Induced Transparency

CHAIRMAN: W. B. McKnight, University of Alabama in Huntsville

4:25 Role of Radiation Trapping on Stimulated VUV Emission in Laser-Produced Plasmas — K. G. Whitney, J. Davis, and J. P. Apruzese, US Naval Research Labs

5:00 Adjourn
(Buses provided to Hilton Inn)

Wednesday Evening, 1 December 1976

Huntsville Hilton, Freedom Plaza (Opposite Von Braun Civic Center)

6:00 Cocktails - Monte Sano Room

6:30 Dinner - Monte Sano Room

Workshop Session - Monte Sano Room

8:00 Panel Discussion: Superradiance, Superfluorescence and Swept Gain Excitation Superradiance in the FIR and near FIR

Panel Members:
H. R. Robl, ARO, Chairman
R. Bonifacio, U. of Milan
R. K. Bullough, U. of Manchester
S. R. Hartmann, Columbia U.
M. S. Feld, MIT

8:30 Open Discussion

Thursday Morning, 2 December 1976, Bldg 5250, Room A115, Redstone Arsenal, AL

SESSION V: Superradiance Theory
CHAIRMAN: J. H. Eberly, U. of Rochester

8:00	Discussion of Recent Numerical and Experimental Results on Superradiance	R. Bonifacio U. of Milan
8:50	Coffee Break	
9:00	Pure Superfluorescence?	R. K. Bullough U. of Manchester
9:50	Coffee Break	

SESSION VI: Swept Gain Superradiance and Amplifier Theory
CHAIRMAN: C. A. Coulter, U. of Alabama in Tuscaloosa

10:00	Superradiance in Swept Gain Systems	Pierre Meystre U. of Arizona
10:35	The Two-Photon Amplifier	L. M. Narducci Drexel U.
11:10	Coffee Break	

SESSION VII: New Concepts in Cooperative Effects
CHAIRMAN: T. A. Barr, Redstone Arsenal, AL

11:20	Theoretical Development of the Free Electron Laser	F. A. Hopf U. of Arizona
11:55	Lunch	
1:00	Workshop Session: Impact of Theory on Experiment and Vice Versa Panel to be announced	
3:00	Adjourn	

DISCUSSIONS

M. S. Feld - FIR Superradiance - An Experimental Point of View

Q - What is the cooperation length in the x-ray range?

Feld - If the excitation is swept, it will be infinite.

Q - Why doesn't the inversion keep building up in the x-ray case?

Feld - In most cases the lower level fills rapidly, so a population inversion becomes more difficult to create.

Q - Where does the loss factor of 2.5 come from in HF, where the Fresnel number is ~ 0.1?

Feld - The loss is due to near-field diffraction. It's only a small correction.

Q - But its value strongly affects the ringing.

Feld - True.

Q - What linewidths are predicted?

Feld - In the x-ray case the pulses (10^{-13} s) are about ten times shorter than those predicted by rate equations. The linewidth is determined by saturation broadening.

Bonifacio - Under what condition could one observe a hyperbolic secant pulse?

Feld - One can only approximate it ...

Q - Is your T_R the same as that of Rehler and Eberly?

Feld - Eberly could give us the best answer.

Eberly - Probably it is.

T. A. DeTemple - Experiments in FIR Superradiance

Feld - What is the area of your output pulses? Less than π?

DeTemple - The area is about 0.2 π.

Feld - So could you say that the pulse tries to evolve, but its buildup is limited by population decay?

DeTemple - Yes. It looks like the higher the gain, the longer the delay. The interesting thing is that it can be described in terms of a simple equation.

Feld - Where did your equations come from?

DeTemple - From the swept gain paper of Bonifacio.

Bonifacio - You have $\alpha L > 1$ so $\alpha L > \phi$ also. You are in the fully nonlinear regime.

Feld - Not so. If population decay is rapid there is an additional requirement to be in the nonlinear regime: $\alpha L > (\phi^2/4)T_2'/T_1$, as explained in our paper. It would be interesting to compare these experiments with Crisp's theory.

DeTemple - Yes, that would be interesting.

Bonifacio - But let me point out that if the results fit this form given, they cannot fit the other (Crisp's).

S. R. Hartmann - Cooperative Effects in Metal Vapors

Feld - I have two points. First, the use of a fast photomultiplier opens the interesting possibility of studying fluctuations in the superradiant output pulse. From this one can get information on the photon statistics. Second, with a magnetic field, there is one more effect - the Hanle effect - which changes the coupling by removing the level degeneracy. Breaking the degeneracy (by the natural width) changes the nonlinear coupling, which enhances the pumping efficiency. Changes in pump efficiency produce corresponding changes in the output. This may allow studying g-factors and linewidths.

Flusberg - Note that there is no saturation in the magnetic field. This is more than just breaking the symmetry by a strong field. One could also use an oscillating magnetic field to produce modulation of the envelope of the electric field.

H. M. Gibbs - Quantum Beat Superfluorescence in Cs

Flusberg - Does the pulse die after 20 nsec?

Gibbs - In the beam the output pulse intensity goes as $1/t_o^2$, i.e., proportional to n^2. In the cell the dependence is $1/t_o^2$ out to $t_o \approx 20$ ns, beyond which the pulse intensity goes down faster and rapidly disappears in the noise.

Flusberg - Did you ever try pumping off-resonance?

Gibbs - We did at first, but not in later stages. In later stages we committed ourselves to longer delays.

Flusberg - I'm also curious about stimulated Raman scattering.

Vrehen - In early experiments we saw SRS with very short delays. It was done at high densities.

Feld - You said that in the beam, T_2^* was not important, but even in the beam, T_2^* is the shortest of the relaxation times.

Gibbs - Yes, but it is longer than t_o. In fact we need to know it only to calculate αL. (NOTE: $\tau_R = 8\pi\tau_o/3n\lambda^2 L$ is calculated from measurements of n and L and known values of τ_o and λ. L is then T_2^*/τ_R.)

Q. H. F. Vrehen - Single-Pulse Superfluorescence in Cs

Bonifacio - First of all, let me say that it makes me happy to see a sech^2 pulse with $\alpha L \gg 1$, and without ringing, which is absent because $\tau_R/\tau_E \gg 2$ implies $L/L_c \gg 1$. Second, the fast increase of τ_D/τ_R at high densities is explained by the fact that τ_R is approaching τ_c. It was predicted by our theory. (NOTE: $\tau_R = 2\tau_E$ is the condition for the onset of ringing.)

Feld - I'm impressed with your careful experiments. We always used a computer program to analyze our detailed experimental results. The analytical expressions are only approximate. Comparison of theory and experiment should really be done with the full computer solution. We'd be happy to run the program for you.

Vrehen - I would be delighted to see such calculations made.

K. G. Whitney - Role of Radiation Trapping on Stimulated VUV Emission in Laser-Produced Plasmas

Bowen - Do you have a uniform radial distribution?

Whitney - Yes.

Bowen - The blast wave theory gives a shock front and blowoff...

Whitney - It's an idealized calculation, but even with gradients in the density, the effect could still be seen.

Bowen - What is the percent population inversion?

Whitney - $\Delta N/N$ is about 10%, based on a fairly simple excited state level structure.

Panel Discussion - Superradiance, Superfluorescence and Swept Gain Excitation Superradiance in the FIR and near FIR - Robl, Bonifacio, Hartmann, Bullough, and Feld

Robl - To begin, we'll have each of the panel members say a few words about the present situation in cooperative effects and the outlook for the future. Let's begin with Dr. Feld.

Feld - Experiments that we currently have in mind will study the evolution of the superradiant pulse. Strong and limited (FID) superradiance are well understood, but the intermediate regime is not. We investigate this in samples of different optical depth by monitoring the free induction decay at one transition by probing a coupled transition with a weak monochromatic field. The frequency profile of the decay as a function of time should have interesting structure. This decay is determined by both collision processes (T_2) and collective effects. The collective effects should become more important as the optical depth of the decaying transition is increased.

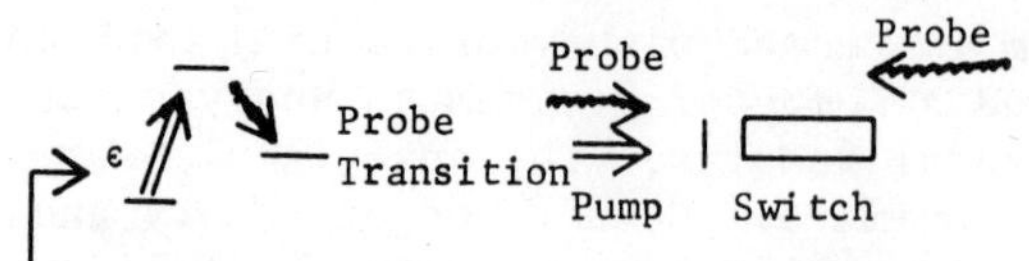

Hopf - This is free induction decay? But there's a field present.

Feld - The free induction decay at the pump transition is monitored by a probe at the coupled transition. Incidentally, the frequency spectrum depends on the relative direction of the probe beam: it will be narrow in the co-propagating direction and wider in the counter-propagating direction.

Bowden - How is this experiment related to collective phenomena?

Feld - As the optical depth of the sample increases (say, by varying the pressure), the rate of free induction decay should increase. At high pressures, it should vary more.

DeTemple - Why isn't it just an amplifier then?

Hartmann - I'm a little confused about it, too. S gets amplified, but it is the probe.

Feld - The pump and probe fields are not at the same wavelength.

Q - How do you distinguish between the two?

Feld - You could use a filter, or slightly misalign the two beams.

Karras - Will the output be observed without the probe?

Feld - The free induction decay will still occur. The advantage of using a probe is that one can get information in the frequency domain, as well as in the time domain.

Flusberg - So you call it superradiance because the polarization is there after the driving field is gone.

Feld - The sample would have to be optically thick to observe collective behavior.

Bonifacio - When you create a macroscopic polarization with an external source inducing coherence, it is called superradiance; this is not the same as just exciting and allowing the correlation to develop spontaneously, which is superfluorescence. They are two radically different things. In superfluorescence, the system starts to radiate by itself as N, and after lethargy, reaches coherence.

Feld - To me, the essential feature of superradiance is the N^2 emission occurring from a phased array of atomic dipoles (coherent state). You can prepare the system in a state of coherent polarization, or else you can start with a high gain inverted medium and let it evolve into a coherent state, as in our experiments. Dicke discussed the two different ways. I called our experiments Dicke superradiance because high gain amplifiers began to be called "superradiant." They will give an output linearly dependent on L, but Dicke superradiance gives output proportional to L^2. There is a basic distinction.

Bonifacio - I called the process superfluorescence because it begins as ordinary fluorescence.

Feld - Seems to me that a laser also starts from fluorescence. Crisp showed that in high gain systems the small-signal growth is more complicated than a simple exponential.

Eberly - The point is, there is utility to having two names.

Bonifacio - But there is a macroscopic difference: one radiates in both directions; the other, only in one.

Robl - Then does the same distinction apply to a small sample?

Bonifacio - Of course.

Q - What's the difference between superradiance and ordinary laser emission?

Feld - First, the condition for superradiance is that the single pass gain $\alpha L \gg 1$, whereas for an ordinary laser oscillator it is only $\alpha L_{eff} > 1$ (the effective gain length $L_{eff} = L/(1-R)$, $1-R$ = mirror loss). Thus, you need high gain for superradiance, but you only need net gain for laser oscillation. However, high gain is not a sufficient condition for superradiance. For example, in an N_2 laser (with or without mirrors) the output is proportional to L and to N, so the output intensity doubles if you double the length or the density. But in a true superradiant system doubling N or L leads to a quadrupling of the output intensity. This distinction arises because superradiance is inherently a transient effect - the inversion process must be rapid compared to the emission time - whereas in a high gain laser medium the response is quasi-steady state. Incidentally, the presence or absence of feedback (mirrors) is not a distinguishing feature. Mirrors only affect the details - the fast turn-on of a high-gain system gives rise to a radiation burst proportional to N^2. Eventually the system settles to a steady state.

Hopf - But that's simply not true. It's even contrary to fact! What about a ruby laser?

Feld - Ruby lasers aren't pumped fast enough. So the gain medium of a ruby laser is always in a quasi-steady state.

Hopf - There are two different regimes, switched on fast and high gain, and then long pumped ones. Superradiance does not have anything to do with the latter.

Feld - And the Maxwell-Bloch equations describe both ...

Hopf - And anything else you like!

Whitney - Isn't a Q-spoiled laser an N^2 effect?

Bonifacio - No, only near threshold.

Hopf - Lamb said: build an array of antennae, match phases, and you can call it superradiance; it radiates like N^2.

Feld - Our HF experiments were first done in a cavity ...

Karras - So what we have is a phased array of dipoles.

DeTemple - What used to be called a superradiant laser is now called a superfluorescent laser; can we come to some agreement on the words we use to describe these phenomena?

Bonifacio - But the physics involved is different: one can distinguish superfluorescence from superradiance by their second moments. In superfluorescence fluctuations are of the order N^2. If you are in a prepared state - no fluctuations.

Eberly - The physics is different? The distinction should be experimentally feasible. I am interested in Mike's opinion. Rodolfo is trying to suggest that we should be able to distinguish the two. You, Mike, advocated homogeneity.

Feld - Let me find the place where we agree.

Bonifacio - Nothing.

Feld - Most interesting thing to study, in the long run, is statistics.

Bonifacio - He cannot measure it, so I object to his statement.

Feld - A fully quantum-mechanical treatment would be interesting, earlier papers laid the groundwork, and now we have experiments to guide the theory, but a fully quantum model will be difficult. The approach to be taken should be to simplify the problem as much as possible without going to a single mode and try to find a solution for small Bloch angle (say $\theta < \pi/10$), then join that solution with the semiclassical one. We should try to get a self-consistent expression for the evolution of the pulse as it starts. By the time you get to $\pi/2$ case you should have a very strong pulse. Thus you are in the semiclassical regime. This would seem to be the path to a statistical understanding.

Hopf - I can write an analytic expression from memory.

Feld - You should publish it.

Hopf - It is published. This can be done — in fact, it was, a year ago. Haroche is starting from ground zero.

Bonifacio - The mean-field model agrees within a factor of 2. So let's look at the present model. Probably nobody realizes that we have a quantum model. No computer fittings, no factorization. We factorize only when we want to compare our results to mean-field or semiclassical ones. The field is incoherent and stays incoherent all the time, even at the peak. We did not expect it. We expected a Glauber coherent state. Also, the ringing does not go to zero. I suggest we check these things.

Robl - Is the replacement of operators with their expectation values justified?

Bonifacio - You don't have to; since $\langle I^2\rangle/\langle I\rangle^2 \sim 1$, there are large fluctuations. This is not a Glauber coherent state. It never vanishes in quantum mechanics because of the finite vacuum expectation.

Gibbs - Are the fluctuations proportional to $\sqrt{N}$ or to $\sqrt{\mu N}$?

Bonifacio - They are as large as 50-100%.

Bullough - The observation of these things is not an indication of the correctness of the Bonifacio model, just that quantum processes are important. What is needed is something more analogous to the semiclassical — something including spatial variations.

Flusberg - Why isn't the phase change of the field reproduced in a quantum treatment?

Vrehen - We are not looking at averages but at single shots so why could it not be zero?

Bonifacio - Quantum mechanics does not describe a single experiment. This is first year quantum mechanics.

Robl - This is in analogy with quantum mechanics not predicting the phase of a laser.

Hopf - Does a varying initial tipping angle result in fluctuations?

Feld - A varying polarization source has a small effect on fluctuations.

Hopf - Is your source Gaussian?

Feld - Yes.

Whitney - The distinction is based on initial conditions? How about the intrinsic difference between the behavior of small and extended samples?

Feld - As far as I know, there has never been a clean experiment of superradiance in a sample small compared to the wavelength. It would be fascinating to do such an experiment.

Hartmann - It would have to be in a cavity, though.

Eberly - The original experiment was done by Dicke in a microwave cavity, where he tried to get diffusion inversion.

Hartmann - But the field in a cavity is uniform. This is not the case with Dicke's system.

Vrehen - Wouldn't the dipole interaction wash out the superradiant effect?

Hartmann - The superradiant condition lasts longest in a sphere - but there are large frequency shifts.

Bullough - Are the frequency shifts from the dipole-dipole interaction?

Hartmann - Yes.

Vrehen - Is there anything to be learned in this?

Hartmann - Yes. It's a textbook case - one has to be careful with geometry.

Whitney - How about the cooperation length? Throw out swept excitation. Then superradiance goes away after a length L_c, but amplified spontaneous emission needs a finite length to be observed.

Feld - It would be a good experiment to pump a long cell transversely and look for length dependence.

Bonifacio - In Vrehen's experiment, we have $\tau_R/\tau_E = 2$, where $\tau_E = L/c$ and $\tau_R = 8\pi T_o/3\rho L\lambda^2 \propto 1/\rho L$. The cooperation time is $\tau_c = \sqrt{8\pi T_o/3c\rho\lambda^2} \propto 1/\sqrt{\rho}$, so that $\tau_c = \sqrt{\tau_E\tau_R}$; now $L_c = c\tau_c$, so $\tau_R/\tau_E = (L_c/L)^2$, so if $L_c/L > \sqrt{2}$, then there is a single pulse - otherwise there is ringing. So this verifies the fundamental role of the cooperation length. Also, Vrehen said that the pulse width stops decreasing as ρ increases, that is, as τ_D approaches τ_c. He has shown the phenomenon of τ_D/τ_R versus τ_R: when $\tau_D/\tau_R \propto \tau_c/\tau_R = \sqrt{\tau_E\tau_R}/\tau_R \propto 1/\sqrt{\tau_R}$. Is τ_R calculated or derived from the pulse width?

Vrehen - It is calculated.

Bonifacio - This is a beautiful proof! Well, an interpretation that makes sense. They observed everything we predicted.

DeTemple - This assumes disc geometry, Fresnel numbers of the order 1. Shape factors should play some role.

Eberly - You do not disagree with each other.

Bonifacio - It is different when Fresnel numbers are much smaller than one.

Vrehen - We keep Fresnel number of the order 1.

Feld - Are you using swept excitation?

Vrehen - Yes.

Bonifacio - But the sample is very short.

Feld - What is your cooperation length?

Bonifacio - Of the order of sample length. This is a perfect verification.

Hopf - So what Rodolfo is saying is right.

Feld - Rodolfo, do you feel that a semi-classical treatment describes average results?

Bonifacio - Roughly, yes. I wasn't completely right. I was log-right; Bullough's results from the Maxwell-Bloch equations agree — and they do not contradict the mean field results.

Bullough - They agree, but at this transition ....

Bonifacio - The HF experiments are strange. $T_o \sim 1$ sec, $\tau_R \sim 10^{-9}$ sec, so $\theta_o \sim 10^{-8}$. Vrehen needs a larger angle. So HF is peculiar. They agree in relation to the experiment.

Feld - At the time of the experiment, there was no theory. What was achieved in the lab was to take an essentially forbidden transition and make it strongly allowed. The agreement we found between the experiments and the semi-classical description is a major advance. We have shown that superradiance is a simple effect — readily describable in terms of the coupled Maxwell-Bloch equations. From the point of view of pulse propagation it can be thought of as the amplification of the tail of a noise pulse. So now there is a common ground of understanding.

Gibbs - There is disagreement with the plane-wave, large Fresnel number predictions.

Feld - My intuition is that the agreement will be close.

Gibbs - What about the initial tipping angle?

Feld - In the detailed model we take a fluctuating polarization source constructed to be consistent with requirements of thermal equilibrium. We use this for data fitting. In the ideal case, it is equivalent to an initial tipping angle θ_o, as given in our paper.

This approximation is reasonable since the earliest fluctuations at the far end of the medium are amplified over the greatest length for the greatest amount of time and therefore dominate.

Bonifacio - Feld's formula gives $\theta_o \sim 1/\sqrt{N}$. The mistake was to solve the Maxwell-Bloch equations with a very small angle. When the angle is not small — the correction is negligible.

Feld - We use a fluctuating polarization source in the Maxwell-Bloch equations, not a θ_o. When $\alpha L \gg 1$ (the limit depends on the source of dephasing), the polarization source may be replaced by θ_o, which gives the delay T_D.

Eberly - Gentlemen, much heat and light has been generated, but we must still coexist, as the proverbial lions and lambs. As an author of one of the prototype papers, I don't have to worry about its interpretation. Rodolfo and single-pulse superradiance have taken a lot of flak, but now we have experiments from the Netherlands and from Illinois showing that single-pulse superradiance does exist. There is no controversy about what is to be done theoretically. The point is that there <u>is</u> a single-pulse regime, in addition to whatever else. The existence of other regimes is not denied.

<u>Hopf</u> - Some Frenchman wrote a strange mean-field theory that fits the data of the experiment. Self-induced transparency theory applied to a dense ($\rho \sim 10^{20-21}$ cm^{-3}) system does not fit the data. There is something wrong with the propagation theory. I do not know what.

<u>Gibbs</u> - For Fresnel numbers close to one, spatial variations, i.e., local variations in phase and intensity, cause self-focusing, etc., and this leads to radical departures from the pulse evolution predicted by uniform plane wave simulations. Inclusion of radial and phase terms may have an equally significant effect on superfluorescence evolution of an inverted medium with Fresnel number close to unity.

<u>Hopf</u> - I am talking about $\alpha L \sim 2$, not like in your case.

<u>Whitney</u> - Maybe it was dipole-dipole interaction.

<u>Hopf</u> - Maybe.

<u>Gibbs</u> - Whose experiment was it?

<u>Eberly and Bullough</u> - Asher's.

<u>Bonifacio</u> - I don't understand. Feld says the number of lobes is $\ln\theta_o/4$; this is ~ 3 in the Netherlands experiment. What is the problem?

<u>Feld</u> - The question is, how much is $\alpha L > 1$? This depends on many things, and only in the limit when all these are negligible does this form occur. Under some experimental conditions, no ringing will occur. In fact, in our first paper we also published some data with no ringing. However, we felt that the observation of

ringing was important because it pointed out that the effects of propagation are inherent in superradiance in extended ($L \gg \lambda$), optically thick samples.

Gibbs - In your experimental paper you state that the Fresnel number is ~ 1, which gives $kL \sim 0.35$, so why do you fit your data with $kL = 2.5$ which corresponds to a Fresnel number of 0.08?

(NOTE: $kL = (1/2)\ell n\ (1 + 1/N^2) : N = A/\lambda L$).

Feld - It's based on experimental measurements, and is due to near field diffraction.

Hopf - Haroche sees ringing - six to ten lobes.

Gibbs - Haroche is looking at the same transition in Cs as we did - he's hitting it hard and looking early.

Karras - We see six or seven lobes in copper vapor lasers with longitudinal discharge excitation and $\rho \sim 10^{14}$ cm^{-3}. As the gain increases, we see pulse breakup and the number of lobes increase. So the basic pattern is the same as what was described here. You can see it with completely incoherent excitation. In the highly oscillating regime the intensity goes like N.

Bullough - Haroche has very short delays and $I \propto N$.

Hopf - The only difference between Bonifacio and Feld are two waves. Mike claims that he is not getting any differences for 2-wave propagation.

Bonifacio - If you want to use Maxwell-Bloch equations you have to choose different θ_o.

Hopf - There is still a theoretical conflict. In the semiclassical theory, the ringing disappears as superradiance goes away. The mean field theory fundamentally disagrees.

Feld - To change the topic: Is superradiance a useful effect? Recent superradiance experiments, such as the quantum beat phenomenon, are interesting, and possibly useful. The possibility of producing ultrashort, superradiant pulses may also be useful. Haroche is exciting high Rdyberg states to observe highly excited transitions, and doing spectroscopy by observing the cascades. Perhaps the most interesting possibility is the application to x-ray lasers where superradiant theory predicts shorter delays, shorter pulses, and lower thresholds than those predicted by rate equations.

Karras - We have metal-vapor lasers operating in these regimes, and would like to know what's going on.

Hopf - One usually thinks of narrow pulses in terms of bandwidth — there may be applications to fusion work.

R. K. Bullough - Pure Superfluorescence?

Vrehen - You replace diffraction loss with an adjustable parameter k - I would argue that F = 1 is basic; by changing the density by a small amount, one goes from a single pulse to full ringing behavior. Will your theory predict this?

Bullough - It has damping, but also has σ_{tz} which provides loss as does the k of Bonifacio. We have not reached as low intensities.

Gibbs - How large are the spatially dependent chirps that are predicted?

Bullough - They can be infinite - this may be a problem.

R. Bonifacio - Discussion of Recent Numerical and Experimental Results on Superradiance.

Q - How about the needle configuration: there $\tau_R = \tau_c = 8T_o/\rho\lambda a \propto 1/\rho a$, which implies $L/c = \tau_R$, so your arguments are invalid.

Bonifacio - As L becomes less than L_c, the time scale goes to τ_c and so intensity goes as N and ringing is due to cooperative oscillation, not propagation. Observe that the minimum pulse width is $2\tau_c$ as the length goes to infinity. We find that we can fit behavior of delay with

$$\frac{\tau_D}{\tau_R} = \left[\frac{L}{L_c} + 2 + \left(\frac{\ell n\theta_o}{4} - 1\right) e^{-L/L_c}\right] \ell n\theta_o \quad .$$

Meystre - Where does this come from?

Bonifacio - I propose it on the basis that it fits the data - it does not come from a theory.

Vrehen - If one did a Maxwell-Bloch analysis for our situation, would one find the same thing?

Bonifacio - Yes, in both directions; the backscatter is uniformizing.

Hartmann - What do you choose for initial conditions?

Bonifacio - Uniform excitation - more classical than Feld - We just put in a random, not fluctuating, source; we want to start from something other than a uniform polarization.

Flusberg - There is no disagreement in the limit of the HF experiment. On that basis, it is likely that the Feld theory is workable.

Bonifacio - But it's not a theory, just a computer fitting.

F. A. Hopf - Superradiance in Swept Gain Systems

Flusberg - Early papers show Raman gain that goes as $e^{\sqrt{z}}$.

Hopf - That's right. It's nothing fundamental.

Bowden - θ_o does not enter. Is this true? Doesn't this demarcate the regimes?

Hopf - Yes, but in a given regime, it has little effect. It tells how far you have to go in an amplifier ....

Eberly - Being the last, rather than the first speaker, you have an integrative, rather than a provocative role so would you give us your view of how these different approaches are related?

Hopf - The difference is in treating a one-way theory.

Whitney - Isn't the main distinction whether the gain is swept or not?

Hopf - That depends on the length of the sample and how it radiates.

Eberly - I'd like to request a comment on the same issue by Rodolfo.

Bonifacio - I find myself somewhat confused.

L. M. Narducci - The Two-Photon Amplifier

Hartmann - If you have enough gain to see two-photon amplification, you will have even more for electric quadrupole radiation.

Narducci - Yes. Competition is expected. As of now we have analyzed for competitive mechanisms.

Flusberg - The order of magnitude of both processes is the same.

Narducci - Electric quadrupole decay starts from noise....

Flusberg - Yes, but with high gain, it makes no difference.

Hartmann - We have never seen this in our experiments.

Narducci - You did not have an external signal of the right frequency.

Hopf - Where does the dynamic Stark effect play a role?

Narducci - In the resonant coherent limit there is a conservation law that links the detuning parameter to the field intensity.

Eberly - Will k_{at} acquire a power width?

Narducci - No, I don't think so.

Hopf - This bothers me - Liao sees the system go through resonance as a function of power.

Narducci - We claim that the detuning remains zero if it is zero initially. There is no frequency pulling in this case.

P. Meystre - Theoretical Development of the Free-Electron Laser

Madey - A few comments on the spread: our calculation for a monoenergetic electron beam gives us the spread and the average energy deposited, hence the gain.

We haven't changed the volume in phase space so there's no reason not to re-narrow the beam. We have had good luck using a short helical magnet after the long magnet. I don't know what problems are included with the angular spread, but in the long wavelength regime, it seems to work.

Workshop Session - Impact of Theory on Experiment and Vice Versa - Panel of Hopf, Bullough, Vrehen, and Gibbs

Bowden - We've heard two papers describing experimental work in the far infrared: the question then is, in the interpretation of these results, what do you regard as the role of the superradiant and swept gain criteria as far as analysis of data goes?

Hopf - I'm not absolutely sure - but I can make a guess; the swept-gain (undirectional) effect is to some extent superradiant pulse formation. The problem is that the sweep isn't for a long enough time.

Bowden - You would suggest that the cell be lengthened, then.

Hopf - Exactly.

Gibbs - What's interesting is just a traveling-wave amplifier.

Hopf - That's where experiments have generally been lacking.

Whitney - What would happen in a physical situation like the one used for a nitrogen laser with one end mirror?

Hopf - Just amplification of the reflected wave.

Whitney - I am talking about a system like the one described by Rodolfo.

Hopf - Two waves present are your source.

Bonifacio - When is an excitation swept, and when is it not? The ideal is a delta function pi pulse. The pulse must be very short with respect to what? Transit time?

Hopf - Is that the right question? Consider under what conditions swept excitation gives radiation in only one direction: 1) pump pulse width less than spontaneous lifetime, 2) pump pulse moves with the wave generated and 3) transit time greater than width of pump pulse.

DeTemple - How does the forward-to-backward intensity ratio compare?

Hopf - Want it to be much greater than 1.

Bonifacio - The pump pulse width is less than spontaneous lifetime, so the gain is swept ....

Hopf - No, pump width is greater than the transit time in the experiment. Gain has to disappear after the pulse is gone. It means $L > L_c$.

Bonifacio - Swept gain and $L > L_c$ give one wave.

Gibbs - No, the superradiant decay time less than transit time is the condition necessary, if not, you have just a traveling-wave amplifier.

Flusberg - T_s less than transit time implies $L > L_c$, so ....

Gibbs - For emission in only one direction one needs both conditions, namely $\tau_R < \tau_E$. In our experiments the whole sample sees the same thing, i.e., the excitation plus duration is long compared to the sample transit time. Therefore emission comes out both ends.

Vrehen - How can one experimentally separate stimulated Raman scattering from this?

Hopf - What if you pump with a large bandwidth pulse - if the pulse is chaotic enough, the correlations will be destroyed. They are not the same. Isn't it true that, in a three-level system, the density matrix is just a sum of two terms, corresponding to the two phenomena.

Whitney - This is a 1-step - 2-step controversy.

Flusberg - The Raman is closely related on resonance to stimulated emission. The two effects blend into one. For resonance, there is a controversy where the border between them is. How about cooperation length - does one need $L \gg L_c$ for ringing?

Bonifacio - $L = \sqrt{2}L_c$ - there is no ℓn factor.

Flusberg - In the Gibbs-Vrehen experiment, the ℓn is negligible - so the experiment doesn't test the ratio. The Feld experiment may test the ratio.

Bonifacio - Ressayre and Tallet show that for delays much greater than T_2^*, it is not superradiance. A factor of 2 density change in the Feld experiment gives a large change in the pulse width, indicating that it's out of the superradiant regime.

Hopf - Like a T_2^* -dominated process. There is still an area theorem, though - the effect is a consequence of the area theorem in inhomogeneously broadened systems.

Bonifacio - The area theorem doesn't give the shape of the pulse.

Hopf - One needs to know the area and the energy.

Vrehen - Let's change the topic to the future. Experimentally, to get further confirmation, what conditions should we seek? How can we prepare an experiment to cover both regions?

Bonifacio - To change the ℓn factor you have to change the experiments or change the density or length by orders of magnitude.

Bullough - The letter on sodium shows the results, Haroche has seen oscillations.

Karras - We see many oscillations.

Bonifacio - Let me suggest that you change your length and try to see more pulses.

Karras - We did it. We started with a shorter laser and had only one pulse. Then we doubled the length with the mirror and observed several oscillations.

Hopf - You should be careful. Relaxation oscillations can be of the same character. Rising pump pulses can additionally obscure results.

Hartmann - What superradiant radiation pattern would one expect from a large spherical sample simultaneously pumped?

Bonifacio - A spherical wave, described as before with L replaced by the radius - because a cylinder emits equally in both directions on each shot.

Bullough - I remember doing the sphere, in a different context, about 20 years ago. It should not be difficult.

Karras - How much different from a sphere would the sample have to be to give a directed output?

Whitney - You might also be talking about diagnostics concerning the homogeneity of the gain medium.

Bullough - That's determined by diffraction. Quantum description is not easy. You have to do correlations.

Vrehen - How about a sample of large Fresnel number - will it radiate all in phase or will different modes go independently?

Bonifacio - Expect it to radiate into independent lobes simultaneously.

Whitney - I disagree. I believe they "communicate".

Vrehen - What would the divergence be?

Whitney - You should go back to Rehler and Eberly.

Flusberg - There is gain in all directions.

Bowden - In this case speaking of N_c is meaningless (à la Dicke).

Vrehen - I am asking about spatial coherence.

Whitney - You may need to define an L_c in all directions.

Bonifacio - $\theta_o \sim \sqrt{a}/L$ ....

Flusberg - Different time structure in θ ....

Bowden - So what is N_c then?

Hopf - What about the possibility of doing statistics?

Vrehen - One would need a very stable experiment. When strongly saturating, one sees small fluctuations $\sim$ 10%.

Bonifacio - That's exactly right! DiGiorgio predicted 10-20% fluctuations in peak pulse intensity.

LIST OF CONTRIBUTORS

Anderson, C. E., General Electric Company

Anderson, R. S., General Electric Company

Apruzese, J. P., Science Applications, Incorporated

Bonifacio, R., Instituto di Scienze Fisiche dell' Universitá, Milano, Italy

Bowden, C. M., Redstone Arsenal, Alabama

Bricks, B. G., General Electric Company

Bullough, R. K., University of Manchester, Manchester, United Kingdom

Davis, J., U.S. Naval Research Laboratories

Deacon, D. A. G., Stanford University

DeTemple, T. A., University of Illinois

Eidson, W. W., Drexel University

Feld, M. S., Massachusetts Institute of Technology

Flusberg, A., Columbia University

Furcinitti, P. S., Pennsylvania State University

Gibbs, H. M., Bell Laboratories

Gilmore, R., University of South Florida

Gronchi, M., Instituto di Scienze Fisiche dell' Universitá, Milano, Italy

Hartmann, S. R., Columbia University

Hopf, F. A., University of Arizona

Johnson, L. G., Drexel University

Karras, T. W., General Electric Company

Konopnicki, M. J., University of Rochester

Lugiato, L. A., Instituto di Scienze Fisiche dell' Universitá, Milano, Italy

MacGillivray, J. C. Massachusetts Institute of Technology

Madey, J. M. J., Stanford University

Mattar, F. P., Laboratory for Laser Energetics, University of Rochester

Meystre, P., University of Arizona

Mossberg, T., Columbia University

Narducci, L. M., Drexel University

Newstein, M. C., Polytechnic Institute of New York

Petuchowski, S. J., University of Illinois

Ricca, A. M., CISE, Milano, Italy

Rosenberger, A. T., University of Illinois

Sanders, R., University of Manchester, Manchester, United Kingdom

Seibert, E. J., Drexel University

Vrehen, Q. H. F., Philips Research Laboratories, Eindhoven, The Netherlands

Whitney, K. G., Science Applications, Incorporated

SUBJECT INDEX